D1525154

The Common Tern
Its Breeding Biology and Social Behavior

The Common Tern
Its Breeding Biology and Social Behavior

Joanna Burger and
Michael Gochfeld

Columbia University Press
New York

COLUMBIA UNIVERSITY PRESS
New York Oxford

Library of Congress Cataloging-in-Publication Data

Burger, Joanna.
The common tern : its breeding biology and social behavior /
Joanna Burger and Michael Gochfeld.
p. cm.
Includes bibliographical references and indexes.
ISBN 0-231-07502-2
1. Common tern—Reproduction. 2. Common tern—Behavior.
3. Social behavior in animals. I. Gochfeld, Michael. II. Title.
QL696.C46B87 1991 91-8578
598′.338—dc20 CIP

Casebound editions of Columbia University Press books
are Smyth-sewn and printed on permanant and durable acid-free paper

♾

Printed in the United States of America

c 10 9 8 7 6 5 4 3 2 1

FOR OUR PARENTS,

OUR BROTHERS AND SISTERS:
E. Melvin Burger Jr., Christina Wiser, John A. Burger, Barbara Kamm,
Roy W. Burger, Robert Gochfeld.

AND THE NEXT GENERATION:
Deborah and David Gochfeld, Kathy and Edward Burger,
Michael, David and Daniel Wiser, Jacob and Andrew Burger,
Benjamin Kamm, Erik and Elizabeth Burger,
Jennifer Wolfson, Douglas Gochfeld

Contents

THREE
Breeding Biology, Phenology, Activity Patterns, and Synchrony 36

FOUR
Habitat Selection and Territoriality 91

FIVE
Aggressive Behavior 119

SIX
Predation, Vigilance, and Antipredator Behavior 168

Preface and Acknowledgments

Dramatic movements of endless hordes of passenger pigeons and bison awed early settlers in North America, yet within a century these species had been reduced to near extinction, largely because the conspicuousness of the large groups rendered them vulnerable to human exploitation and profitable to harvest. Less conspicuous, but no less certain, were the extinction of the once abundant great auk by fishermen and sailors and the near extermination of populations of egrets and terns by insatiable hunters—all for the market and the millinery trade.

To some extent these species, particularly the pigeon and the auk, may have required large social groupings to breed successfully. Among other group social phenomena, impressive evidence of massive synchrony of behavior is seen in the simultaneous aerial maneuvers of flocks of starlings and in the breeding of immense colonies of an African weaver finch, the quelea. The passenger pigeon and great auk are a missed opportunity, but the starling and quelea, at least—as well as boobies, terns, and other seabirds—are still with us in great numbers. They afford the opportunity to study the breeding behavior and social interactions in large groups of birds.

Animals form groups as an antipredator strategy and to facilitate detection and exploitation of food (Alexander 1974). Group members benefit from collective vigilance, reduction in time devoted to alertness, and increased opportunities for mate selection and social stimulation. Feeding flocks often attract other individuals to ephemeral or patchy food sources, and group behavior may increase or decrease foraging success. However, members of groups also suffer disadvantages such as interference, aggression, disease, piracy, and competition.

Coloniality is one form of group living whereby many individuals nest in close proximity. Colonial nesting occurs in 29 of the 129 avian families and is characteristic of many seabirds. Many birds that breed in colonies also feed in large flocks.

The adaptive advantages of coloniality relate to predation and to resource utilization of food, space, nest sites, nest material, and mates. Social facilitation and information transfer are two phenomena that theoretically enhance reproductive success in groups with regard to predation and food acquisition respectively. The costs of groupings such as enhanced conspicuousness are readily apparent. Birds nesting solitarily may be quiet and cryptic, escaping detection by sitting tight or slipping away surreptitiously while an intruder is a long way off, whereas large groupings of boldly patterned and often noisy seabirds are conspicuous. However, in large groups an individual may be protected by the sheer force of numbers or by the collective defenses of the group. Competition for territories, mates, nesting material, and even eggs or chicks occurs. In some species, adults frequently steal eggs or chicks from neighbors. In others, adults kill neighboring chicks that wander into their territory.

The variability in these phenomena across species has attracted the attention of ecologists and ethologists for decades. Indeed studies of seabirds have resulted in many fundamental contributions to the development of behavioral, ecological, and evolutionary thought. Yet each new "answer" raises additional questions and promotes the search for new manifestations and interpretations (see Darling 1938; Tinbergen 1953; Wittenberger and Hunt 1985; Kharitonov and Siegel-Causey 1988).

Our book focuses on behavioral, ecological, and sociobiological studies of colony dynamics, breeding biology, and interspecific interactions of common terns. Our overall aim is to examine the evolution and adaptive significance of components of breeding behavior and coloniality. For more than twenty years we have had the opportunity to study common terns and related colonial birds, to observe some of these phenomena at work, and to test theories about social and breeding behavior. The increasing emphasis on systematic data collection and quantitative analysis has enhanced our ability to progress from one hypothesis to another, and to investigate some general questions that intrigue biologists. This book addresses many of the questions that have lured us through innumerable seabird colonies in varied habitats for many years. The data presented in this book have not hitherto been published unless specific citations are given.

Common terns have a wide geographical distribution, occurring at one season or another on most continents and occupying a variety of habitats from rocky islets, to sandy beaches, dunes, marshes, and artificial structures. Of the approximately 44 species of terns in the world (depending on taxonomy, appendix B), this species is probably the best known because of its relative abundance in Europe and North America. During the course of our study, "our" colony at Cedar Beach, Long Island, New York, increased to over six thousand pairs, becoming one of the largest common tern colonies currently known in the world. Yet early descriptions suggest that much larger terneries existed in North America within the past three centuries. Colloquially referred to as sea swallows, the buoyant and graceful terns are an intrinsic part of coastal landscapes, inspiring those lured by the beach and the sea, regardless of their ecological interests. For the biologist, however, the common tern has been an object of intensive study and an abundant literature, and there are few seabirds which are better known.

We studied birds nesting on sandy beaches and on salt marshes, for the problems faced by the birds in the two habitats are very different. Although both habitats experience avian predation, the salt marsh islands offer little opportunity for mammalian predators, but are more vulnerable to storm tides and flooding. Mammalian predators pose a threat to terns nesting on dry land colonies, and these include not only wild predators, but human commensals (dogs, cats, rats), not to mention humans themselves. The physiognomic features of the landscape, historical factors, and current social factors influence where the birds will nest in any season. Common terns also breed on artificial "spoil" islands created by dredging, on islets, shell bars, and in other habitats where we have not yet studied them.

The role of breeding synchrony and its relation to group size and reproductive success has been a challenging concept. Although many studies have offered alternative explanations for apparent synchrony, a comprehensive study of the problem offers immense logistical problems, since one has to be in many colonies at the same time. Nonetheless, this remains a fruitful concept for study; and because we were able to study a number of colonies during many seasons, our data help to illuminate the issue.

The persistence of terns and other colonial species within the New York megalopolis is a tribute to their adaptability. At present, populations of common terns in New York and New Jersey appear stable. Although many colonies have persisted in close proximity to human activity for decades, their nesting sites are by no means secured for posterity. The birds and man have an uneasy peace at present, and there is no basis for complacency. Perhaps development of shore properties has proceeded as far as it can go in some places, granting the birds amnesty from further habitat usurpation. In any case, whatever transpires in North America may pale to insignificance beside habitat destruction, chemical contamination, and human exploitation that threaten these same tern populations on their tropical wintering grounds.

Among the topics germane to coloniality but not examined in this book are questions related to feeding behavior and ecology (see reviews in Croxall 1987), information transfer, cooperative feeding, and the influence of these factors on social behavior. Some of our current research activities are focused along these lines, and while recognizing their importance in the discussion of coloniality, we confine ourselves in this book primarily to the social behavior of this colonial species.

There is an aesthetic dimension to our studies of terns as well. Patient observation of a nesting tern, 4 ounces (120 grams) of fluff, that has completed a 15,000 kilometer round trip journey to its tropical wintering grounds, affords opportunity for insight, contemplation, and a bit of wonder. Learning that a tern one banded as a chick last spring has been recovered in a remote part of South America punctuates our understanding of their migration. The discovery of interspecific hybridization on the one hand or a novel developmental defect on the other are sobering reminders of biologic variability and frailty. Just walking through a thriving tern colony provides an aura of excitement even if by so doing one becomes a target for the noisy attacking adults, and for their repeated strikes.

Over the past two decades we have come to understand many aspects of common tern behavior and breeding biology, but we have also come to recognize that no season is truly typical, and that observations vary from colony to colony and year to year (in fact, 1990, our most recent year, presented us with unprecedented weather conditions and a much delayed breeding season). Tern behavior in large colonies differs from that in small ones and behavior in growing populations differs from that in declining ones. As it is very likely that common terns do things differently in other parts of their range, this book affords investigators elsewhere the opportunity to validate the generality of some findings or demonstrate the uniqueness of others.

During our studies of the breeding behavior and ecology of the common terns we also studied the black skimmers nesting among them. Although we originally intended to publish the two studies in a single integrated form, several readers advised strongly against that. Accordingly, our study of the black skimmer has been published separately (Burger and Gochfeld 1990a), and forms a companion to this volume.

Our earliest field work with common terns dates back to 1964, although annual work was not begun in New York until 1969 and in New Jersey until 1976. Many of the specific projects have been completed at different times during this period, and a few are ongoing. This is reflected in the data tables, some of which end in 1985, others (e.g., productivity) in 1988, while others (e.g., population dynamics, colony selection, and turnover) have been updated through 1990. We expect to continue these and other studies of terns for many years.

ACKNOWLEDGMENTS

Over the years, many people have helped us with our studies of colonial birds, some for only a few hours, others persistently for over a decade; we thank them all. We particularly thank Fred Lesser, Betsy Jones, Jim Jones, Carl Safina, Darrel Ford, and Robert Paxton for many hours of stimulating discussions, field companionship, and abundant logistical support. Robert Paxton and Sarah Plimpton generously extended to us use of their cottage close to Cedar Beach, which made access to the terns convenient and enjoyable. Carl Safina, who began working with us as an undergraduate and has blossomed into a valued colleague, also shared his abode with us on many a night. Our children, Deborah and David, grew up during this research with noisy terns swooping overhead, and they participated in marking nests, banding chicks, trapping adults, and eventually analyzing data. Our parents sparked our interest as children and encouraged us throughout, although they sometimes wondered if this was a real occupation for grown people.

We also thank Francine Buckley, Paul Buckley, Darrel Ford, Ann Galli, Joan Galli, Robert Gochfeld, Laurie Goodrich, Joy Grafton, Betsy Jones,

Jim Jones, Fred Lesser, Maria Mikovsky, Bertram Murray Jr., Patti Murray, Danielle Ponsolle, Diane Riska, and particularly Carl Safina for field assistance, help with censusing, and field companionship.

Our understanding of colony dynamics would be impossible without the extensive banding of adult and young terns carried on by numerous individuals over many years. We particularly appreciate the labors of Tom David, Darrell Ford, Robert Gochfeld, Helen Hays and her colleagues, Fred Heath, Richard Kremer, Fred Lesser, Ian C. T. Nisbet, Robert Paxton, Peter Post, Gilbert Raynor, Carl Safina, Jeffrey Spendelow, and LeRoy Wilcox, some of whom are sadly no longer with us.

Ronald Cody and Richard Trout provided essential statistical advice and help with programming. Deborah Gochfeld and Maria Mikovsky helped with data entry, and Alex Gochfeld and Brook Lauro printed many of the photographs. Jorge Saliva provided line drawings of tern behavior, and Betty Green and Bobbe Philip helped in ways too numerous to mention. Extensive help with the bibliography and index was provided by Marilyn Arciszewski, Dale Bertrand, Deborah Gochfeld, Betty Green, Kevin Staine, and Kirk Waterstripe. Alex Gochfeld translated important Russian literature that greatly expanded our horizons. Ralph Morris, Raymond Pierotti, and Carl Safina provided careful and extensive comments on the manuscript. Finally, Deborah Gochfeld not only helped collect and analyze data, but worked on several versions of the manuscript, and we thank her now.

We thank the managers of the Edwin B. Forsyth [Brigantine] National Wildlife Refuge, the Endangered and Non-Game Species Program of the New Jersey Division of Fish, Game, and Wildlife (particularly JoAnn Frier-Murza, Larry Niles, and David Jenkins), the New York Department of Environmental Conservation, the Long Island State Parks and Recreation Commission, the Town of Babylon, the National Audubon Society, and the U. S. Fish and Wildlife Service for the necessary permits to conduct this research.

Over the years, several organizations provided logistical support or funding and we thank the Endangered and Non-Game Species Program of the New Jersey Department of Environmental Protection, the New York Department of Environmental Conservation, the New Jersey State Mosquito Commission, the Ocean County Mosquito Commission, the Research Council and the Charles and Johanna Busch Fund of Rutgers University, the Society of Sigma Xi, the National Institute of Mental Health, and the U. S. Environmental Protection Agency. The Scully Science Center of the National Audubon Society provided logistical support in many ways.

Tico shared with us the many hours of manuscript preparation and review, and cheerfully consumed several chapters thereof.

Finally, we have discussed seabird breeding biology and social behavior and the adaptive significance of coloniality of terns with many people and have benefited from these discussions and from their encouragement. Among the many are Sandy Bartle, Keith Bildstein, William Boarman, R. G. B. Brown, John Brzorad, Charles Collins, Kendall Corbin, John Coulson, Malcolm Coulter, Jack Cowie, Michael Erwin, Darrell Ford, Sidney Gautreaux, Helen Hays, Thomas Howell, Kees Hulsman, George Hunt, Michel Kleinbaum, James Kushlan, John Krebs, Brook Lauro, Charles Leck, Fred Lesser, Mary LeCroy, Martin McNicholl, Douglas Mock, Bill Montevecchi, Ralph Morris, Bertram Murray, Jr., David Nettleship, Ian Nisbet, Ken Parkes, Kathy Parsons, Ray Pierotti, Peter Post, Iola Price, Jim Rodgers, John Ryder, Philip Regal, Carl Safina, Jorge Saliva, Dave Shealer, Gary Shugart, Jeffrey Spendelow, Kevin Staine, Niko Tinbergen, H. B. Tordoff, Greg Transue, Guy Tudor, Kees Vermeer, Dick Veitch, David Wingate, and—last but by no means least—Victor Zubakin.

ONE

Common Terns, Other Terns, and Coloniality

Many species of birds breed along our coasts, nesting solitarily or in groups or colonies. With the increasing development of coastal areas for homes, recreation, and commercial interests, it is critical to understand the fragile ecosystem of beach, estuary, and salt marsh, as well as the species living within them. One or more species of terns are characteristic of most coastlines, and among these the common tern (*Sterna hirundo*; henceforth, scientific names appear in appendix A) is certainly the best known in both eastern North America and Europe.

In this book, we examine several aspects of the breeding biology and behavior of common terns nesting in New York and New Jersey. In addition to basic breeding biology such as habitat selection, phenology, activity patterns, and synchrony of egg-laying, we are particularly interested in social interactions such as aggressive behavior, and how the terns are affected by and respond to adversities such as predation, piracy, and flooding. We explore the effects of these factors on the terns' reproductive success. Since common terns in our area (Long Island in New York and Barnegat Bay in New Jersey) nest on salt marshes (Burger and Lesser 1978, 1979; Safina et al. 1989) as well as on sandy beaches, this book focuses on contrasting their behavior in these two habitats. We did not study terns in other habitats which figure prominently elsewhere in their range and which may afford different constraints.

Recently Nisbet (1989) emphasized the challenge of studying the population ecology of long-lived seabirds, noting that among these the common tern requires more than a decade of study. Indeed, as the lifespan of a bird approaches the effective field lifespan of the student, the limitations on what one can hope to accomplish become painfully apparent. Nisbet (1989) identified the need for synthesis of long-term study data, whereas his analysis showed the contrary: of nearly twenty long-term seabird studies, virtually all publications (including most of ours to date) have focused on short-term substudies. This book and its companion work on the black skimmer (Burger and Gochfeld 1990a) attempt to rectify that aspect, bringing together data obtained over a twenty-year period. Our work with the common tern did not focus on individually marked birds, since our population expanded too rapidly for us to keep up with the banding, and other studies are utilizing individually banded birds in long-term studies. In our work, we have focused on colony dynamics and behavioral patterns compared in different habitats over a period of years, particularly such topics as colony occupation, habitat and nest site selection, territoriality and aggression, and reproductive success.

1.1 STUDY OBJECTIVES

Our overall goal is to describe the breeding behavior of common terns and to assess the advantages and disadvantages of colonial nesting in common terns. Our study objectives include comparisons of breeding behavior and reproductive success in different habitats, at different densities, and in colonies of varying composition. In our area of the northeastern United States, common terns often nest in mixed species colonies on sandy barrier beaches and on salt marsh islands. Since the constraints imposed by sandy beach and salt marsh habitats differ, we were particularly interested in how the terns respond to and use each habitat type. We examine several aspects of tern breeding behavior that in the short term affect reproductive success and in the long term shape their adaptations and the evolution of their colonial habits.

We will discuss breeding phenology, colony and nest site selection, colonization, colony turnover, species interactions, territoriality, social facilitation, courtship behavior, incubation and chick rearing, predation, antipredator behavior, and reproductive success. We concur with Wittenberger and

Hunt's (1985:57) statement that the significance of avian coloniality is not yet completely understood, and no single hypothesis is likely to provide a general explanation. Our work tests many of the hypotheses adduced to "explain" coloniality and clarifies those patterns which appear important for our populations of common terns. Whereas we point out generalizations likely to apply to many species in many circumstances, we agree that a quest for a single hypothesis is too simplistic to explain behavior which varies across taxa and regions.

This book covers field studies begun in 1964 and conducted annually from 1969 through 1990. The common terns we studied rarely nested solitarily, but nested in colonies of up to six thousand pairs. Particularly for marsh-nesting common terns, there appears to be abundant available habitat close to sites used as colonies. One could infer that if coloniality were not advantageous, the birds would disperse more widely into this available habitat. Therefore we examined habitat selection and focused on environmental factors (tides and flooding) and biotic factors (predation, piracy, and aggression), which influence the social and colonial behavior of common terns.

1.2 STUDY SPECIES: THE COMMON TERN

The common tern (fig 1.1) is a nearly cosmopolitan seabird. In North America, it breeds from the Atlantic Coast through much of the interior of North America, but not on the Pacific Coast. Common terns breed southward to North Carolina, with another small population on the Gulf Coast and in parts of the Caribbean, where its distribution is somewhat confused due to its similarity to the roseate tern. Common terns also breed across Eurasia south to the Mediterranean and the north side of the Himalayas. In the Western Hemisphere, it winters in the Caribbean and South America south to Argentina (AOU. 1983).

Along the Atlantic coast of North America, common terns nest on a variety of substrates including sandy beaches, dredged islands, rocky islets, shell bars, and on salt marshes. They form both monospecific and mixed species colonies, often with other tern species and with black skimmers (fig. 1.2). The species is highly colonial and rarely nests solitarily. Its colonies range in size from a few pairs to thousands of pairs. One of the colonies we studied, Cedar Beach, grew to about 6000 pairs in the mid-1980s, becoming one of the world's largest known common tern colonies in this century.

The common tern (see fig. 1.1) weighs about 100 to 130 grams. The males and females are virtually identical in appearance and size, although subtle differences in bill size have been detected statistically (Coulter 1986). They are long-lived and show delayed reproduction (Austin and Austin 1956). Banding studies show that many live for 15 years or more, meaning that they probably breed at least a dozen times in their lives. Austin (1942) estimated a life span of 8+ years, and an annual adult mortality of about 25 percent (Austin and Austin 1956), although both figures are biased because of significant levels of band loss and trap shyness, and probable intercolony movement (Nisbet 1978; DiCostanzo 1980). Nisbet (1978) estimated annual adult mortality at 7.5 to 11 percent and DiCostanzo (1980) reported an estimate of 8 percent. Most common terns do not breed before they are three years old, although a proportion breed at age two, and rarely even at age one. Most of these younger breeders are in adult plumage, although occasionally birds with non-breeding plumage are found nesting. Figure 1.3 illustrates an incubating bird in the nonbreeding, white-faced *portlandica* plumage (Palmer 1941b) originally described as a distinct species.

Evolutionary and Taxonomic Relationships

Terns belong to the avian order Charadriiformes which contains some of the most familiar of birds. In general, the order has been divided into three major groups, the shorebirds (Charadrii), the auks (Alcae) and the gulls and their allies (Lari). Elsewhere we have summarized the classification of the suborder Charadrii (Gochfeld et al. 1984), and have discussed the biology and relationships of the skimmers (Burger and Gochfeld 1990a). The auks, puffins, murrelets, and their relatives (family Alcidae) form a naturally and closely linked assemblage of species (Nettleship and Birkhead 1985). The relationships in the suborder Lari, containing gulls, terns, skimmers, and skuas have been examined with emphasis on morphology (Schnell 1970), behavior (Moynihan 1958, 1959a), biochemistry (Sibley and Ahlquist 1972) and DNA hybridization (Sibley et al. 1988). The data are open to various interpretations. Most authors agree that the gulls and terns are closely related (family Laridae), while the skuas and skimmers are often sorted into discrete families. Most recently four groups have been lumped into the family Laridae (AOU. 1983), each being relegated to subfamilial rank, namely the skuas and jaegers (Stercorariinae), gulls (Larinae), terns (Sterninae), and skimmers (Rynchopinae).

Figure 1.1. Common tern in flight (top) and incubating at a nest in the Cedar Beach colony (bottom).

Figure 1.2. (top) Common tern attempting to land in the midst of a cluster of black skimmers nesting at Cedar Beach and another (bottom) that has already established its territory and begun to nest. This shows the intimate relationship between the two species.

Figure 1.3. Adult-plumaged common tern mated with a bird in a non-breeding or subadult plumage (note white forehead) often called a *portlandica* plumage.

1.3 WORLD DISTRIBUTION OF TERNS

Terns are characteristic of most of the world's shorelines. Of the 44 species of terns in the world, only ten breed regularly inland, while few are restricted to inland habitats. Most of the species breed along continental coasts or on offshore or oceanic islands. As far as is known, with the exception of the insectivorous and lizard eating gull-billed tern, all rely primarily on aquatic organisms (mainly fish) for their food. Most nest in very close proximity to water, some actually nesting on floating substrate.

Many species of terns are very widespread in range. The common, Arctic, roseate, Caspian, and gull-billed terns, for example, occur on virtually all major land masses except Antarctica, although they do not breed in South America, and the sooty tern and brown noddy occur virtually throughout the tropical oceans (pan-tropical distribution).

Although many species of terns are abundant, some are threatened, endangered, or perhaps extinct. Two North American species, the roseate tern (Nisbet 1981; Gochfeld 1983a) and least tern are in jeopardy (Burger 1984a, 1989). The northeastern North American population of the roseate tern is endangered and the Caribbean population is threatened. The common tern is currently listed as a threatened species in New York, primarily because of threats to its breeding beaches; while, in New Jersey, common terns have largely abandoned beach habitat and nest mainly in marshes.

1.4 TERNS AS COLONIAL SPECIES

Most species of terns nest colonially in a variety of habitats (appendix B). The largest tern colonies are those of sooty terns which approach a million pairs (Ashmole 1963), and this species is among the most abundant of seabirds. The "crested" terns include a number of species that form extremely dense colonies, often with neighboring pairs able to jab at one another while incubating. This represents the maximum packing among seabirds. Good accounts of these species are provided by various writers (Buckley and Buckley 1972; Veen 1977; Langham 1974; Langham and Hulsman 1986). Although Sandwich, royal, and crested terns form very dense groups, rarely do they match in size the number of pairs found in many sooty tern colonies.

Further along the density spectrum are species like the common and roseate terns. These nest in assemblages ranging from dozens of pairs to several thousands of pairs. The characteristic density is reflected by internest distances ranging from 45 cm to 300 cm. The least gregarious of the terns are species such as the black-bellied, the Damara, and the Peruvian tern that may form clusters of two to ten pairs, but typically have distances of 10 to 100 m or even more between nests. Some of these birds may be characteristically solitary.

Even in a colonial species one may find isolated pairs nesting with no conspecifics nearby (Bergman 1980). We occasionally find one pair of common terns nesting alone, far from conspecifics, but sometimes in association with another species. It is this spectrum of colony sizes from one or two pair to millions, and of densities ranging from touching distance to hundreds of meters, which render the species of terns intriguing subjects for studies of social behavior and coloniality. Their diversity also suggests plasticity with respect to coloniality, and that the degree of coloniality within a species may be facultative rather than obligate.

1.5 WHY BIRDS FORM COLONIES

The spatial pattern of nesting in birds ranges from solitary nesters who may have no conspecifics within visual or acoustic range to densely packed colonies of thousands of pairs where individuals can touch their neighbors. Even within a species there may be substantial variation in the dispersion pattern. Colonial nesting occurs in 29 of the 129 avian families (Lack 1968), but is particularly common in seabirds. Species that feed in flocks are more likely to be colonial (26 percent compared to only 1 percent of solitary feeders). Birds might nest in close proximity because space is limited, but truly colonial species will still show a preference for nesting with conspecifics, even when there is no evidence that space is limited.

The adaptive advantages of nesting in colonies include avoiding predation and optimal resource utilization (Horn 1968; Burger 1981d, 1984d; Wittenberger and Hunt 1985; Kharitonov and Siegel-Causey 1988; Siegel-Causey and Kharitonov 1990). Antipredator behavior, social facilitation, and information transfer are proposed mechanisms that could increase the fitness of birds breeding in groups (Ward 1965; Ward and Zahavi 1973; Waltz 1982, 1983). Social facilitation, leading to increased breeding synchrony and reduced predation due to predator swamping, was proposed by Darling (1938) as an important advantage to colonial living, and has been discussed in several contexts (Burger 1974a, 1979a; Wilson 1975; Gochfeld 1980a,b; Hogstad 1983; Nisbet and Welton 1984). This relationship, often called the Darling Effect, remains controversial, but proves useful heuristically in examining colonial behavior.

The increased antipredator behavior of colony members can deter predators (Tinbergen 1963, 1967; Kruuk 1964; Patterson 1965; Andersson 1976; Montevecchi 1977, 1978a; Krebs 1978; Wiklund and Andersson 1980). Similarly, colonies can serve as information centers where individuals can learn foraging sites or discover patchily distributed and ephemeral food resources (Phillips 1962; Ward 1965; Murton 1971a,b; Krebs et al. 1972; Krebs 1974; Kushlan 1974; Gochfeld and Burger 1982; Ward and Zahavi 1973; Emlen and Demong 1975; Bayer 1982; Waltz 1982).

There are, however, disadvantages to breeding in colonies, including competition for resources such as nest sites and materials (Cullen 1957; Fisher and Lockley 1954; Sladen 1958; Crook 1964; Tenaza 1971; Siegfried 1972; Burger 1978a,b,), mates (Mayr 1935; French 1959; MacRoberts 1973;

Kushlan 1973; Schoener 1974; Mock 1976a,b; Minaeu and Cooke 1979), food (Wittenberger and Hunt 1985), and brood parasitism (Friedman 1960; Brown 1984). Other disadvantages include increased disease transmission potential and increased detectability by predators (Weller 1959; see discussion in Shields and Crook 1987; Shields et al. 1988).

The balance of these factors should influence the propensity of individuals to seek or shun other individuals, and should be reflected in the dispersion or degree of coloniality manifested by a population (Siegel-Causey and Kharitinov 1990). In some cases, species that usually nest solitarily will sometimes nest colonially (Pienkowski and Evans 1982). For species where coloniality is facultative and a result of limited safe nesting sites, selection pressure for coloniality itself would be lower. It is likely that the selection pressures for and against coloniality operate over a time frame much longer than that available in a field biologist's working lifetime. Nonetheless, we can gain insight into these factors from long term studies.

1.6 COMMON TERNS' FOOD AND FEEDING BEHAVIOR

Notably absent from this book are data on feeding behavior and food. We do not underestimate the importance of food, or of cooperation or competition for food in common tern biology. Our studies of food and feeding are ongoing and have been published separately, mainly in conjunction with our colleague and former student, Carl Safina, who is taking the lead role in studying the terns at sea (Safina and Burger 1985, 1988a,c; Safina et al. 1989; Safina 1990a,b).

There is a large literature on seabirds and their food and social feeding behavior (e.g., Sealy 1973; Davis 1973; Andersson 1976; Erwin 1978; Baird and Moe 1978; Andersson and Gotmark 1980; Hoffman et al. 1981; Bayer 1982; Evans 1982b; Nettleship et al. 1984; Khleboselov 1986; Birt et al. 1987; Croxall 1987) as well as on the feeding biology of terns in general (e.g., Ashmole and Ashmole 1967; Gochfeld and Burger 1982; Kaverkina 1989), and common terns in particular (Safina and Burger 1985, 1988a,c; Safina 1990a,b). The feeding of terns has resulted in such notable and widely generalized studies as Nisbet's (1973) model on the relationship of food availability, courtship feeding rates, egg size, and mate preference in common terns, and Willard and Salt's (1971) study of foraging behavior in Forster's terns. Safina and Burger (1988a) provide an extensive review of the

Figure 1.4. Early dawn foraging flock of common terns outside of Fire Island Inlet, Long Island, New York. Birds are foraging over schools of small fish driven toward the surface by predatory bluefish.

literature on the relationship of food availability to breeding of common terns.

The work of Safina and Burger (1985, 1988a,b, 1989a,b) has involved extensive sonar sampling of fish available to, and exploited by, terns in the vicinity of their Long Island colonies. They have shown that common terns laid eggs earlier, fed chicks more frequently, and had faster chick growth and better survival, in a year when the sonar documented higher fish density compared with years of lower density (Safina and Burger 1988a,b). Safina and Burger (1988a) showed that potential prey fish were very patchily distributed and that tern flocks were closely associated with the more dense aggregations of prey within 3 m of the water surface (fig 1.4). They found a correlation of tern flock size with prey fish school size as printed by the sonar, but not with the number of predatory blue fish driving the smaller fish to the surface. Safina is continuing these studies, comparing the feeding behavior of the common tern with its close relative, the roseate tern, near several colonies (Safina 1990a,b).

Common terns take a wide variety of fish, yet usually at any one time only one or two species predominate. In most years at Cedar Beach, two species, sand lance and bay anchovy, have been dominant, while baby bluefish, silversides, killifish, butterfish, and a variety of herrings are taken in most years (Safina and Burger 1988a; Safina et al. 1990). In a few years shrimp have been a major part of the diet in some colonies. One year, we noted a large number of squid being brought to chicks at Cedar Beach. When available, the sand lance (also called sand eel) is one of the preferred foods. On many days, it constitutes 100 percent of the fish brought back to the colony. Its long slender body makes it ideal baby food since even day-old terns can swallow a substantial sand lance. By contrast, the wide-bodied butterfish are too large for most chicks to swallow. On occasion, one finds several discarded butterfish lying about a nest, indicating that at least some adult terns do not adjust their fishing in response to a chick's inability to swallow. The most dramatic example of this is the rather uncommon observation of 20 to 80 or more desiccated pipefish scattered about a nest. Some parents apparently find these easy prey, while chicks may struggle vainly for hours trying to swallow these extremely long and not very meaty morsels, which remain alive and struggle and twist themselves around the chick's bill. Yet hour after hour the adults return with yet another pipefish. This is clearly an unusual event since one finds the evidence at only a few nests in any year.

Common terns characteristically feed by coursing or hovering 3 to 7 m over the surface and plunging almost vertically to seize small fish with their bill. At times they may submerge completely, but they probably only take fish from the top 20 to 30 cm of water. They usually capture only a single fish, but at times take two or more (Gochfeld 1978). When prey fish were dense (as visualized by the sonar), the terns hovered less and dove more often and had fewer abortive dives, but there was no apparent effect on the percentage of successful dives (Safina and Burger 1988a). The presence of bluefish driving small prey fish to the surface had an important impact on the terns' feeding behavior, and they hovered over the bluefish waiting for them to drive prey within reach.

The dense flocks of terns that form over a school of bluefish are extremely fluid. A thousand birds may form a swirling mass in one location (fig 1.4), yet they may suddenly disband only to regroup a few minutes later over another promising patch, as the fish—which humans can see only on the sonar screen—shift their location. After a few minutes in a dense flock, the terns' success rate may decline, and some individuals become discouraged and disperse. Eventually, one of these discovers new prey and dives, and

Figure 1.5. A three-week-old common tern chick (top) struggles to swallow a large sand eel after a tug-of-war with its sibling. It is unable to complete the process at one sitting. The fish protrudes from its bill until digestion proceeds. The bottom chick attempted to swallow a baby bluefish tail first, but in the process the bill slipped through the gill preventing further progress.

almost immediately large numbers of terns congregate at this new site (Gochfeld and Burger 1982).

Near Cedar Beach there is generally a significant decline of prey availability with the advancing season, and small fish become scarce in mid- or late summer, after most of the young terns have fledged. Late in the season, terns will exploit low density patches that are ignored earlier in the season (Safina and Burger 1985). However, not all years follow the same pattern, with variation in average flock size and fish-capture success (Safina and Burger 1989).

Most common terns in the northeastern United States feed in flocks over local and ephemeral patches of high prey fish density. However, Nisbet (1983) found that some individuals defend exclusive shoreline territories, and that such behavior may be profitable in those locations where there is a consistent and predictable source of fish. Erwin (1977) provided some evidence that inshore waters of the Virginia coast tend to yield more consistent fish samples than offshore waters. But meeting a profitability threshold for defense requires adequacy as well as consistency.

Having captured a fish, the tern may swallow it on the spot or may fly directly back to the colony to feed it to its mate or chicks. Chicks seize the fish and with a few vigorous tosses of their heads work the fish headfirst into their gullet and swallow it (Gochfeld 1980d). If the fish is large, they may not be able to swallow it entirely at first (fig 1.5 top). Occasionally a chick attempts to swallow a fish tail first, but this is a much slower process, increasing the risk of losing the fish to pirates. Figure 1.5 illustrates a rare accident in which a chick has impaled a fish through the gill and is unable to either swallow or disgorge it (Gochfeld 1975c).

The work in this book is focused on the breeding colonies themselves. Thus, many other important aspects of tern biology, such as foraging behavior, though vital to the reproductive success of the birds, will not be found in these pages. However, one of the dramatic feeding strategies of the terns, namely piracy, does take place at the colony, and this behavior figures prominently in this book. Many studies of piracy have focused on strategies of the pirates or the benefits of engaging in piracy, while our work has focused on the strategies available to the potential victims to minimize their risk of losing a fish.

TWO

Study Sites, Methods, Samples, and Definitions

Common terns provide excellent opportunities for examining breeding and social behavior, behavioral adaptations to varying habitat availability and suitability, and the significance of coloniality. They nest together in accessible mixed-species colonies of various sizes in both salt marshes and on dry land (beaches and fill). The adversities affecting dry land colonies are mainly predation (both mammalian and avian) and human disturbance, since these colonies are readily accessible. By contrast, the main adversities befalling the salt marsh colonies are flooding and avian predation. These adversities contribute to the selection pressures facing tern populations.

In this chapter, we describe the colony sites we studied and outline the techniques of observation and recording of behavior. We describe the methods we used to survey and census them, and the methods we used to study colony and nest site selection and for determining reproductive success. We explain our behavioral observations, statistical analytical techniques, and experimental procedures which added to our understanding of the phenomena we observed. Specific methods are described in more detail, where warranted, in the appropriate chapters. Below, we provide operational definitions of some terms.

Colony: Usually refers to an aggregation of breeding of birds. However, for our purposes a single nesting pair of common terns constitutes a colony. By this definition, a colony cannot be unoccupied.

Site or Colony Site: A place where a group of birds breeds. A site is a geographical unit and it may be unoccupied in a particular year.

Colony-year or Occupancy: Each time a site is occupied in a particular year, it counts as a colony-year. In a matrix of colonies by years, each cell which has some birds nesting would constitute a colony-year of occupancy.

Subcolony: A discrete portion of a colony identified either by a habitat change or by a discontinuity or gap among the nesting birds.

Neighborhood or Cluster: A group of nests in close proximity that form an active social grouping, rather than just a spatial cluster.

Colony Size: The number of breeding pairs of birds in a colony.

Synchrony: A measure of the temporal clustering of a phenomenon. We employ the standard deviation of the egg-laying date as the measure of nesting synchrony (after Emlen and Demong 1975; Gochfeld 1980b).

Productivity: The number of young birds fledged per pair nesting in the colony.

2.1 COLONIES STUDIED

We studied terns in sandy beach and salt marsh colonies in New York from 1969 to 1990 and in New Jersey from 1976 to 1990 (fig 2.1; appendix C). During this period, we gathered data on the distribution of the populations, colony size, density, and phenology; made observations on various aspects of behavior; and examined reproductive success.

Data on phenology, colony occupancy, habitat and nest site selection, predation, antipredator behavior, and reproductive success were gathered insofar as possible in both habitats in most years. Some species-specific behaviors (i.e., display sequences or some aspects of parental behavior) should not vary among habitats, and we gathered such data at only one or two colonies. When colony size was a critical variable (i.e., mobbing and colony defense), we examined behavior at many colonies over a wide range of sizes.

Over the years, we have visited and studied many dry land and marsh colonies along the coast of Long Island (New York) and New Jersey. Some sites have been occupied for only a few seasons, others continuously. In New York our principal study sites were two large dry land (sandy beach) colonies (West End II and Cedar Beach) and one salt marsh colony (Seganus Thatch), all on the south shore of Long Island. In New Jersey, we studied 34 salt marsh colonies (one of which, Tow Island, was visited infrequently) and two sand colonies (Pelican and Holgate). Our study of colony dynamics did not include Tow Island or Holgate, but did include Pelican. For productivity we studied 33 salt marsh islands (not including Tow), nor were Holgate and Pelican included. The two states are immediately adjacent, and the distribution of tern colonies along the south shore of Long Island, New York, and the northern coast of New Jersey is virtually continuous (fig. 2.1). Additional colonies in both states were censused periodically to determine patterns of colonization and intercolony movements. All colonies contained predominantly common terns; most colonies contained black skimmers; other associated species included roseate and least terns, herring and laughing gulls, and various shore and marsh birds. The three dry land colonies (West End, Cedar Beach, Holgate) were occupied in almost all years, while the occupancy of salt marsh islands varied, some being occupied in all years, others only once.

Natural Vegetation of the Barrier Beach and Salt Marsh

The natural vegetation of the barrier beach is a continuum extending from the ocean berm with scattered sea rocket and outer dune dominated by beach grass, across the interdune area dominated by beach grass and seaside goldenrod, to the back dune area dominated by bayberry, to the inner marsh dominated by common reed or phragmites and the bushes *Iva* and *Baccharis*, and finally onto the saltmarsh dominated by cordgrass *(Spartina alterniflora)* and salt hay *(Spartina patens)*.

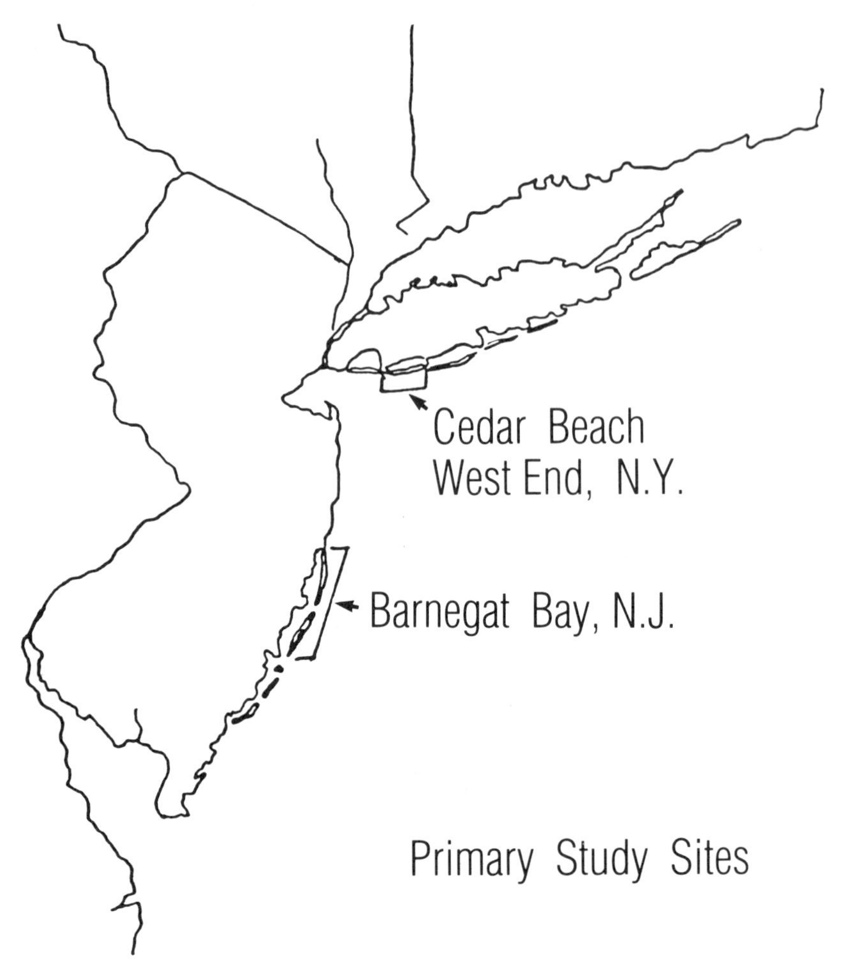

Figure 2.1. Map of the New York (Cedar Beach and West End) and New Jersey (Barnegat Bay) common tern colonies.

Beach Colonies

The dry land colonies were West End Beach II (in Jones Beach State Park, 5 km south of Freeport in southeastern Nassau County), Cedar Beach (5 km south of Babylon in southwestern Suffolk County, New York, see Gochfeld 1976), and Holgate (in the Edwin B. Forsyth National Wildlife Refuge at the southern tip of Long Beach Island, Ocean County, New Jersey). All three are on the oceanic barrier beach. We also examined colony dynamics at

Figure 2.2. Aerial view of West End II colony. Distance between the two roads at widest point of loop is 150 m.

Breezy Point (Post and Gochfeld 1978), a 5.5 ha site in Brooklyn, New York, for this colony was abandoned and then recolonized during the study. These colonies are all on sandy beach, mostly in interdune areas which in some cases have been artificially stabilized. The vegetation is dominated by beach grass, seaside goldenrod, and sea rocket which are important determinants of where the terns nest. All are adjacent to beaches that are heavily used for recreation, particularly swimming and fishing. The potential for human disturbance is high, but all three are patrolled by park police. Management practices have varied from deliberate attempts to discourage tern nesting at West End II, to nearly total fencing at Cedar Beach, to beach closure during the breeding season at Holgate.

West End Beach II: This site (8 ha), designated as West End II, is part of Jones Beach State Park (fig. 2.2), and has been occupied by common terns continuously from the mid-1950s to 1987. The substrate is sand mixed generously with road-fill, and between 1963 and 1970, the colony was entirely inside the loop of the access road which serves the large recreational beach facility immediately south of the colony. Since 1971, as the colony

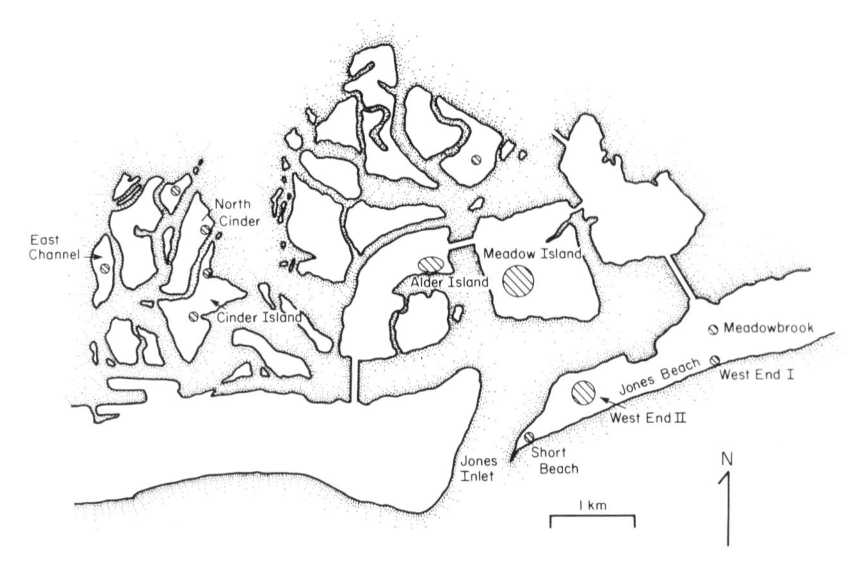

Figure 2.3. View of the Jones Inlet area with the location and relative size of common tern colonies indicated. Small circles indicate < 100 pairs. The oval indicates 100–200 pairs. Larger circles indicate > 1000 pairs.

expanded, birds have nested outside the loop as well, until by 1980 the majority of birds nested outside, some as close as 10 m to the large and busy parking lot. From 1984 onward, the colony began to decline, and in 1988 it was abandoned altogether, the terns nesting in a new colony 1.5 km to the east at West End Beach I. About 150 pairs of black skimmers nested at West End in most years, and prior to 1970 there were about 25 pairs of roseate terns as well. Figure 2.3 shows the location of the colonies in the vicinity of Jones Inlet, where there was frequent interchange of birds.

Human disturbance was a frequent and sometimes daily event during the entire nesting season, increasing after the Memorial Day (May 30th) holiday and peaking on the Independence Day (July 4th) weekend. In the late 1970s, park officials, anxious to dissuade the terns from nesting in close proximity to the road and recreational beach, made additional plantings of beach grass, but this did not appear successful in discouraging nesting. In addition to the natural vegetative cover of beach grass and goldenrod, there are numerous exotic weed species as well as horticultural planting of rose and Japanese black pine.

Cedar Beach: On the barrier beach 18 km east of West End, the Cedar Beach site (10.5 ha) lies 300 m west of the Town of Babylon's recreational beach. Human intrusion is regular, but less frequent than at West End. The colony is about 800 m long and up to 150 m wide and lies between two dunes both artificially stabilized with snow fences. The colony is bordered on the north by a nearly continuous row of bayberry and black pine; otherwise most of the vegetation is natural. The construction of a sewage outfall pipeline through the center of the colony in 1977–78 has left a scar, but has not permanently affected the colony. In 1984 the foredune along the eastern half of the colony was breached by winter storms, increasing public access to the colony. In 1985 most of the colony was fenced, thereby restricting access. In 1988 beach contamination resulted in closure of the beach several times during the season and discouraged recreational use in general, thereby nearly eliminating unwarranted human intrusion.

In the mid-1980s the Cedar Beach common tern population exceeded 6,000 pairs of common terns. Up to 200 pairs of black skimmers, 150 pairs of roseate terns, 75 pairs of least terns and several pair of piping plover nest there as well.

Holgate: This is a sandy beach site (5 ha) just north of Beach Haven Inlet, with scattered low dunes and 10 to 15 percent beach grass cover. Although flat, the colony is infrequently flooded by high tides. The breeding population ranges from 100 to 200 pairs of common terns, 100 to 400 pairs of skimmers, and up to 350 pairs of least terns. The main vegetation is beach grass and seaside goldenrod. Since 1986 the beach has been closed to people during the tern breeding season.

Salt March Colonies

Salt marsh colonies are listed in Appendix C along with their characteristics. A typical salt marsh colony may be approximately circular, and most range in diameter from 50 to 300 m, although the size range was 0.4 to 45 ha. The 34 salt marsh colonies in Barnegat Bay (fig. 2.4) extend from the Lavallette group in the north (39 57′ N) to Hester Sedge and Tow (39 31′ N) (approximately 55 km), and—with Holgate and Pelican—account for all of the tern colonies in Ocean County, New Jersey.

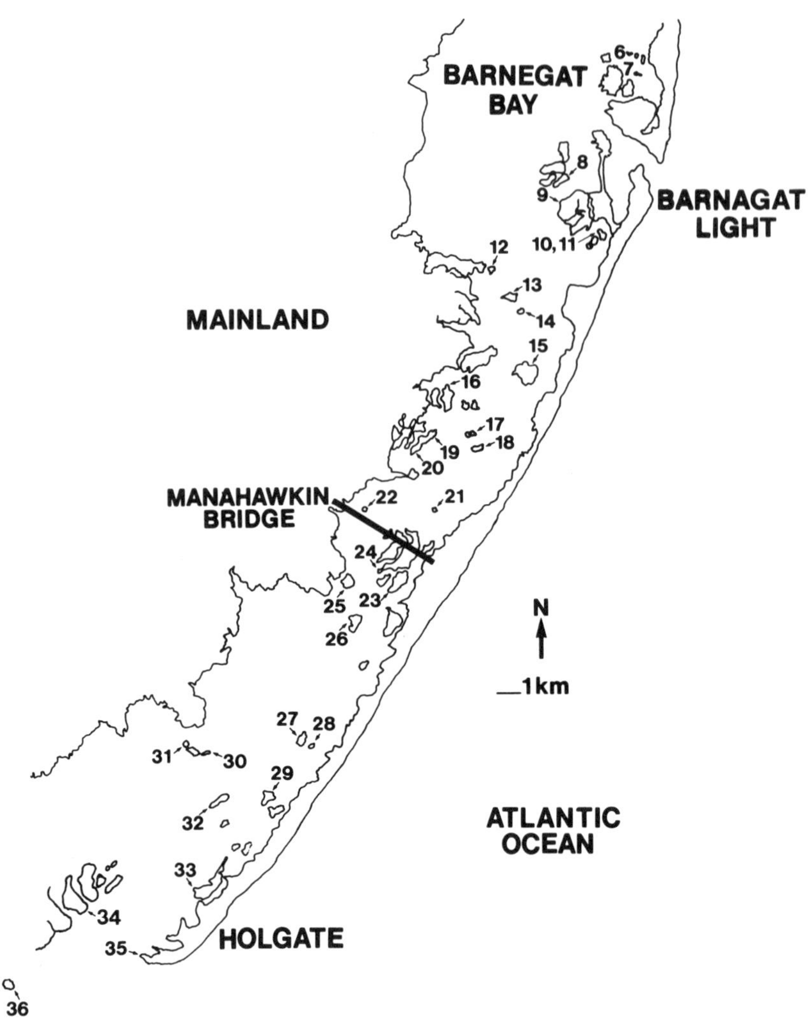

Figure 2.4. Map of Barnegat Bay, New Jersey, colonies. Note that the Lavallette colonies (numbers 1 to 5) in the extreme north are not shown. The bold line indicates the Manahawkin Bridge.

Tow Island lies just to the south of Beach Haven Inlet in Atlantic County. It provides both sand and marsh on which terns and skimmers often nest. On most islands the predominant vegetation is salt marsh cordgrass and salt hay, with stands of common reed and bushes on slightly higher ground. The reeds and bushes make up as much as 15 to 20 percent of the habitat on 11 of the islands (appendix C). Herring gulls nest among or adjacent to the terns and skimmers on 19 of the islands (appendix C). Laughing gulls nest or have nested on Clam, High Bar, East Carvel, East Vol, Thorofare, Egg, West Ham, and the Long Point islands. Barnegat Bay has only two inlets (one in the south and one in the middle). Thus, most areas of the bay have a very small daily tide swing (usually less than 0.5 m).

2.2 COLONY SURVEY AND CENSUS METHODS

We located and censused colonies by car, boat, or helicopter (Buckley et al. 1978). The frequency of comprehensive surveys varied from every 3 to 8 days to twice per month, depending on the year and location. Salt marsh colonies were also censused immediately after severe high tides, in order to determine flood damage. Ground surveys consisted of surveying all available habitat and sites known to be used previously (Burger and Lesser 1978, 1979). This involved driving along the barrier beach (New York) and taking boats from island to island (New York and New Jersey). Once colonies were located, we observed from the edge, and then entered to estimate population size and phenology by making partial or complete nest counts. In some years, aerial surveys by helicopter were made throughout the season in New Jersey (Burger 1978) and at the peak of nesting (early June, Buckley et al. 1978) and midsummer in New York.

Aerial surveys allowed estimates of available habitat, and assured location of new and small colonies. Without landing, we were able to census terns with some practice. We could land and ground-truth our aerial estimates, as well as assess the stage of the breeding effort. Determination of causes of nest-failure and productivity required ground surveys.

2.3 COLONY AND NEST SITE SELECTION

We examined colony selection by describing and quantifying the physical characteristics of occupied sites and comparing these with the characteristics of unoccupied sites which represent the available but unused habitat (Burger and Gochfeld 1986). We contrasted colony site selection on beaches and in the salt marshes. We studied nest site selection by comparing occupied nest

Figure 2.5. Researchers Michael Gochfeld (right) and David Gochfeld (left), taking data on nest site characteristics and contents, are being dive-bombed by common terns.

sites with unoccupied sites chosen at random (fig. 2.5). We located random points using random number tables to determine the rectangular coordinates of points within a colony or quadrat. The number of random points in each area equalled the number of nests studied.

For each colony, we computed the mean clutch initiation date for all nests and again for all initial nests (minus late outliers). Taking May 1st as day 1, we estimated the date of clutch initiation for each nest based on dates at which the 1st, 2d, and 3rd eggs were found or extrapolating backwards from the date on which hatching occurred. We then computed the mean date and standard deviation of clutch initiation. We used the standard deviation of laying as the measure of colony synchrony (Emlen and Demong 1975; Gochfeld 1980b). A low standard deviation (SD) indicates high synchrony.

2.4 NEST CHECKS

We performed two types of nest site checks: routine nest checks and in-depth nest checks. Routine nest checks involved walking though the colony and recording the following information for all nests: contents, status (evidence of flooding, predation, or abandonment), and chicks present (alive, or dead from starvation, exposure, predation, flooding, or adult aggression). These nest checks were designed to provide data on colony size, clutch size, and fledging success for each colony.

In-depth nest checks involved marking each nest individually with a numbered tongue-depressor (dry land colony) or flag (marsh colonies), banding all chicks with U.S. Fish and Wildlife numbered metal bands, and following on a regular basis the fate of all eggs and chicks. Adults were trapped at the nest using Potter-type treadle traps (hardware cloth boxes, 23 × 25 cm × 23 cm high; Post and Gochfeld 1979). Adults were not overly disrupted by this procedure since after being examined, banded, weighed, and measured, they were quickly released, and usually returned to incubate within minutes (fig. 2.6). Moreover, as the terns became accustomed to our visits they attacked more vigorously encouraging some of us to use helmets as protection from their dives and strikes, and we often needed scarfs to protect against their well-aimed defecations.

In salt marsh colonies (usually no larger than 500 pairs), it was almost always possible to perform a complete nest check on every visit. The large size of dry land colonies (1000 to 6000 pairs) precluded complete nest checks. The New York colonies were divided into a grid of 20 m × 20 m quadrats, and randomly selected study quadrats were monitored throughout the season. These were delineated by permanent, numbered, wooden posts which had to be sufficiently inconspicuous to avoid detection by picnickers searching for firewood. The rigors of the maritime climate required refurbishing the quadrat markers at least every other year.

We assessed the condition of nests in a variety of ways. Eggs that were flooded by high tides had a characteristic chalky white film which they retained for about five days. Abandoned eggs were cold to the touch and, after a few days, lost their sheen, and became paler. When lifted, they left a characteristic indentation in the sand which does not occur with eggs that are being turned and rearranged during normal incubation. We tested whether eggs were abandoned by standing them vertically in the nest or by separating the eggs in a clutch by about 3 cm. Incubating birds quickly gathered the

Figure 2.6. Incubating terns (top left) were trapped by placing a treadle trap over their nest. Birds readily entered traps and attempted to incubate. Joanna Burger (bottom left) removes a trapped tern. Michael Gochfeld (top right) examines a captured common tern. Burger, Gochfeld, and Carl Safina (bottom right) process an adult tern. This involved examining birds for oiling or injuries, and measuring, weighing, and banding the birds.

eggs together into a clutch or rolled the vertical eggs onto their sides. Eggs remaining in the displaced position for 24 hours were judged abandoned.

Eggs with slits characterized predation by American oystercatchers whereas ruddy turnstones, red-winged blackbirds or common grackles left small peck-holes in the eggs, and squirrels left a jagged edge (see fig. 6.1). Fish crows and gulls usually carried the eggs away. If, on our in-depth nest checks we found that eggs were missing, we considered this evidence of predation, unless there was evidence of alternative disturbance, such as signs of human intrusion.

We could often ascertain the cause of chick death. Starved chicks were underweight for their age or size. Drowned chicks were usually wet and bedraggled, and were sprawled at a distance from the nest, often in the high tide line with scattered eggs. Bedraggled chicks found huddled in vegetation or dead on nests were considered to have died of exposure, particularly from cold rains. Death from territorial adults on adjacent territories were characterized by peck marks on the head and back. It was more difficult to ascertain the fate of "missing" chicks than of missing eggs. Some of the former were certainly victims of predation, but others may have simply moved to cover and escaped detection or may have run and eluded capture.

2.5 DETERMINATION OF REPRODUCTIVE SUCCESS

Reproductive success or productivity was determined from 1976 to 1988 for the 33 New Jersey salt marsh islands and for selected years for the two New York dry land colonies. In the New Jersey colonies, productivity could be determined in each colony; but in New York, success was determined for sample plots. Individual nests were fenced so that young birds could be followed to fledging (fig. 2.7). For all years, we determined the mean number of young fledged per pair as the overall measure of success. Fledging refers to flying young although some workers use the number of birds reaching 10–14 days of age as a reasonable estimate of fledging. We estimated the number of pairs present in the colony, sometimes by taking the maximum number of nests present at any one time.

Figure 2.7. A random sample of individual nests was fenced so that young terns could be located until they fledged. Note trapped adult tern in foreground.

2.6 BEHAVIORAL OBSERVATIONS

We made prolonged behavioral observations from blinds (fig. 2.8) to collect data on courtship behavior, territoriality, social facilitation and information transfer, parental behavior, predation and antipredator behavior, and activity patterns (both diurnal and nocturnal, Palmer 1941a; Ewald et al 1980; Burger 1984c). Canvas blinds were used at both salt marsh and dry land colonies, while at the West End colony it was often most convenient to use a parked car as a blind since the birds nested close to a beach-access road (fig. 2.2). Observations were made for two to five days per week from early May until late July for periods of two to twelve hours per day.

Territorial and Courtship Behavior

Early in the birds' pre-nesting phase, we observed from a blind located on the edge of colonies to determine where the first birds established territories

Figure 2.8. Field assistants (David and Debbie Gochfeld) in front of a blind used for behavioral observations. Helmets and neck coverings provide protection from the diving birds.

and the pattern of initial settlement. We usually watched from 06:00 to 10:00 and from 15:00 to 20:00. We used outline maps of each study area and recorded the number and location of all birds that landed, interbird distances, and aggressive interactions. Interbird distances were estimated by the number of body lengths (one tern body length = 35 cm).

We counted the frequency and type of territorial activities in a given time period. We used time blocks as short as 5 sec. (social facilitation of courtship) or as long as 15 min. (parental investment). At the beginning of each observation period, and at 15-min. intervals thereafter, we recorded general data such as time of day, weather variables (temperature, wind velocity and direction, cloud cover, precipitation), and number of birds present (in singles and pairs).

Sex Differences in Behavior

Unlike the sexually dimorphic black skimmers, common terns are virtually monomorphic in size. We could only sex members of a pair by observing courtship feeding (only males bring fish to feed their mates; Nisbet 1973) and copulation behavior of color-marked individuals. Adults were trapped at the nest (fig. 2.6), banded, and marked with a felt marker which lasted for about two to three weeks, allowing us to distinguish the members of a pair.

Social Facilitation

We examined social facilitation (Darling 1938; Burger 1979b; Gochfeld 1980b) directly by recording the frequency of particular courtship displays among neighboring pairs of birds. We usually watched groups of birds of varying size and density and recorded the number of events occurring in time blocks that ranged from 5 sec. to 15 min.

Social facilitation of courtship is most dramatic during the pre–egg-laying period, and we gathered data on it during May and early June. We recorded the date, time, weather conditions, species present, numbers of pairs and single birds present in 15-min. intervals. we then recorded such courtship behaviors as landing, takeoff, and aerial displays, courtship feeding, circling, mounting, copulation, and nest scraping. Where feasible, we counted the numbers of birds present each minute. Early in the season this often changed dramatically and contributed to both the interest and complexity of our study of social interactions.

Parental Behavior

When we used blinds we had an assistant walk us to the blind and then depart, thus encouraging the terns to settle more quickly (fig. 2.8). Observations on parental behavior were usually conducted in four- to twelve-hour time blocks, and were made at all times of day, and occasionally at night. We recorded time, species composition and weather variables, as well as which parents were present and which were brooding or incubating. We noted the time of feeding of chicks, aggression against intruders, and distance of chick(s) from nests. We nest trapped some terns and marked them for individual identification (fig. 2.6).

Activity Patterns

Common terns are generally diurnal. In 1982 we studied the terns from 5:00 a.m. until midnight to gather information on daily cycles of behavior.

During the day we conducted continuous watches for four- to six-hour time blocks, recording behavior in 1-min. samples. At night we used an image-intensifying "starlight" scope (4×, Smith and Wesson), and watched birds for 30-min. intervals. This is a time-consuming activity, since the field of view is necessarily very restricted and a second observer is required to record data on data sheets illuminated by a small pocket flashlight. Moreover, constant viewing through the scope proved very fatiguing.

Data were recorded directly on computer sheets for subsequent analysis. The data were used to determine general activity patterns, frequency of events at different densities or stages of the cycle, and as a function of time of day. In addition, specific aggressive and courtship activities were analyzed to detect temporal clustering of events.

Piracy

A common cost of coloniality is the loss of food through piracy or kleptoparasitism, either by conspecifics or other species. We examined this aspect of behavior in common terns by observing subsections of the colony and recording data on fish delivery, attempted and successful piracy, as well as fish size estimated with reference to the bill length. Although many studies focus on strategies of pirates, we were most interested in the strategies of potential victims.

Predators and Antipredator Behavior

Throughout the breeding season we recorded data on the presence and activity of potential predators near or in the colony, and we examined the responses of the terns (fig. 2.9). Predators included hawks, gulls, crows, oystercatchers, blackbirds, cats, dogs, and humans. These data were collected while studying other aspects of breeding behavior as well as systematically. For each time block, we noted whether any other bird, mammal, or aircraft appeared in or over the colony. We recorded the following information: species, age class (for gulls), number, distance from colony edge, height above ground, number of terns responding, type of response (escape, mobbing, or attack), duration of response, and duration of absence from territory. Responses ranged from no response to head up (alert), to standing, to brief up flights or panic flights, to escape or attack. We distinguish "mobbing," a group of birds circling noisily and threateningly overhead, from "attack,"

Figure 2.9. Joanna Burger observing piracy attempts on male common terns returning to courtship feed females.

where the birds swoop close to an intruder and may actually strike it. The two grade into one another and, in the literature, both are subsumed under the term mobbing. We found it useful to make the distinction. During nest checks, we could count the number of times we were struck on the head as a measure of the attack tendency of a group. In addition, we employed selected experimental methods to further examine predation and antipredator behavior.

Since vigilance time is a cost all birds must bear to assure adequate detection of approaching predators, we also noted the behavior of the colony in general: whole colony takes flight, or small percentage takes flight. We also examined individual birds, noting what proportion of the time was spent being alert. Alertness was devoted to territorial neighbors as well as approaching predators, and these could often not be distinguished. We also noted whether the whole colony flew up, or only those individuals whose nests were immediately threatened by an intruder.

2.7 EXPERIMENTAL PROCEDURES

Many aspects of breeding behavior are best examined by combining observation and experiment. Descriptive observations can be systematized. One's design can include making observations at different times of the day, in different habitats, at different seasons, and in areas of different species composition and density.

Although observations allow for examination of many aspects of coloniality, we found it necessary to develop experimental or manipulative methods to study other aspects. We studied antipredator behavior experimentally, employing two methods: 1) we introduced potential predators to test the responses of birds, and 2) we used a human as an intruder to examine variation in mobbing response.

Mobbing Behavior Experiments

We performed two experiments on the mobbing behavior of common terns. In the Cedar Beach colony, we designed experiments to measure mobbing as a function of the intruders' location (center versus edge), nest density, and date. We used live predators as well as human intruders. In salt marsh colonies, we examined mobbing as a function of colony size and stage of breeding cycle.

In salt marsh colonies in 1979, we examined mobbing behavior as a function of colony size. From 1 May to 30 July 1979, we visited 30 colonies (ranging from 6 to 500 pairs) every 4 to 5 days. We approached each colony by boat, and one person walked slowly from the edge of the island into the center of the colony. We took care not to wave arms or deliberately frighten the birds, and we wore the same clothing each time. The second observer remained in the boat recording data in 1-min. samples. We were able to obtain ratios of the number of birds striking and diving to the number of birds overhead. We used the data to estimate the costs of mobbing to an individual nesting in the colony. We related all of our measures to colony size, density, and stage of cycle.

2.8 DATA ANALYSIS

Descriptive statistics will usually be presented as the mean ± one standard deviation (±1 SD), unless otherwise specified. Many of the biological phenomena we investigated are not normally distributed and do not lend themselves to parametric analysis. We used a variety of nonparametric statistical procedures as appropriate (Siegel 1956). We used chi square tests on contingency tables and to estimate goodness of fit (Sokal and Rohlf 1982). For correlational analysis, we preferred the Kendall tau rank correlation method. When data transformations were appropriate, yielding approximately normal distributions, we used *t* tests to determine differences between samples and occasionally a Pearson product moment correlation. A variety of analysis of variance methods were employed. Most of our procedures employed the Statistical Analysis System (SAS 1982; Cody and Smith 1985) although we wrote some programs for unique applications.

For multivariate analysis we initially explored our independent variables using a stepwise multiple regression, and then used those variables that entered significantly in a general linear models procedure which gives a more valid estimation of the coefficients and significance level for each variable. We usually present the actual probability of *P* value associated with our analyses, since in evaluating the robustness of tests a reader will be better persuaded by a $P = .004$ than a $P = .04$. Recognizing that much of our data does not fit a normal or log-normal distribution, we show a strong preference for nonparametric statistics.

THREE

Breeding Biology, Phenology, Activity Patterns, and Synchrony

3.1 GENERAL BREEDING BIOLOGY

There are many aspects of common tern breeding biology and behavior which have been described in substantial detail by earlier authors (Jones 1906; Bent 1921; Marples and Marples 1934; Southern 1938; Palmer 1941a; Austin and Austin 1956; Roselaar 1985). Some of the major longterm studies include Marples and Marples (1934), the Austins (1929, 1932, 1949, 1956), Nisbet (1978a), Nisbet et al. (1984), Kaverkina (1986a), and Dicostanzo (1980). Our study did not attempt to repeat the excellent observations of many of these authors.

The breeding season for common terns along the New York–New Jersey coasts lasts from late April into September. During this time, the birds arrive from their winter quarters, settle on colony sites, establish territories, mate, lay eggs, incubate, and care for chicks. Unless disturbed, terns will rear their young within the nesting territory, and even after fledging the young return to the territory for up to several weeks.

Incubation requires three to four weeks depending on the degree of disturbance, and the chicks first fly at about four weeks of age. Most common terns do not begin breeding until they are three years of age (Austin and Austin 1956), although a few return in their third year (at age 2).

Migration and Arrival

Through extensive banding studies much is known about the migration and wintering quarters of the common terns that breed in the northeastern United States. There are few United States recovery records during the winter, from which we surmise that they migrate mainly offshore to their winter quarters in the Caribbean and South America. Common terns return to New York and New Jersey usually during the second or third week of April when, on warm afternoons, their cries can be heard faintly as they circle high over the colonies. In the following week, they are seen in increasing numbers at the colony itself. During these early days, they spend much of their time in aerial courtship and are easily disturbed. Palmer (1941a) has described this phase in detail.

The propensity of residents of Guyana to capture terns for food (Trull 1983) has resulted in a disproportionate number of banding returns from northeastern South America. Large flocks of terns have indeed been seen in the southern Caribbean during migration and smaller flocks in winter (Blokpoel et al. 1982, 1984). The birds apparently winter mainly from northern South America to central Brazil, although we have seen flocks of hundreds of common terns on the coast of southern Argentina (Chubut Province). Winter mortality (Blokpoel et al 1982, 1984, Trull 1983) may be a critical factor in the population dynamics of this species.

Preincubation, Courtship, and Mating

During the preincubation and courtship phase, male terns arrive at the colony and establish territories, display with females (figs. 3.1, 3.2, and 3.3), and engage in courtship feeding (Palmer 1941a; Brown 1967; Nisbet 1973a; Lobkov and Golovina 1978; Wiggins and Morris 1988). Terns disperse throughout the available habitat and select and defend territories. The females spend much time on the territory while males are away foraging. Males return with fish, calling as they arrive, and feed the fish to the female; often a fish flight display ensues (Palmer 1941a). Males soon leave to find another fish (Nisbet 1973a, fig. 3.1). Although in some larids courtship feeding immediately induces a copulatory response from the female, this appears not to be the case for common terns (our observations; Palmer 1941a; Wiggins and Morris 1988). Nisbet (1973a, 1978b) pointed out that courtship feeding, traditionally viewed as a pair-bonding behavior, conveys the impor-

Figure 3.1. Pair-bonding displays: (top) Nest-showing display. Male elevates body and engages in vigorous sand-kicking to produce scrape. (bottom) Courtship feeding: female crouches and begs from male who will usually leave to procure a fish. See also figure 3.11.

Figure 3.2. Precopulatory displays: (top) Bird on right is in full "bent" posture, bird on left is in full upright (also used in aggression). These often progress to the Upright parade (bottom) in which male and female circle with bodies parallel and head and necks stretched upward. Female is just beginning to lower head.

Figure 3.3. (top) Female begins to crouch and male assumes a bent position prior to mounting. (bottom) Copulatory stance immediately prior to copulation. Note female elevating her tail to facilitate cloacal contact, male wing flags vigorously (facing page top) accompanied by cloacal contact (facing page bottom).

tant message of male competence which the female uses to assess the suitability of her mate. Smith (1980) suggests that courtship feeding is a primary way that a male can contribute directly to the quality of his offspring through higher quality eggs.

Females, as well as males guard the territory and chase intruders that land. Occasionally unmated males with fish land in search of females and these are vigorously chased by females or by the males (see chapter 5). As the pre-laying period advances, courtship feeding intensifies, continues during egg-laying, and decreases thereafter (Nisbet 1973a).

Among many papers on tern behavior Marples and Marples (1934), Southern (1938) and Palmer (1941a) describe courtship displays and mating in detail (see also Roselaar 1985 for recent summary). The premating display involves ritualized postures and circling (fig 3.2), and these displays are often synchronized among neighboring pairs (Gochfeld 1980b). At first, males mount and stand for a minute or two before stepping or being shaken off. Later mounting lasts longer (fig. 3.3) and assuming that the pair is not interrupted by neighbors or intruders, the mount usually culminates in copulation, accompanied by conspicuous wing-flagging.

Egg–Laying

The egg-laying period for any pair of terns lasts only about four days. Eggs may be laid on successive days or on alternate days. During this time the territorial boundaries are solidified, courtship feeding activity is at its peak, and birds mate frequently. Nisbet and Cohen (1975) found an average of 3.6 days between laying of the first and third eggs in a three-egg clutch.

After the first egg is laid, the birds may or may not incubate. One bird—usually the female—attends the nest, usually standing beside it or shading the egg, but incubation is sporadic at best. Judging from the difficulty of nest-trapping individuals with one egg, it is clear that their motivation to incubate is low. When the second egg has been laid the incubation tendency becomes stronger, and most pairs initiate incubation, but the constancy of incubation is not established until the third egg is laid. Thus, although the eggs hatch asynchronously (Nisbet and Cohen 1975), the hatching interval is shorter than the laying interval.

In our study area, the typical common tern clutch is three eggs. Variation in clutch size will be discussed below. Nests may be sparse collections of shells or debris (Marples and Marples 1934), or may be well-formed from dead vegetation (fig. 3.4 top). Nests on salt marsh mat are often built up of vegetation (fig. 3.4 bottom). Many observers have been impressed by the intra- and inter-clutch variability in the size, shape, and coloration of common tern eggs, and there have been several extensive biometrical studies of tern eggs (Rowan et al. 1917; Gemperle and Preston 1955; Gochfeld 1977a).

Early biologists suspected that terns might raise two broods in a season, but this belief was denigrated until very recently. Hays (1984) and Wiggins et al. (1984), however, documented cases of common terns raising young from a first brood and, at the same time, initiating a second brood. Such events must be very infrequent and are probably rarely successful.

Incubation

There have been numerous studies of the behavior and physiology of birds, particularly larids, during the incubation phase. Studies by Tinbergen (1959, 1953, 1967), Baerends and Drent (1970), Drent (1967) and Beer (1961, 1962) have documented nest-attentiveness and mate roles. For common terns the incubation period has been reported as 23–28 days (Austin 1933).

Austin (1932) found a modal incubation period of 26 days in a partially disturbed colony. Nisbet and Cohen (1975) found a range of 21–22 days in undisturbed nests and 27–28 days in areas with disturbance. Nisbet and Welton (1984) found that the incubation period at the Monomoy, Massachusetts, common tern colony varied from 22 days in years with little or no nocturnal predation to 29 days in years with heavy owl predation. Our modal values for different colonies ranged from 22 to 25 days, with a maximum of 28 days at West End Beach in a year (1971) with heavy nocturnal predation. Both male and female terns incubate in shifts of three to five hours, with the female performing slightly more than half of the incubation. During the day, while one member of the pair incubates, the other is off foraging. At night, the nonincubating tern is usually present at the nest site.

Chick Stage

Adult terns attend and guard their chicks from hatching to fledging, although usually there is only one adult at the nest at any one time. Aggression increases at hatching when young chicks presumably become mobile and can easily be killed by conspecifics or eaten by predators. Some adult common terns vigorously attack and kill unprotected chicks that approach their

Figure 3.4. (top) A well-formed common tern nest with a complete clutch of three eggs in a beach colony. The nest is formed mainly during nest-exchanges, when one adult collects bits of shell or vegetation and tosses it toward the nest. A day-old tern chick and newly hatched chick are shown on their nest (bottom). The open beak represents gular fluttering, a means of thermoregulating. Well-formed nests are more characteristic of salt marsh tern nests (facing).

territory (Austin 1932). Skimmers may kill and even eat tern chicks that wander into their territories. Zubakin (1973) reported cannibalism for gull-billed terns, but it is not known to occur in common terns. Both sexes bring back food to feed the young, although males may do more feeding of very young chicks. Although chicks remain in their nests for the first day or two of life (fig. 3.4 bottom), young chicks often hide in the vegetation when not being fed, while chicks older than 10 days wander about their territories more. In the absence of human disturbance they remain on their territories until after fledging. The mobile chicks of gulls and terns can run from their nest or territory at an early age. Increasing interest has focused on the problem of adoption, for chicks that are getting meager care at home (the youngest of a brood), may benefit from imposing on a neighboring nest where it may suddenly become the oldest and largest chick (Graves and Whitten 1980; Pierotti and Murphy 1987). Adults faced with the raising of foster young could be aggressive in driving away such chicks, although Carter and Spear (1986) argue that occasionally raising a foster chick does not necessarily harm the adult's fitness in the long run.

Fledging and Post-Fledging

Fledging takes place at 27 to 30 days of age, by which time the wings are long and strong enough for flight. For several days the chicks can be seen jumping up and down and flapping vigorously, and at some point when the wind is favorable they suddenly become airborne. Often, when we have chased older chicks to capture them for banding or when we have released them, we have helped them discover their flight capabilities.

At fledging, young terns are suddenly liberated from some of the most serious hazards they have faced, i.e., flooding and predation. Yet they face a new hazard, for although capable of taking off and landing, the young frequently crash-land in undesirable locations, such as the territories of neighboring birds or near loafing gulls on the beach. Such risks are reduced in colonies that are not disturbed.

Even flying chicks tend to return to their territories for several days before moving to the shoreline (Nisbet 1976). In midsummer, we have seen them gathering on open areas (beaches, parking lots, large expanses of mat). Here they are attended and fed by parents, and at this stage it is easy to estimate the colony's productivity, particularly because adults and young are together and conspicuous. After about two weeks, however, chicks may start to disperse (Nisbet 1976), and within a few weeks, flocks with young from several colonies may aggregate at a single locality, and these flocks increase in size into September. By mid-September, most common terns have departed for their wintering grounds in the southern Caribbean and South America.

3.2 BREEDING PHENOLOGY OF TERNS

Salt Marsh Colonies

Most colonial seabirds migrate into the general area where they will breed days or weeks before they take up residence on their nesting colonies. They range over a wide area around the colony, gradually focusing more of their activity on a particular colony site, and spending more and more of their time there. We examined breeding phenology in detail in salt marsh colonies during three years, by censusing several available colony sites every two to four days throughout the breeding cycle (early April to early September).

The breeding phenology for several colonies is shown in figure 3.5.

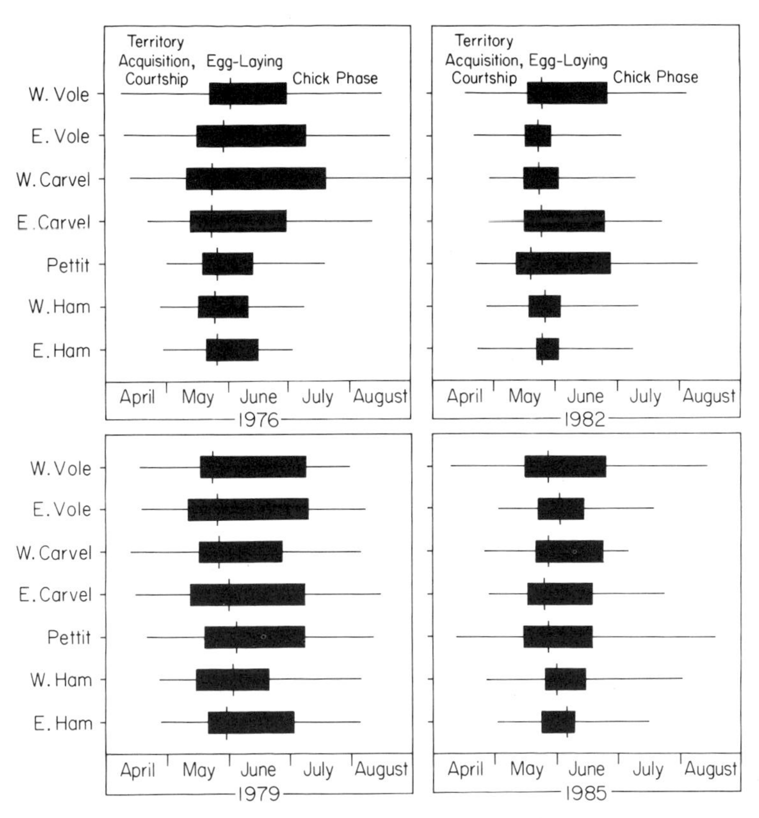

Figure 3.5. Phenology of breeding activities of common terns nesting on seven salt marsh islands in Barnegat Bay in four years. Shown are territory acquisition stage, egg-laying period (black bar), mean date of egg laying (vertical line), and duration of chick phase. Note greater synchrony of egg laying in 1982 and 1985, and peak falling toward the beginning of the egg-laying period in most cases.

Several conclusions can be drawn: 1) among years, the length of all phases (preincubation, egg-laying, chick phase) varied; 2) within years, the lengths of the phases varied among colonies; and 3) for any given colony, the length of the phases varied among years.

Among years, the earliest date of arrival of common terns on Barnegat Bay salt marsh islands ranged from April 5 to April 10, and the latest date of arrival on islands ranged from April 29 to May 1. Earliest egg-laying dates among years ranged from May 10 to May 24. Thus there is less variability

Table 3.1. Mean date of first egg-laying for common terns nesting in salt marshes.

	1976	*1977*	*1978*	*1979*	*1980*	*1981*	*1982*	*1983*	*1984*	*1985*
Buster			18 M		20 M					
Clam	9 M			12 M						
East Vol	16 M	15 M	18 M	10 M	13 M	14 M	16 M			20 M
West Vol	22 M	18 M	23 M	16 M	14 M	12 M	19 M	16 M	12 M	
Gulf Point				14 M						
Flat Creek			1 J							
West Carvel	10 M	12 M	20 M	14 M	15 M		15 M			18
East Carvel	11 M	13 M	12 M	10 M	11 M	10 M	15 M	10 M	12 M	14 M
W. Log Creek			19 M		15 M					
Log Creek			14 M	18 M			21 M			16 J
Pettit	19 M	16 M	18 M	16 M	12 M	13 M	12 M	9 M	11 M	12 M
Cedar Creek					17 M		21 M		17 M	
S.W. Cedar Bonnet	16 M	17 M	15 M	16 M	18 M	20 M	21 M	20 M		
Thorofare				16 M			18 M			
Egg						15 M				12 M
East Ham	19 M	18 M	18 M	16 M	19 M	13 M	21 M	19 M	18 M	22 M
West Ham	16 M		16 M	16 M		12 M	19 M		16 M	24 M
Marshelder				1 J						
Mordecai				20 M						26 M

Note: Mean date = arithmetic mean of all nests initiated during the first wave of nesting. M = May; J = June.

among years in date of first egg-laying than in date of arrival on the colony site.

Data on first egg-laying date are available for many more salt marsh colonies than are included in figure 3.5, and table 3.1 lists the mean date of clutch initiation (first wave nests only) for New Jersey salt marsh colonies arranged from north to south. In only a few cases were we unable to document clutch initiation. Several colonies such as East and West Vol, East Carvel, Pettit, Cedar Bonnet, and East Ham showed a very narrow spread of clutch initiation over the ten-year period with only a six-day span for East Carvel. Other colonies, particularly those that tended to be later, were more variable. There were some years when we could not ascertain the date of clutch initiation. Table 3.1 shows some interesting patterns. For example, East and West Vol, although only 100 m apart, showed a consistent relationship with West Vol lagging 1–6 days behind East Vol in 6 out of 7 years when both were occupied. There is no clear interyear pattern of which colonies had the earliest eggs. In some years northern colonies were earlier (e.g., 1976), in most years central colonies were earlier (e.g., 1977, 1980), and generally the southern colonies (Cedar Bonnet to Mordecai) were later. This trend, however, also reflects colony size distribution since southern colonies averaged fewer pairs of terns.

Among years the total maximum period of egg-laying (including re-laying) usually spanned 45 to 70 days, and the shortest egg-laying periods spanned 15 to 40 days. Taking the first egg-laying date for each of the colonies, we found that among years the mean date of clutch initiation occurred from May 22 to 28, a remarkably short span. For any year, the mean date of egg-laying among years was always closer to the beginning of egg-laying than to the end. That is, the egg-laying period is skewed and extends well beyond the peak of egg-laying; most birds lay early, but a small proportion breed later.

The date of completion of the chick phase varied substantially among years (August 6 in 1979, August 31 1976). This phase is affected by the timing of all other phases and reflects the sum total of adverse factors (floods, predators, food shortages) during the entire season.

Within-year variations among colonies are interesting because the birds are all exposed to the same weather variables (wind, temperature, precipitation, cloud cover), food resources and, to an extent tidal regimes, although island height, mat height, and island location influence tidal effects. In 1976, the length of the egg-laying period varied among colonies, with the northern colonies having longer periods than the southern colonies. There was a concommitant difference in the length of the chick phase.

In 1979, there was much less variability among colonies with respect to the length of the phases. In 1982, the egg-laying periods were very variable among colonies, and were generally terminated by floods (see below). In some colonies, some terns renested after the early June floods. In 1985, the length of the egg-laying period was variable among colonies, and the resultant chick phases were very variable in length.

Comparisons Among Years and Colonies in Salt Marshes

In salt marshes the phenologies of terns differed dramatically with respect to several factors: 1) within and among years there was variability in phenology, 2) terns rarely abandoned a colony site completely, 3) terns were not persistent in re-laying after loss of nests. Some of these differences are caused by differential colony size. Some tern colonies had up to 500 pairs of breeding terns, whereas others were under 50 pairs. Larger colonies tended to have longer laying periods. Moreover, the social stimulation to be derived from conspecifics varied markedly with colony size and density.

Beach Colonies

On the beaches the overall phenology of common terns was similar to the phenology on the nearby salt marshes, but with some consistent differences. Common terns arrived at the Long Island colonies in late April. They were first heard flying high above the colonies, and settled on the nearby beaches after dark. Within 3 to 4 days, the first birds were seen on the colony site and shortly the males began to defend territories. They were joined almost immediately by females. Egg-laying has been initiated between May 8 and 26. Initiation has occurred much earlier in the 1975–1985 period than at the start of the study (Mann Whitney $U = 1$, $P < 0.001$; table 3.2), but was delayed to late May in 1990. Prior to 1970 only a few nests were initiated before May 31. In 1969, for example, 2 nests were found on May 26, whereas in 1984 and 1985 clutches were initiated on May 8 and 10, and 85 percent of first wave nests were initiated before May 31.

Because the beach tern colonies we studied were so large (usually larger than the largest marsh colonies), there was continual recruitment to the colony, and the egg-laying period was extended to mid-June in all years and to early July in some years, as birds that failed in early attempts renested. Late waves of nesting occurred in some years, with new clutches being found up to August 10. Most terns departed from the colony by late August, and only those feeding injured young remained until mid-September, at which time chicks that were unable to fly were abandoned (Gochfeld 1971, 1973a).

3.3 BREEDING PHENOLOGIES OF ASSOCIATED SPECIES

Several other species nest on sandy beaches or in salt marshes with the common terns including black skimmers, roseate terns, least terns, laughing gulls, herring gulls, willets, and oystercatchers. In general, gulls and oystercatchers nest earlier than the terns, while skimmers and piping plover nest at about the same time. Since some of these species potentially compete with

Table 3.2. Earliest egg-laying dates at Cedar Beach and West End Beach tern colonies. Dates are presented as days after May 1. For both colonies, egg-laying was earlier in 1975–1988 than in 1970–1974 (Mann-Whitney U test, $p<0.001$).

Year	*Cedar Beach*	*West End*
1970	—[a]	26
1971	—[a]	19
1972	18	22
1973	16	21
1974	16	18
1975	12	14
1976	13	11
1977	7	8
1978	12	13
1979	9	10
1980	8	12
1981	11	13
1982	12	12
1983	11	13
1984	10	12
1985	12	12
1986	13	19
1987	19	—[a]
1988	22	18

[a] No early nesting in this colony.

terns for nesting space (gulls), or prey upon their eggs and chicks (gulls, oystercatchers) we describe them briefly below.

Salt Marshes

Since flooding is a major threat to a salt marsh colony, there is a premium on nesting in the higher and dryer parts of the marsh, areas usually dominated by bushes rather than by *Spartina*. Gulls also prefer these sites and, since they start nesting in early April before the terns have arrived at the colony, they can completely exclude common terns from the higher habitat. Terns frequently seek the mat or wrack, piled high on the marsh, as a suitable breeding substrate. Although herring gulls do not nest on the mat, laughing gulls nest on or along the edge of mat.

Common terns generally arrive much later than the gulls or oystercatchers. Not only does this timing preclude their obtaining sites on the higher parts of the marsh, but the terns are nesting at a time when herring gull chicks are hatching. Hence, their eggs or young offer attractive prey to the gulls.

Beaches

In the Long Island beach colonies there is abundant nesting space, and the tern nests are not densely packed. The gull colonies are 1 to 8 km from the tern colonies, and human disturbance has kept gulls from becoming established in these tern colonies. However, in New England, the tendency for herring gulls to establish colonies at sites traditionally occupied by terns has had a great impact on tern populations (Crowell and Crowell 1946). Willets and piping plover nest within tern colonies, and gulls, oystercatchers and marsh hawks (Northern harriers) nest nearby. The gulls are present for two months before the terns and skimmers, and have chicks while the terns and skimmers are still laying eggs and incubating. At Cedar Beach, the harrier is a consistent predator that nests in the nearby salt marshes. Harriers also are feeding chicks when the tern chicks are vulnerable.

3.4 CLUTCH SIZE

Clutch size is an important component of fitness because it directly affects reproductive success in any breeding season, and ultimately the lifetime fitness of any individual. Several studies of gulls and terns have shown that the last egg of a clutch is smaller (Gochfeld 1977b) and its resulting chick is at a disadvantage and usually does not survive (e.g., Nisbet 1973a; Richter 1983; Pieroti and Bellrose 1986). Factors influencing clutch size can be divided into proximate and ultimate (Klomp 1970). Proximate factors are those internal or external factors controlling the number of eggs laid in any breeding season, and ultimate factors are the selective factors that influence the evolution of clutch size.

Lack (1968), Klomp (1970), Hussell (1972), and Murray (1979) review factors affecting clutch size, and these can be divided into geographic factors (latitude, longitude, altitude), demographic factors (adult age, survivorship), temporal factors (interyear and seasonal), and ecologic factors (population density, habitat, and food availability). Some of these effects may result from phenotypic modification, but others may be due to evolutionary, adaptive responses affecting genotypes. Lack's (1968) hypothesis that clutch size reflects the clutch that produces the most surviving young is still being debated and contradictory results from brood addition experiments do not clarify the matter (Ydenberg and Bertram 1989).

Lack's hypothesis can be contrasted with Moreau's (1944) hypothesis, further amplified by Murray (1985). This hypothesis states that individuals may produce more offspring in a season or on a lifetime basis, by reducing the number of chicks in a brood relative to the maximum number of young that could be reared under prevailing conditions. Murray (1985) uses this hypothesis to explain apparent paradoxes in clutch size, for example the lack of geographic variation in clutch size of some taxa (e.g., Procellariiformes) compared with others (e.g., Phasianidae). This is predicated on the hypothesis that birds which attempt to rear their theoretical maximum are at greater risk themselves, and hence may reduce rather than increase their fitness.

In this section, we discuss the clutch size of common terns in salt marsh and beach colonies and consider factors affecting clutch size. We compare our data with previous studies of common terns. This species has been extensively studied throughout temperate regions, and clutch size data are sometimes reported, although often only a single value or a range of mean values are given, without any indication of the actual distribution or variance of clutch size, or the extent of spatial or temporal variability.

Methods

From 1976 to 1985, we visited New Jersey salt marsh colonies every three to four days during the egg-laying period to determine the number of nests containing 1, 2, 3, or more eggs. We calculated clutch size of those nests initiated during the peak period of egg laying. Where possible we censused all nests, but in large colonies we sampled along transects through all habitats in the colony. We then examined differences in clutch size as a function of year, colony, and colony size.

On Long Island, most data were obtained for selected quadrats at the large West End and Cedar Beach colonies, with occasional sampling of other beach and salt marsh colonies during the peak of the breeding season. Data for West End and Cedar Beach were obtained in most years from 1970 to 1985. All nests in sample plots were marked and the contents recorded one or more times during the season.

Table 3.3. Mean clutch size by colony and year for New Jersey common tern colonies (salt marsh). Sample sizes are equal to number of nests in colony, and are given in table 9.7.

Colony	*1976*	*1977*	*1978*	*1979*	*1980*
Buster	2.84	2.00	2.21	1.86	
Cedar Creek	2.84	1.80		2.17	2.61
Clam (small)	2.78				
E. Carvel	2.66	2.09	1.85	2.52	2.08
E. Cedar Bonnet	2.11				
E. Long Point	2.66				
E. Sloop Sedge	2.75				
E. Vol		2.00	2.69	2.59	
East Ham		1.80		2.48	2.13
Egg (small)	2.46				
Flat Creek		1.88	1.67		2.60
Gulf Point		2.29		2.24	
Hester Sedge	2.67	2.50			
High Bar (small)		2.13			
Little	2.95	1.73	2.31	1.89	2.32
Log Creek	2.68	1.50		2.00	2.37
Marshelder				2.50	
Mordecai		2.07		2.13	2.26
N. Lavallette	2.80	2.24	2.68	2.00	2.53
N.W. Lavellette	2.74	2.27	2.33	2.28	
Pettit	2.31	1.98			2.01
S. Lavallette	2.82	2.46	2.82	2.60	
S.W. Cedar Bonnet	1.75	1.68		2.46	
S.W. Lavalette	2.60	2.27	2.47	2.24	2.00
Thorofare		2.23		2.24	
W. Carvel	2.89	1.97	1.96	2.18	2.11
W. Log Creek	2.50	2.10			
W. Long Point	2.69				
W. Sloop Sedge	2.71				
W. Vol	2.55	2.24	2.41		1.89
West Ham	2.25	2.63		2.78	2.13

Clutch Size in New Jersey Salt Marsh Colonies

Mean clutch size for marsh colonies over a ten year period is shown in table 3.3, while the actual clutch size distributions for each colony appear in appendix D. We used a general linear models procedure to determine the contributions of several factors (e.g., year, colony, colony size) to variations

1981	1982	1983	1984	1985	1987	1988
					1.67	2.83
		2.76	2.35			
2.73	2.18	2.66	2.40			
2.66		2.60		2.44		
2.25		2.55		2.01		
		2.90		2.05		
	2.22	2.83				
					1.97	
2.22			1.91			
1.96	2.59	2.44	2.33			
1.73	2.33		2.21	2.24		
2.50	2.41	2.69	2.23	1.97		
2.38	2.75		2.29	2.31		
2.43						
2.52	2.44	2.45	2.35			
		2.54				
	2.42			2.56		
2.49	2.09	2.27	2.44	2.49		
2.58		2.57	2.25	2.07	2.10	

in clutch size. Overall, for the ten years of data, the best model explained 53 percent of the variability ($F = 2.70$, $P < 0.0001$, $df = 42, 143$) by year ($F = 5.85$, $P < .0001$), colony ($F = 1.52$, $P < 0.06$) and number of nests in the colony ($F = 3.15$, $P < 0.06$). We also computed the best model explaining variation in the percent of three egg clutches. Over 57 percent of the variability was explained ($F = 3.23$, $P < 0.0001$, $df = 42, 143$) by colony ($F = 1.73$, $P < 0.02$), year ($F = 7.20$, $P < 0.0001$) and number of nests in the colony ($F = 4.41$, $P < 0.03$).

Larger colonies generally had higher mean clutch sizes. For example, the Lavallette Islands consistently had more breeding pairs, and usually had a higher percentage of three-egg clutches, while traditionally small colonies such as Flat Creek and Log Creek had relatively low clutch sizes (table 3.3).

There were yearly differences in clutch size (table 3.4), which might reflect differences in food availability. Overall, clutch sizes were high for 1976, 1978, and 1983, and were low in 1977, 1980, and 1985. Similarly, in some years there were more colonies where three-egg clutches were most common (1976, 1980, 1983), and other years where two-egg clutches were most common (1977, 1984, 1985). Four-egg clutches never exceeded 0.4 percent, and were not found at all in some years. Overall, some colonies (e.g., Pettit) tended to have higher clutch size and others (e.g., Flat Creek, Log Creek, and Mordecai) had smaller clutch size regardless of year.

Clutch Sizes in New York Beach and Marsh Colonies

Table 3.5 summarizes nine years of clutch size data from West End and twelve years from Cedar Beach, supplemented by data from seven other beach and nine salt marsh colonies. In 1976 and 1978, we recorded clutch size at seven colonies on the south shore of Long Island. In 1976, clutch sizes ranged from 2.19 to 2.53 at five small and medium-sized colonies (table 3.6), while the large Cedar Beach colony had a clutch size of 2.09. In 1978, two salt marsh colonies (Seganus, South Line) had the lowest clutch sizes while the two large beach colonies (West End and Cedar Beach) were among the highest (table 3.5).

In the years when adequate clutch size data were obtained at both Cedar Beach and West End, there was a positive correlation (tau $= 0.62$, $P = 0.05$) in clutch size, both colonies showing high clutches in some years (e.g., 1974, 1975, 1977) and low clutches in others (1971, 1982). Overall the mean clutch size was 2.43 for Cedar Beach and 2.40 for West End. There was more correlation between the two large beach colonies than between beach and New Jersey marsh colonies, suggesting that food, as a factor influencing clutch size, varied differently in Long Island and New Jersey.

Table 3.4. Clutch sizes in New Jersey salt marsh common tern colonies. Shown are the data from colonies with clutch size data, total number of 1-, 2-, 3- and 4-egg clutches, percent of 3-egg clutches summed over all the colonies and number of colonies where there were more 2-egg clutches and where there were more 3-egg clutches.

					Number of Clutches				Number of Colonies	
Year	*Colonies (N)*	*Clutches (N)*	*Mean Clutch*	*% c/3*	*1-egg*	*2-egg*	*3-egg*	*4-egg*	*3-egg dominant*	*2-egg dominant*
1976	23	2,536	2.77	81%	111	359	2,063	3	22	0
1977	23	1,562	2.23	40%	256	685	621	0	6	17
1978	11	1,465	2.53	63%	146	400	917	2	6	5
1979	18	1,346	2.44	52%	116	525	705	0	9	8
1980	14	774	2.17	39%	173	296	304	1	9	5
1981	12	997	2.41	52%	118	355	522	2	8	4
1982	9	997	2.40	51%	113	365	495	4	6	3
1983	12	1,004	2.55	62%	67	314	623	0	12	0
1984	10	982	2.30	43%	130	427	425	0	4	5
1985	9	815	2.21	37%	134	374	305	2	3	6

Table 3.5. Mean clutch size in New York common tern colonies. Sample size is equal to number of nests in colony for all but West End and Cedar, where a sample of nests was examined in each year.

	1970	*1971*	*1972*	*1973*	*1974*	*1975*	*1976*	*1977*	*1978*	*1979*	*1980*	*1982*	*1983*	*1984*	*1985*
Primary Beach colonies															
West End	2.13	2.14	2.29		2.68	2.66		2.65	2.40			2.15	2.78		
Cedar Beach		2.08	2.53		2.51	2.72	2.09	2.69	2.62	2.51	2.69	2.15		2.55	2.11
Other Beach colonies															
Short Beach		2.17	2.23		2.57										
Meadowbrook			2.44												
Cartwright							2.53								
Shinnecock Bar							2.28								
Wantagh			2.39												
Bostwick									2.64						
Hicks Is									2.89						
Salt Marsh Colonies															
South Line									2.22						
North Line							2.19								
Shinnecock Island							2.41								
Seganus									2.10						
Lanes Island							2.45	2.55	2.54						
Loop						2.76									
Islet Z								2.30							
West Moriches						2.76	2.52	2.38							

Table 3.6. Clutch size of common terns on several islands on the South Shore of Long Island in 1976.

Clutch Size	*Cartwright Island (June 7)*	*Shinnecock Bar (June 7)*	*Lane Island (June 8)*	*Moriches, West Inlet Island (June 8)*	*South Line (June 9)*
Habitat	*Sand Bar*	*Sand Bar and Salt March*	*Mat and* S. Patens *Total*	*Sand*	*Spoil Island*
One egg	10 13%	6 16%	21 14%	16 10%	5 24%
Two eggs	16 21%	10 27%	44 29%	47 29%	7 33%
Three eggs	51 66%	21 57%	87 57%	101 61%	9 43%
Mean clutch	2.53	2.41	2.43	2.51	2.19

Discussion

Biases in Estimating Clutch Size

Before comparing our findings with previously published information on common terns, we point out several problems in estimating clutch size for this species. On a single visit to a colony some nests may have disappeared (through predation or hatching), while others are yet to be initiated. For those nests encountered, some clutches may be incomplete while others may have been reduced by partial predation or by hatching. To some extent, the impact of these phenomena can be estimated. Very early in the season, there is a high likelihood that clutches are incomplete and no likelihood that hatching has begun. Also, early in the season, the presence of empty nests (nest material present but no eggs) indicates that egg predation is probably occurring.

The best way to estimate clutch size is to mark nests individually and record the contents on visits over a period of at least ten days. During this period clutches will certainly be completed, and the tendency for clutches to be reduced (by predation) can be estimated. One can then make corrections and come up with a "true clutch size estimate." If only a single visit is possible, the ideal time is about three weeks after the earliest clutch initiation, which should cover the peak period when most nests in the colony will be active but hatching will not have begun.

Two important biases must be realized. Since some one- and two-egg clutches will increase to three-egg clutches, there is a general tendency to underestimate clutch size. Also some two-egg or one-egg clutches may have been three-egg clutches; hence, a three-egg clutch is almost certainly a true three-egg clutch, while a one- or two-egg clutch may not be. Thus, there is a strong likelihood that all clutch size distribution reports underestimate true clutch size.

Large Clutches

Supernormal clutches of eggs in gulls have attracted dramatic attention in recent years (Hunt and Hunt 1977; Ryder and Somppi 1979; Fitch 1979), and have been attributed to polygynous matings or female-female pairs. Conover (1984) considered that for species normally laying a three-egg clutch, five or six eggs would constitute a supernormal clutch, but since four-egg clutches proved rare in our colonies we include these as well. Very occasionally we found nests with four or more eggs (0–5 per season, maximum = 7 eggs), and it was difficult to determine if these resulted from two females depositing eggs, or reflected one female laying four or more eggs. Often, but not always, there were two distinct color patterns, suggesting two different females.

Large clutches were more common in salt marsh (0.4 percent) than in sandy beach colonies (<0.1 percent), and were perhaps correlated with an increased density of nests. This is consistent with the observations of Richards and Morris (1984) who found 2.8 percent of clutches larger than three eggs, but had much higher nest densities than we found in our colonies. We suspect that egg retrieval from a neighboring nest may have accounted for the extra eggs, particularly when high tides may have floated eggs out of some nests allowing them to be retrieved by nearby incubating birds. In some cases intraspecific brood parasitism may account for supernormal clutches. Despite repeated attempts we never trapped three birds on one of these nests.

In addition we note that some of the early reports of common tern colonies indicated a higher proportion of four- and five-egg clutches than is typically found today. Mackay (1898) reported frequencies of 1–1.5 percent four-egg clutches for New England common terns, more than 10 times the rate we currently find in our beach colonies. This is of particular interest in view of early suggestions that supernormal clutches in gulls might be a new phenomenon related to the effects of chemical pollutants (Fry and Toone 1981).

Variation

Spatial and temporal variation in common tern clutch size is constrained by the fact that two- and three-egg clutches account for more than 80 percent of all clutches. Table 3.7 reviews literature reports of common tern clutches, mainly in the eastern United States, with representative reports from elsewhere. The table is not exhaustive, but reveals the kinds of data that are reported and some of the limitations. Only Bird Island has reported values covering a 50-year period, while for most colonies there are only one to three years of data on clutch size. Conversely, most studies report data for fewer than four colonies (except for Stelfox and Brewster 1979; Erwin and Smith 1985). In addition, since mean clutch size may not be as sensitive an

Table 3.7. Representative clutch size of common tern from published data.

Colony Site	*Year*	*Mean Clutch Size*	*Reference*
Northeastern United States			
Bird Island, Mass.	1929	1.9–2.7	Austin 1932
(sandy island)	1932	1.9	Austin 1933
	1933	2.3	Austin 1933
	1971	2.95	Nisbet 1977
	1975	2.78	Nisbet 1977
	1979	2.04–2.96	Nisbet & Welton 1984
	1983	2.14–3.00	Nisbet et al. 1984
Great Gull Island, N.Y.	1969	2.54	LeCroy & LeCroy 1974
(rocky island)	1970	2.16	LeCroy & LeCroy 1974
Cedar Beach, N.Y.	1984	2.52 (SD=0.58)	Safina et al. 1988
(sandy beach)	1985	2.12 (SD=0.59)	Safina et al. 1988
	1986–1987	2.05 (SD=0.6)	Safina et al. 1989
Eastern Long Island beach colonies	1986	2.13 (SD=0.6)	Safina et al. 1989
	1987	2.07 (SD=0.6)	Safina et al. 1989
Eastern Long Island salt marsh colony	1986	2.29 (SD=0.7)	Safina et al. 1989
	1987	2.21 (SD=0.6)	Safina et al. 1989
Shinnecock Bay colonies (salt marsh)	1981	1.9 (SE=0.2)–2.5 (SE=0.1)	Buckley & Buckley 1982
Southern New Jersey	1980	1.67 (SD=0.8)–2.59 (SD=0.6)	Erwin & Smith 1985
(salt marsh colonies)	1981	2.23 (SD=0.6)–2.79 (SD=0.4)	
	1982	1.60 (SD=0.7)–2.50 (SD=0.7)	
Holgate (sandy beach)	1982	2.74 (SD=0.5)	Erwin & Smith 1985

Elsewhere			
Maryland-Virginia beach	1980	2.67 (SD = 0.6)–2.76 (SD = 0.4)	Erwin & Smith 1985
	1981	2.00 (SD = 0.4); 2.24 (SD = 0.5)	Erwin & Smith 1985
	1982	2.40 (SD = 0.6)–2.58 (0.6)	Erwin & Smith 1985
Maryland-Virginia salt marsh	1980	2.40 (SD = 0.8)	Erwin & Smith 1985
	1981	1.84 (SD = 0.5); 2.04 (SD = 0.5)	Erwin & Smith 1985
	1982	2.12 (SD = 0.6); 2.19 (SD = 0.7)	Erwin & Smith 1985
Toledo, Ohio			
(Lake Erie Island)	1980	2.36 (SD = 0.6); 2.67 (SD = 0.6)	Shields 1985
Northern Saskatchewan			
(27 river colonies)	1973–1974	1.0–2.8	Stelfox and Brewster 1979
Britain			
Coquete Island, England	1965	2.59	Langham 1972
	1966	2.37	
	1967	2.45	
Orkney and Shetlands		2.55	Cramp et al. 1985
West Germany			
Wangervage	1982	2.4–2.7	Becker et al. 1983
Angustgrodon		2.8, 2.9	
West Germany		2.84	Cramp et al. 1985
USSR		2.88	Cramp et al. 1985

indicator of food availability as the proportion of three-egg clutches, it would be useful to have some indication of the actual distribution of clutches or at least the variance.

Many papers do not report variance or standard deviation, and Langham (1972) is one of the few recent papers to give the actual distributions. Since early or late sampling tends to bias estimates downwards in a nonrandom fashion, it is not clear whether lower clutch sizes (below 2.25) reflect actual clutch size or sampling error. Several studies that report a range of values (Austin 1932; Buckley and Buckley 1982; Erwin and Smith 1985), show a rather large range (from less than 2 to more than 2.5), while other studies (including Erwin and Smith 1985 for Maryland and Virginia) show a narrower range. Of 43 New York colony occupancies, 95 percent had mean clutch sizes greater than two, while this was true for 86 percent of New Jersey colony occupancies.

Of greater interest are consistent inter-year differences that we found in both New Jersey and New York, and that others have reported (e.g., Langham 1972; Nisbet 1977; Erwin and Smith 1985). Assuming that within any study, methods are comparable from year to year, we interpret such interyear variation as a reflection of food availability, consistent with Nisbet's (1973a) findings on male courtship-feeding rates in relation to egg size and ultimate chick survival. Food availability preceeding the egg-laying period may be the proximate cue by which birds reduce their clutch size and reproductive risk as predicted by Murray (1986).

Table 3.8 compares the clutch sizes for 50 occupancies (individual colonies × years occupied) reported in the literature, 157 reported here from New Jersey salt marshes, and 43 reported here from New York colonies. The grand means for the three data sets range from 2.34 to 2.43, and medians range from 2.34 to 2.45. However, if occupancies with a mean clutch < 2.0 are excluded, the means are virtually identical (2.42–2.46), which we offer as a realistic mean value for temperate zone common terns.

It is evident from our literature analysis that it would be critical in the future to give colony size, sample size (where it differs from colony size), timing of estimate relative to peak, and the actual distribution of clutch sizes. With these data an investigator can make a number of comparisons among or within colonies, and across species. Even though many of these studies had different objectives, it would be useful to have the above raw data.

Further, some of the studies examined more than one colony, but lumped the colonies when presenting the data. Thus Stelfox and Brewster (1979)

Table 3.8. Comparison of average clutch sizes for New York and New Jersey common tern colonies with those reported in the literature.

	Years Occupancy[a]	*Mean for All*	*Mean*[c] *>2.0*	*Median*[c]	*%>2.0*	*95%*[d] *Above*
New Jersey	157	2.34	2.42	2.34	0.86	1.75
New York	43	2.43	2.44	2.45	0.95	2.08
Literature	50	2.40	2.46	2.43	0.92	1.84

[a]Each year a colony is occupied by breeding terns constitutes 1 occupancy. Thus 5 occupancies may be 1 colony occupied for five years or 5 colonies occupied for one year.

[b]The mean of all studies reporting clutch size 2.0.

[c]These are the mean and median values from the different data sets.

[d]95% of mean values were above the listed value.

studied 27 common tern colonies, but only reported clutch size as ranging from 1.00 to 2.80. Although this gives a general picture, it is impossible to know the degree of variation or even the mean clutch weighted appropriately by colony size. In contrast, Erwin and Smith (1985) did present mean clutch size (with standard deviations) for the 14 colonies they studied for three years. Most of the reported clutch sizes are between 2.00 and 2.80, indicating that in most years three-egg clutches predominate over one-egg.

Some of these studies were particularly designed to examine for variations in clutch size as a function of age, habitat, or year. Austin (1933) reported an increase in clutch size (from 1.9 to 2.3) when a dense, grassy habitat was burned, and also showed differences in clutch size as a function of degree of habitat openness (Austin 1932). At this same Massachusetts colony, Nisbet and his colleagues (Nisbet and Welton 1984, Nisbet et al. 1984) have demonstrated that older parents have larger clutches than younger parents, and that earlier-laid clutches are larger than later-laid clutches. Our data show that within a season the earliest nests do not always have the largest clutch size, but that nests initiated after July 15 invariably show lower mean clutch sizes.

Our clutch size data for ten years from New Jersey salt marsh clearly showed yearly, colony, and colony size differences. As one might expect, overall mean clutch size was lower in some years, probably due to food shortages (see Safina and Burger 1985, 1988a,b). Presumably, the differences in available food they reported for Long Island might have occurred in adjacent New Jersey waters as well. Even with the yearly differences, there was an effect of colony size: larger colonies had larger clutch sizes than smaller colonies. This could be due to the presence of a larger percentage of older birds (see Ryder 1980), or to the concentration of more experienced birds near areas of abundant food.

Even with yearly and colony size differences, some colonies consistently had higher clutch sizes than others. The Lavalette Islands generally had higher clutch sizes than islands such as Flat Creek. In their study, Erwin and Smith (1985) failed to find any significant differences in clutch size as a function of year or colony size (all $P > 0.06$). Partly this resulted from their study lasting only three years, and partly it resulted from the fact that many of their colony sites were used only one or two years. Further, they were comparing colonies in New Jersey, Maryland, and Virginia, and so might have encountered inapparent geographic variation.

Comparison of Salt Marsh and Beach Colonies

Using parametric statistics, Erwin and Smith (1985) reported no difference in mean clutch size between five beach and nine marsh colonies from Maryland to New Jersey. Inspection of their data shows that using a one-tailed Mann-Whitney test, the marsh colonies had a marginally significant lower clutch size ($P = 0.05$), however, they were not testing that hypothesis. By contrast, Safina et al. (1989) compared marsh and beach nesting colonies for two years, and found that salt marsh terns had significantly higher clutch sizes in both years, although they did not fledge significantly more young.

We could not compare clutch sizes on sand and salt marsh colonies in New Jersey because in this study area the terns mostly nested on salt marsh. However, in New York (see table 3.6), there were no differences in 1976 when we compared clutches on islands (sand versus salt marsh).

Interpretation

Clutch size within a population or species is subject to strong evolutionary selection (Klomp 1970; Murray 1979), while the clutch produced by an individual female is likely to be influenced by her current physical condition as well as social factors. Our data show a rather tight distribution of mean values between 2.34 and 2.43 for different data sets, and we suspect that values below 2.0 for a given colony in the northeastern United States reflect incomplete clutches. If clutch size is sensitive to food availability, terns may benefit by having fewer chicks to raise in a year when it will be difficult to feed them. By laying and hatching two eggs they may do better at raising one or both of the chicks, than they would if they had hatched three chicks, and worn themselves out in a vain parental attempt to feed all three.

3.5 ACTIVITY PATTERNS

Daily and seasonal activity patterns in colonial seabirds reflect varying stimuli and responses associated with reproduction, feeding, weather, and tides (Cullen 1954; Delius 1970; Burger 1976; Conover and Miller 1980; Bernstein and Maxson 1984). Generally the frequency of most activities is highest in the early morning and late afternoon (Cullen 1954). This holds as well for Antarctic species exposed to continuous daylight (Muller-Schwarze 1968).

Activity patterns may, however, be influenced by tidal cycles (Delius 1970), which in turn influence food availability. Tidal patterns have been shown for most shorebirds (Evans 1976; see review in Burger 1984b); some species even feed at night to take advantage of favorable tide conditions (Heppleston 1971; Dugan 1981; Pienkowski 1982). The foraging activities of colonially-nesting seabirds are also affected by tidal rhythms (Delius 1970, Burger 1976; Galusha and Amlaner 1978). The daily activity patterns of terns have not previously been examined in detail. In this section we examine activity patterns for terns to describe the seasonal and daily variation that affects other behavior.

Methods

In 1982, we sat outside the West End colony and watched a group of about 20 to 25 pairs of terns. We noted data in one-minute time blocks, recording the date, time, number of single and paired terns present, the number of intruders, and a tally of various behavior patterns. We recorded the various stages of courtship activity leading up to copulation as well as any interruption of copulation, aggressive behaviors such as wing-up displays, walk toward, fly toward, aerial chases, fights, total aggressive actions, the number of fish brought in, and the number of birds preening and sleeping. Two people were required to obtain these data. We made observations between 0500 and 2400. Observations at night were made with a Smith and Wesson image-intensifying scope.

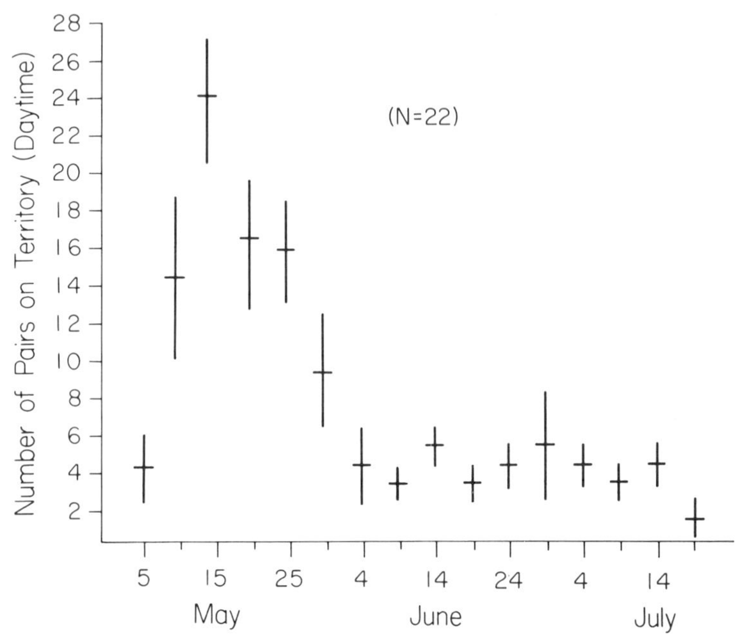

Figure 3.6. Seasonal variation in the number of nests with both members of the pair present for 24 common tern territories (1982). Given are mean (horizontal line) ± 1 SD (vertical line). Note much higher proportion of territories with both pairs present prior to egg laying period. Later in season, the average number of adults present is about 1.1 per territory.

Seasonal Patterns

Activity patterns in many species of birds vary across season. Courtship and aggressive activities are most frequent early in the season, parental activities late in the season. Once terns have begun incubation there is usually only one adult present on the territory, at least until late afternoon. We found clear seasonal variation in presence on the colony (fig. 3.6) and in aggression (fig. 3.7). Aggression was high early in May and again early in June, corresponding to courtship and to hatching.

Daily Patterns

During most of the day only one member of an incubating pair of terns remains on the territory (fig. 3.8). During exchanges, one of the mates may

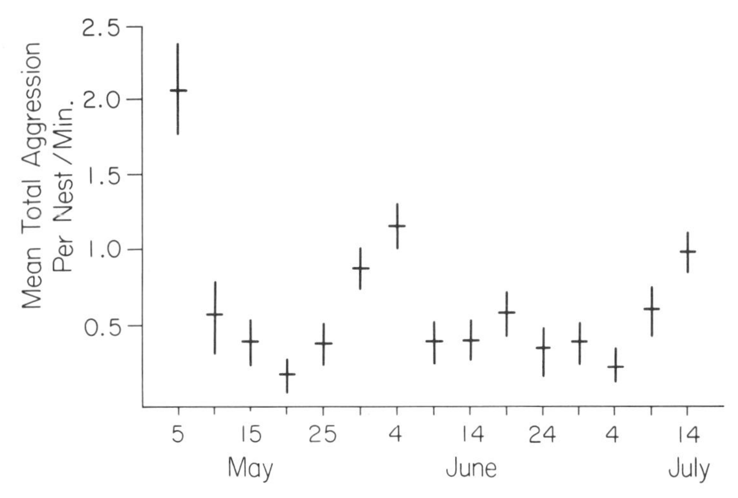

Figure 3.7. Seasonal variation in mean (± 1 SD) rate of total (inter- and intra-specific) aggressive interactions for common terns per pair per minute. (1982 data).

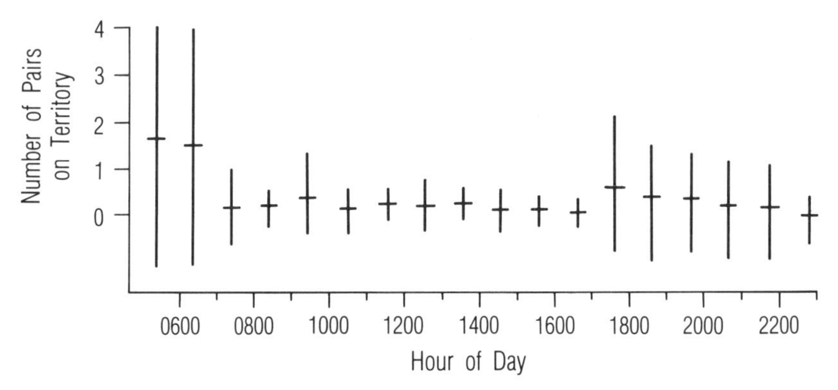

Figure 3.8. Daily variations in presence on territory of 24 pairs of common terns. Shows mean (± 1 SD) number of pairs with both members present, during daytime observations (1982).

linger for a few minutes, facing away from the nest. It will frequently toss bits of shell or straw over its shoulder toward the nest, before flying off to feed. This behavior associated with incubation reliefs accounts for the accumulation of nest material around many nests as incubation progresses. On the average the ratio of adult terns to nests is about 1.1 during the early

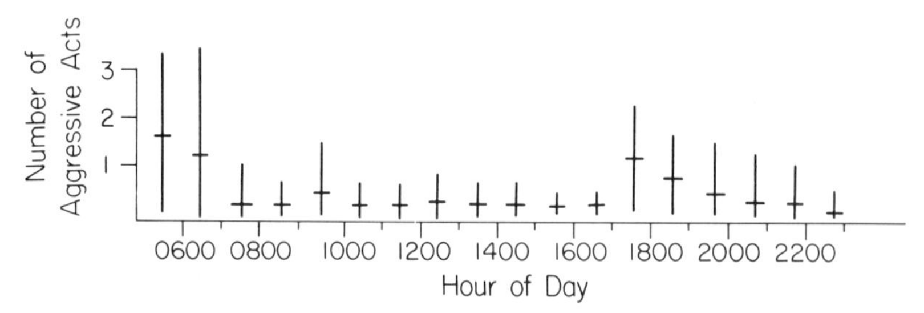

Figure 3.9. Daily variations in mean (± 1 SD) number of aggressive interactions per hour for 24 pairs of common terns.

morning, reaching 1.6 or more by evening as the mates return for the night. Aggression rates varied significantly as a function of time of day (fig. 3.9). Terns engaged in much more aggression from 0600–0800 and from 1600–1800 than at other times.

Discussion

Clearly activity patterns show daily and seasonal variation throughout the breeding cycle. Unlike the black skimmers they nest with, the terns are clearly diurnal, and most activity declines dramatically at dusk. No birds returned with fish after dark. Since most of the time only one member of the pair is present during the day, there is a tendency for chicks to remain close to the nest site, usually hiding in nearby vegetation, where the parents can find and defend them.

3.6 SYNCHRONY

Synchrony of certain behaviors occurs in many groups of animals. For example, in insects there is synchronous flashing in fireflies (Case and Buck 1963; Buck and Buck 1978), synchronous emergence of some cicadas (Hoppensteadt and Keller 1976), synchronous appearances and migration of *Urania* moths (Smith 1972), and synchronous spawning of some corals (Harrison et al. 1984). Under laboratory conditions, deer mice will synchronize activity rhythms (Crowley and Bovet 1980). Human females living in small groups often synchronize menstrual cycles (McClintock 1971, 1981; Quadagno et al. 1979). Knowlton (1979) examined the theoretical implications of synchrony particularly as a factor increasing the payoff of parental care where there is a disparity of investment.

Some activities of terns such as the upflights or "panics" (Marples and Marples 1934; Marshall 1942), show synchrony on a very short time frame (seconds). Reproductive synchrony, however, occurs on the order of days or weeks. Before beginning our study of behavioral and reproductive synchrony in common terns we had studied synchrony in a variety of behaviors such as diving in grebes and waterfowl and song bouts in meadowlarks and skylarks (Burger 1974c; Gochfeld 1978a,b).

Synchrony and the Darling Effect

One of the unifying theories of breeding behavior in colonial birds, was set forth by the Scottish naturalist Frank Fraser Darling (1938) who developed a theory involving behavioral, physiologic, ecologic, and evolutionary phenomena which became known as the Darling Effect. Although the Darling Effect has been criticized by various researchers, we were impressed that many of its features were "robust" in the sense that they apply even when not all their assumptions are met (Gochfeld 1980b).

Darling studied herring and lesser black-backed gulls on Priest Island off the coast of Scotland. The island harbored four gull colonies ranging in size up to a few hundred pairs. He found that in the larger colonies the birds tended to breed earlier, more synchronously, and lost fewer chicks to crow predation. He postulated that in the larger groups there was more opportunity for social interaction, which lead to acceleration of their physiologic cycles, which readied the birds for ovulation. The more social stimulation, the earlier and more synchronously the birds bred. In colonies where breeding was synchronous, most eggs hatched at about the same time. Hence there was a large group of young produced at one time, and even a busy predator could take only a few of these. The rest of the young survived. He concluded that evolution must favor synchrony as an antipredator adaptation, and hence living in larger colonies is also adaptively advantageous.

Since Darling (1938) first proposed that breeding synchrony increases with colony size in gulls, several authors have found supporting (for review see Burger, 1974a, 1979b; Gochfeld 1980a,b) or contradictory results (Orians, 1961a; MacRoberts and MacRoberts 1972; Snapp, 1976). Wittenberger and Hunt (1985) concluded that there were many negative studies of the Darling Effect and few convincing ones. The mechanism proposed for enhancing breeding synchrony is social facilitation or social stimulation. The effects of social stimulation have clearly been demonstrated in weaverbirds (Hall 1970; Collias et al. 1971). Nest building and male displays attract females to village weaverbird colonies, and the number of females attracted is related to the number of males displaying in the colony (Collias and Collias 1969).

To us this seemed like a challenging idea. The process of social facilitation (a term Darling did not use), was amenable to study. We knew that other researchers had failed to find evidence for part of the Darling Effect, particularly because by whatever measure one uses, larger colonies of birds in general are less synchronous than smaller ones. This was not because social stimulation was ineffectual in achieving synchrony, but because the process of social attraction (Gochfeld 1980b), resulted in the continual recruitment of late arriving birds to larger colonies, thereby swamping any tendency to synchrony.

On the other hand, there is positive evidence for social interactions as a synchronizing mechanism. Courting ring-billed gulls (Southern 1974) and Common and Caspian terns (Gochfeld 1980b) on neighboring territories tend to mount and copulate at about the same time. The synchronizing stimulus may be high visibility behaviors such as wing-flagging in gulls (Southern 1974). During copulation the males elevate and flap their wings while balancing on the female. This copulatory wing-flagging is readily apparent to neighboring birds, and one can see them initiate precopulatory displays.

The accelerating effect of these synchronizing behaviors would advance the breeding season to an earlier date were it not for the constraints imposed by cold, wet spring weather. Reduced reproductive success on the part of late-nesting breeders has been found for many species, including Laysan albatross (Fisher 1971), thick-billed murres (Gaston and Nettleship 1981), Atlantic puffin (Nettleship 1972), white pelicans (Knopf 1979), eiders (Milne 1974), storm petrels (Harris, 1969), black-headed gulls (Weidmann, 1956), herring gulls (Paludan 1951; Parsons 1971, 1975), lesser black-backed gulls (Brown 1967a), red-winged blackbirds (Robertson 1973), and bank swallows (Emlen and Demong, 1975). Yet differences in reproductive success can be due to differences in the age or experience of birds (Crawford 1977; Ryder 1980; younger birds tend to breed later in the season), rather than to differences in synchrony. Veen (1977) reported that late nesting sandwich terns had lower synchrony and lower reproductive success.

Several of the authors cited showed that early breeding birds were the most successful. Breeding earlier than the peak is often disadvantageous, as has been documented for great-horned owl predation on pre-peak common terns (Nisbet 1975), for cat predation on early-hatching and late-hatching young sooty terns compared to midseason-hatched young (Ashmole, 1963), and for crow predation on black-headed gulls (Patterson 1965). In cliff swallows, relative predator density decreased five-fold compared to a twenty-fold increase in colony size (Wilkinson and English-Loeb 1982). Yom-Tov (1975) also found that losses due to cannibalism were higher in asynchronous nests in colonial carrion crows. Herbert and Barclay (1986) reported differences in growth rates as a function of the degree of synchronous hatching within nests.

Daan and Tinbergen (1979) showed that young common murres tend to fledge from their cliffs en masse, which has the effect of swamping the gulls that gather there to feed on the inexperienced young as they hit the water. Similarly, great black-backed gulls wait on the edges of nesting cliffs to take young kittiwakes first learning to fly (Burger and Gochfeld 1984). There may be other mechanisms favoring synchrony. For example, Emlen and Demong (1975) showed that bank swallows that hatch at or after the peak of breeding fare better than those hatching before, because of the greater availability of feeding information from adults flying out to favorable feeding areas. Synchrony can also result from external factors such as inclement weather causing prior nesting failure of large numbers of birds or colony desertions (Evans 1982), or to genetic components (Findley and Cooke 1982).

There is another dimension to synchrony—hatching synchrony, which has received extensive study by others. In birds that begin incubation before they have completed their clutch, one sees hatching asynchrony. Earlier eggs hatch hours or days before later ones. The earlier hatching chicks are fed earlier and, particularly since they come from slightly heavier and more nutrient-rich eggs, they grow more rapidly. In a year of food shortage, they may survive at the expense of their younger and weaker siblings (Lack 1968). Nisbet and Cohen (1975) showed that this is an important phenomenon in terns as well.

In our studies of terns, we recognized several related phenomena. *Social stimulation* is the general term accounting for the increase in frequency and/or intensity of behavior when an individual is in the presence of other individuals. The assumption is that it is the social setting or social interaction which leads to the increased behavior. *Social facilitation* is a special case of social stimulation. An individual begins to perform or increases its performance of a particular behavior pattern in the presence of another individual performing the *same* behavior. In a classic paper, Tolman (1964) noted that pecking rates in domestic chicks increased in the presence of another chick that was pecking. Harlow (1932) was perhaps the first to use the term social facilitation when he described feeding rates in rats under social situations. From an operational viewpoint, we counted the number of behaviors (i.e., calls, displays, fish delivered, nest changes) per pair per time period, to obtain evidence of social facilitation.

Synchrony is a measure of the temporal clustering of events. The term is relative, and it is always necessary to define a time scale of interest. In the twentieth century, two events occurring on the same day would be synchrojnous; but on a scale of days, two events occurring in one day would not be considered synchronous. More importantly, we distinguish between synchrony which is a measure of the timing of events, and synchronization which refers to the mechanisms whereby any observed degree of synchrony arises. Synchronizing (or desynchronizing) factors can include weather, food, predation, or social interactions. Veen (1977) suggested that sandwich tern nesting was synchronized by food availability.

One of the important paradoxes is that synchronizing events do not always increase synchrony. Thus the social interactions of breeding birds which may synchronize reproductive cycles (Darling 1938), also serve as social attractants to late-arriving birds which may reduce the overall breeding synchrony of a colony (Gochfeld 1980b).

Methods for Studying Synchrony

Ecologists have developed an extensive literature on the spatial sampling of organisms (e.g., Cain and Castro 1971; Poole 1974; Pielou 1977; Ripley 1981), some of which involve the use of quadrats. We used an analogous approach, the use of time blocks to study temporal patterns. Just as there has been abundant discussion of appropriate quadrat size, so one must be careful that one's time blocks are appropriate to the phenomenon of interest (Gochfeld 1978b). Selection of a time block depends in part upon the frequency of the event being sampled, on the rapidity with which it varies over time, and on the kind of analysis being performed. Ideally, some time blocks will have no events, others may have several. A disproportionate number of time blocks with several events, tested against a Poisson distribution, is evidence of temporal clustering or synchrony.

Empirically, we have settled on time blocks varying from 5 seconds to 15 minutes for analyzing the social facilitation of behavior reported in this chapter. One underlying principle is that if one records data in very small time blocks, it is subsequently possible to aggregate the data into larger blocks while the converse is not possible.

There are several ways of measuring or describing the synchrony of an event (see Gochfeld 1980b). In our discussion of breeding synchrony, we use the standard deviation of timing which seems to provide the single best description, one which agrees with graphic presentations and intuition. Elsewhere, we have provided detailed information on synchronous behavior and apparent social facilitation in the breeding behavior of black skimmers (Burger and Gochfeld 1990a). Copulation among neighboring pairs was clearly synchronized, as was sand-kicking and fish presentations to females.

Synchrony of Behavior in Common Terns

Common terns nesting on salt marshes usually nest in groups of 25 to 500 pairs, and those nesting on beaches often nest in much larger colonies. In large colonies, density varies spatially, resulting in subcolonies. Thus it is more difficult logistically to examine the effects of social facilitation in beach colonies, because the lack of a discrete boundary to groups makes it difficult to determine the effective size of interacting units. Nonetheless, some behavior patterns that could be isolated were studied in 1982 to 1985. We studied aggression and copulation behavior in small groups of terns nesting on salt marshes, and fish transport at a beach colony.

During territorial establishment and the courtship phase common terns engage in frequent aggressive interactions with neighbors and intruders. We observed a group of 24 pairs from 0600 to 0800 and recorded all aggressive interactions in one-minute samples. Aggressive encounters showed a strongly contagious distribution compared with a uniform distribution ($X^2 = 5918$, $df = 8$, $P < 0.0001$), nor did they fit a Poisson. There was an excess of minutes with no aggressive encounters and with seven to ten encounters (fig. 3.10 top).

We examined the frequency of copulations and found that more often than expected, two or more pairs would be mounted simultaneously (fig. 3.10, bottom, $X^2 = 650$, $df = 5$, $P < 0.0001$). Common terns bring back fish

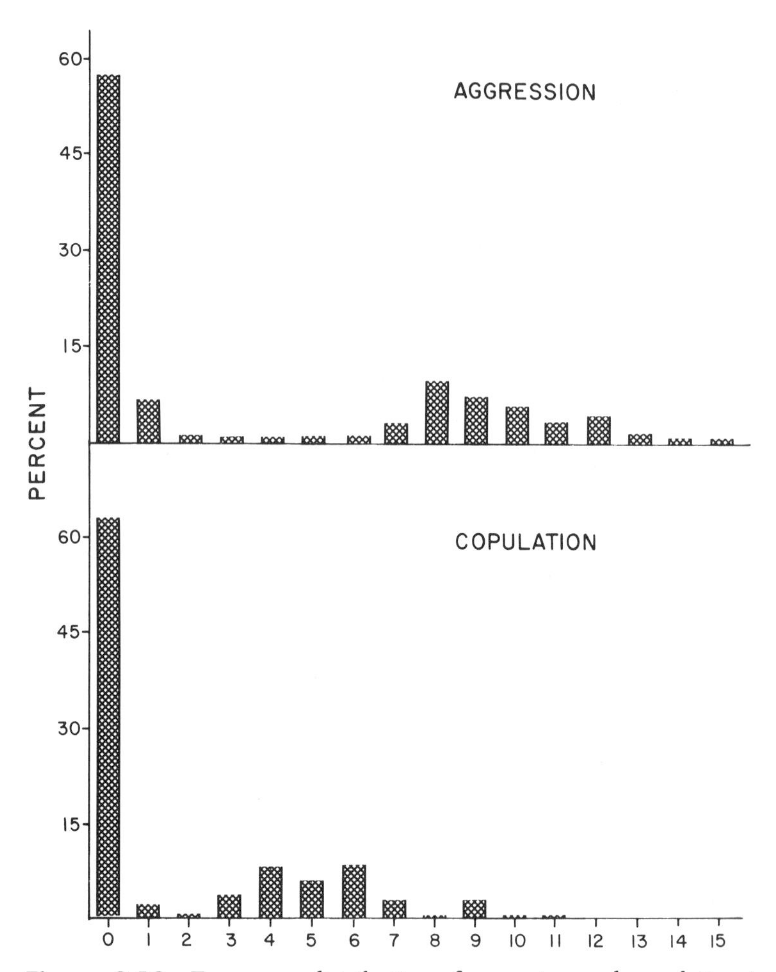

Figure 3.10. Frequency distribution of aggression and copulation in common terns. Shown are the percent of one-minute time blocks representing each frequency of events.

throughout the reproductive period to courtship feed females and to feed young. Some of the courtship feeding visits are followed by copulation (fig. 3.11). We counted the number of fish carried into the Cedar Beach colony in 5-second time blocks. Often there is an apparently steady stream of birds returning to the colony with fish, but we found the pattern did not fit either a uniform or a Poisson distribution. There was an excess of time blocks in which seven or more individuals returned with a fish and an excess with no

Figure 3.11. Male common tern returning with fish for female (top) and pair of terns copulating (bottom). Note pair of black skimmers on adjacent territory.

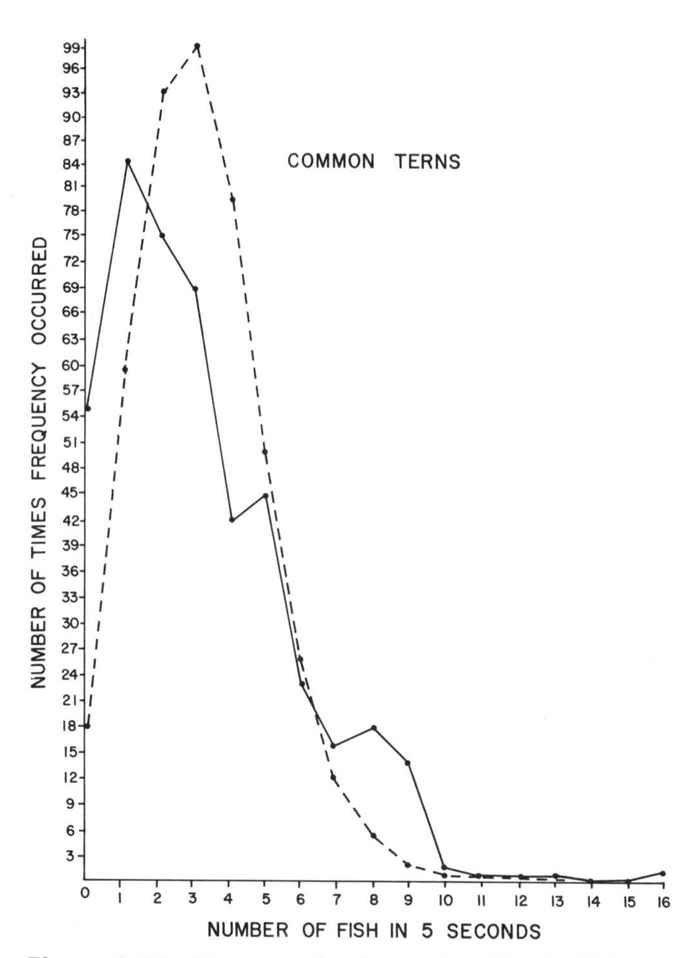

Figure 3.12. Frequency distribution (solid line) of fish transport events per 5-second time interval for common terns. Dashed line is the corresponding Poisson distribution if terns were returning at random intervals. Distribution shows excess temporal clustering of birds returning with fish.

returnees (fig. 3.12, $X^2 = 274$, $df = 8$, $P < 0.0001$). This pattern was less pronounced than the pattern for either aggression or copulation. Fish transport differs from the other behaviors, however, in that the terns leave the colony to obtain fish. Even if they left synchronously, it would be difficult to completely synchronize their arrival, since factors such as fish availability and individual fishing ability affect foraging time.

Discussion

Group Versus Colony Behavior

Patterns of aggressive behavior and copulations were both consistent with social facilitation. The appearance of an intruder stimulated group activity in a neighborhood. Copulations among neighboring pairs occurred together, as Gochfeld (1980b) reported previously. The reason for many pairs engaging in copulation at the same time is unclear. One could implicate a stimulating effect of precopulatory circling behavior. If one male is copulating, he cannot interrupt another's copulation. Our observations indicate that when multiple copulations occur in one time block, it is more likely than chance that the pairs will be in close proximity.

Fish Transport

Courtship feeding often plays a directly stimulating role in the breeding of larids, particularly for such species as black skimmers and least terns. In common terns, Nisbet (1973a) pointed out that it also provides females an opportunity to assess male quality and commitment and contributes directly to formation of larger, energy-rich eggs. Morris (1986) described the relationship between courtship feeding rates and phenology in common terns.

We had originally thought that fish transport behaviors might be clumped, particularly if information transfer occurs. Ward and Zahavi (1973) first proposed that birds congregate in groups to exchange information about patchily distributed food sources. This theory requires that birds somehow transfer information about the food sources. Various authors (e.g., Gotmark 1982) have examined this hypothesis. However, information transfer could be passive in that unsuccessful foragers could simply follow successful foragers (those that returned with fish) from the colony to the food source (Krebs 1978). Alternatively, they could feed by social enhancement; that is, they could cruise the potential foraging habitat looking for successful foragers that are plunge diving for fish (Gochfeld and Burger 1982). In either case, birds should return to the colony in groups, and our data confirmed this. The temporal distribution of fish transport trips showed an excess of periods with no fish, and with seven or more fish. This suggests that social facilitation either affects terns leaving on foraging trips, while actively foraging, or on their return trips. There is a clear advantage to returning in groups, for as the terns approach the colony they are often intercepted by conspecific pirates (chapter 7). A pirate could, however, only intercept one bird out of the group. Hence, there appeared to be safety in numbers.

Comparison of Common Terns and Black Skimmers

Where we could compare behavior patterns (aggression, copulation), both species showed effects consistent with synchronization. There were more skimmer behaviors in which social facilitation was apparent (Burger and Gochfeld 1990a). Partially, this is a logistical problem; it proved more difficult to sample terns (because they don't form discrete, isolated groups). However, we suggest that it is precisely the small groups that make social facilitation stronger. In a small group all birds can interact, identify each other, respond as individuals, and develop particular modes of interacting. These relationships may be long term for both terns and skimmers, since both species show a high degree of group adherence (Austin 1951; Gochfeld 1978a).

By applying two conditions to social facilitation (disproportionate increase in intensity or frequency, nonrandom temporal clumping of events), we have developed a method to measure an effect we can define as social facilitation. The method allowed us to categorize different behavior patterns as consistent with "social facilitation." The alternative would be behaviors whose frequency is uniform or fits a normal or Poisson distribution.

The method has difficulties in that it is still not possible to prove that merely increasing the number of birds causes the change in behavior. Further, granting that these behavioral shifts are due to social facilitation, one cannot prove the link between social facilitation and synchrony first enunciated by Darling (1938). We feel, however, that in its own right it is important to show that birds change their behavior in direct response to that of their neighbors.

Breeding Synchrony in Common Terns

Traditionally, the date on which a clutch is initiated has been used most as the unit of analysis for breeding synchrony (Gochfeld 1980b). There are several measures of synchrony; however, we feel the best measure of breeding synchrony is the standard deviation of clutch-initiation date (see Emlen and

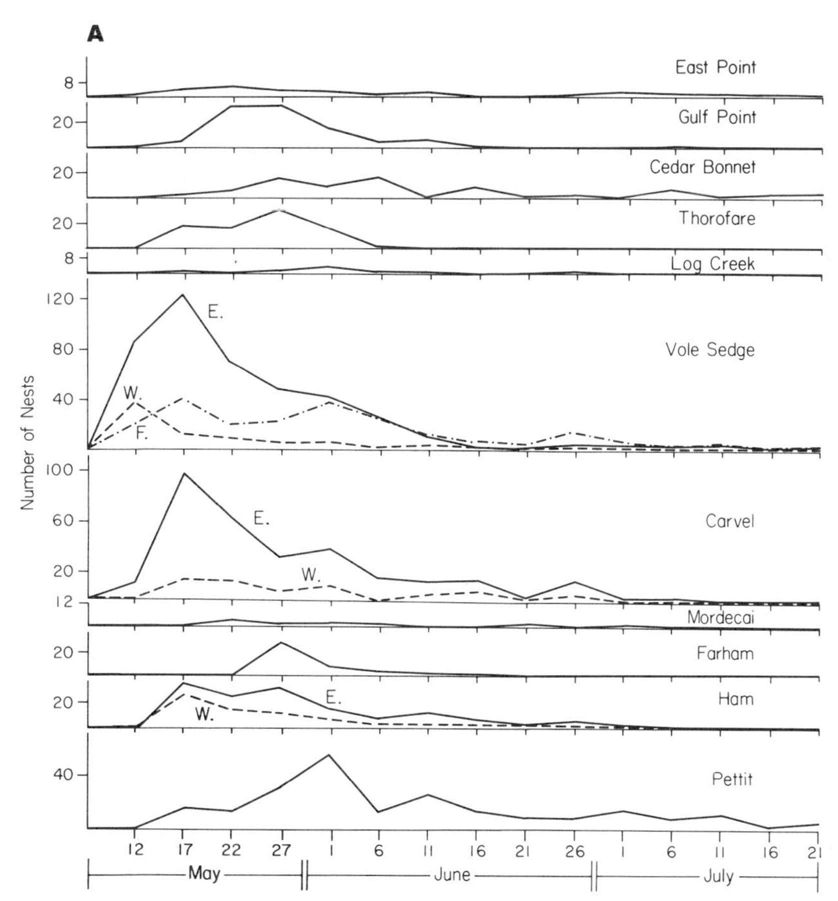

Figure 3.13. Frequency distribution of clutch initiation dates for common tern nests in salt marsh colonies. Shown are the number of nests initiated (first egg laid) in five-day time blocks for each of the three years a = 1979, b = 1982, c = 1985; for Vol Sedge, Carvel, and Ham Islands, E = East, W = West, F = Far.

Demong 1975; Burger 1979a; Gochfeld 1980b) where a low standard deviation indicates high synchrony. Breeding chronologies of egg-laying for several salt marsh colonies are shown in figure 3.13. Not all colonies were occupied in all years, and some colonies monitored less frequently during egg-laying were not included. Several features are immediately obvious from the chronologies: 1) colony size varied from year to year within the same colony, 2)

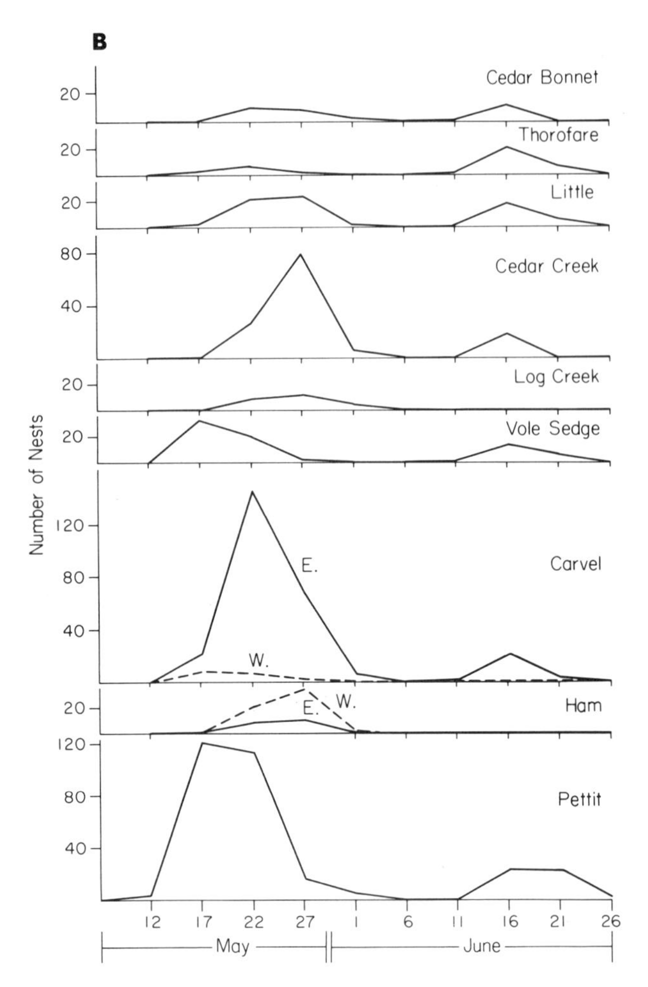

the peak in the first wave of egg-laying varied from year to year and among colonies, 3) the colonies with early peaks of egg-laying one year were not necessarily early in other years, 4) not all colonies experienced second waves of nesting, 5) second nesting waves generally involved fewer birds than initial waves of nesting. Table 3.9 compared synchrony (SD) among colonies and years based on nest initiation dates for the entire season. Colonies that appear synchronous (e.g., East Carvel in 1979) may show high standard deviations for the season as a whole.

In 1979, the mean date of clutch initiation varied across colonies, from

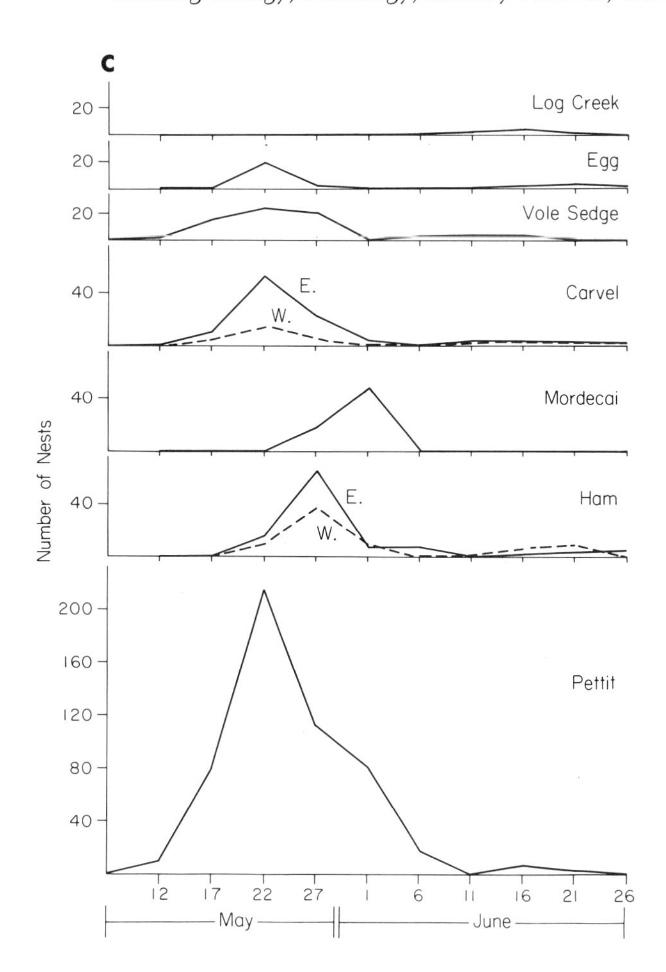

May 21 to June 2. However, in 1982 and 1985, the mean date of egg-laying varied from May 19–26 (except for two late colonies, Mordecai and Log Creek in 1985).

We also examined the relationship of colony size, laying date, and synchrony in common terns (table 3.10). For colonies, the first date of egg-laying was positively correlated with the mean date of clutch initiation; (i.e., when the first tern lays early in the season, the mean date is also early). Also, the first day of egg-laying was negatively correlated with the number of pairs, suggesting that large colonies begin to lay earlier than small colonies; however, the mean date of egg-laying is not correlated with the number of terns.

Table 3.9. Colony size, mean date of clutch initiation, and synchrony of egg-laying for common terns nesting on salt marshes. Synchrony is measured by the standard deviation of egg-laying.

Island	*Number of Pairs of Terns*	*Mean Date of Egg-laying*[a]	*SD of Date of Egg-laying*[a]
1979			
West Vol	110	May 24.5	9.9
Far West Vol	70	May 26.9	10.1
East Vol	300	May 21.4	8.4
East Carvel	210	May 27.1	12.2
West Carvel	47	May 23.1	5.9
Pettit	90	June 1.9	8.9
West Ham	30	May 24.7	9.2
Far Ham	19	May 30.7	4.8
East Ham	90	May 27.7	10.5
Mordecai	10	May 28.0	5.8
Log Creek	18	May 29.4	7.3
Thorofare	77	May 22.2	8.8
Cedar Bonnet	26	May 30.1	6.0
Gulf Point	110	May 27.8	6.7
East Point	52	May 26.8	9.0
1982			
West Vol	61	May 19.1	2.5
East Carvel	300	May 22.8	3.2
West Carvel	19	May 20.4	3.4
Pettit	353	May 20.1	3.2
West Ham	51	May 24.7	2.2
East Ham	20	May 24.7	2.0
Log Creek	26	May 26.6	3.7
Cedar Creek	99	May 26.0	2.1
Thorofare	33	May 22.2	3.9
Little	85	May 25.3	2.7
Cedar Bonnet	26	May 25.9	3.8
1985			
West Vol	64	May 21.7	4.0
East Carvel	90	May 21.8	7.1
West Carvel	26	May 21.5	2.6
Pettit	505	May 23.7	5.4
Egg	21	May 22.1	1.5
West Ham	55	May 26.5	2.4
East Ham	83	May 26.2	2.0
Mordecai	65	May 30.5	1.5
Log Creek	20	June 15.9	5.8

[a] First wave only; does not include nests initiated after most of colony was eliminated by tidal flooding.

Table 3.10. Colony size, synchrony, and first laying date in common tern colonies (1979, 1982, 1985).

Island	*Number of Nesting Pairs*	*Synchrony of Terns*[a]	*Date of 1st Egg (in May)*	*Mean Initiation Date (post May 1)*[b]	*Productivity Young/Pair*
1979					
East Point	52	9.0	12	26.8	0.04
East Vol	300	8.4	10	21.4	0.58
West Vol	110	9.9	16	24.5	0.14
Far West Vol	70	10.1	11	26.9	0.02
Gulf Point	110	6.7	14	27.8	0
East Carvel	210	12.2	10	27.1	1.00
West Carvel	47	5.9	14	23.1	0.04
Log Creek	18	7.3	18	29.4	0.11
Pettit	90	8.9	16	32.9	0.16
Cedar Bonnet	26	6.0	16	30.1	0.12
Thorofare	77	8.8	16	22.2	0
East Ham	90	10.5	16	27.7	0.40
West Ham	30	9.2	16	24.7	0.18
Far West Ham	19	4.8	26	30.7	0.02
Marshelder	250	11.3	32	39.0	0.32
Mordecai	10	5.8	20	28.0	0.20
1982					
West Vol	61	2.5	15	19.1	0
East Carvel	300	3.2	15	22.8	0.23
West Carvel	19	3.4	15	20.4	0
Log Creek	26	3.7	21	26.6	0
Pettit	353	3.2	12	20.1	0
Cedar Creek	99	2.1	21	26.0	0
Cedar Bonnet	26	3.8	21	25.9	0.26
Thorofare	33	3.9	18	22.2	0
East Ham	20	2.0	21	24.7	0
West Ham	51	2.2	19	24.7	0
Little	85	2.7	19	25.3	0
1985					
West Vol	64	4.0	13	21.7	0.39
East Carvel	90	7.1	14	21.8	0
West Carvel	26	2.6	18	21.5	0.23
Log Creek	20	5.8	37	46.9	1.25
Pettit	505	5.4	12	23.7	1.05
Egg	45	1.5	20	22.1	0.40
East Ham	83	1.4	22	36.3	0.36
West Ham	55	2.4	24	26.5	0.45
Mordecai	65	1.5	26	30.5	0

[a] Synchrony measured as the standard deviation of the egg-laying date.
[b] Mean initiation date listed as days after May 1st.

Table 3.11. Relationship of colony size and laying date to synchrony in salt-marsh–nesting common terns (1979, 1982, 1985). Correlation coefficient is shown above the diagonal and the probability that this differs from 0 is shown below diagonal.

Variable	*Breeding Pairs*	*Egg Laying Date*		*Synchrony*
		First	*Mean*	
Mean ± SD	98.2±111	May 21	May 26.4	5.5±3.2
Breeding pairs N	———	−0.36	−0.14	−0.08
First date of egg-laying	0.05	———	0.73	−0.26
Mean date of egg-laying	NS	0.001	———	+0.21
Synchrony[a]	NS	NS	NS	———

[a] Standard deviation of egg-laying date.

This confirms our earlier hypothesis that large colonies attract later breeders, thus prolonging the breeding season and perhaps masking synchrony. Moreover, for the three years combined no single variable was significantly correlated with clutch-initiation synchrony (table 3.11). However, when the years are considered separately, the standard deviation was correlated positively with the number of tern pairs for 1979 ($r = +\ 0.45$, $P < 0.01$, $N = 16$ colonies) but not for 1982 ($N = 11$) or 1985 ($N = 9$); and standard deviation was negatively correlated with the number of skimmer pairs in 1985 ($r = -54$, $P < 0.04$). Using regression models to determine the two variables that most influence the variability in synchrony indicated that the dates of first laying and mean laying explained 35 percent of the variance. Although these variables are related, the statistical procedures separate out their independent contribution.

Synchrony in Second Waves of Nesting

In all the three years, there were multiple waves of nesting (fig. 3.13). In 1979, all colonies experienced second waves of nesting, and half of the colonies had third waves of nesting. Secondary nesting waves, however, were considerably smaller. In 1982 and 1985, every colony had a second wave of nesting, but only one colony had a third wave of nesting.

In general, second waves of nesting involved only a small proportion of the birds breeding in the initial waves. Only Cedar Bonnet and Thorofare (in 1982) had second waves that involved the same or more birds than initial

Table 3.12. Variations in breeding synchrony as a function of breeding wave, island, and cause of failure in common terns. Given are standard deviations of clutch initiation dates (in days). Causes of nest failure included: flood tides (F), heavy rain (R) and predation (P).

	Egg-laying Wave		
Island	*First*	*Second*	*Third*
1979			
West Vol	9.9	6.7	
Far West Vol	10.1	6.0	0.0
East Vol	8.4	6.4	0.0
Pettit	8.9	4.9	1.2
West Carvel	5.9	5.0	
Cedar Bonnet	6.0 (P)	9.6	0.5
Log Creek	7.3	0.0	
East Point	9.0	5.0	0.7
Gulf Point	6.7	0.0	
Mordecai	5.8	5.7	
1982			
West Vol	2.5 (F)	2.4	
Pettit	3.2 (F)	2.8	
East Carvel	3.2 (F)	2.0	
Cedar Creek	2.1 (F)	1.2	
Thorofare	3.9 (F)	1.9	
Little	2.7 (F)	1.9	
Cedar Bonnet	3.8 (F)	1.4	
1985			
West Vol	4.0 (P)	5.0	
East Carvel	7.1 (R,P)	4.7	
West Carvel	2.6 (P)	4.6	
Pettit	5.4 (R)	1.0	
Egg	1.5 (P)	3.7	
East Ham	2.0 (R)	1.4 (P)	3.6
West Ham	2.4 (R)	1.5	

Note: Breeding waves are separated by several days or weeks. Numbers are SD of clutch initiation dates.

waves, as birds moved in from elsewhere later in the season. Breeding synchrony of subsequent waves was less than in initial nestings in 84 percent of the colonies in 1979, 1982, and 1985 (table 3.12). This was true regardless of the degree of synchrony of initial waves.

Causes of Failures and Synchrony

We had initially predicted that causes of reproductive failure would influence subsequent synchrony and timing of subsequent waves of nesting. Nesting failure leading to renesting resulted from flood tides (67 percent), predators (18 percent), and heavy rains (15 percent). Flood tides, however, varied from severe (in 1982) when significant numbers of terns were flooded out, to mild (in 1979) when flood tides generally destroyed 20 percent or fewer of the nests at any one time. In 1982, flood tide losses accounted for 50 percent or more of initial nests in most colonies (see discussion).

In 1979, there was little synchrony when renesting occurred among colonies. High storm tides destroyed some nests in all colonies. Several high tides inundated different colonies at different times, desynchronizing second waves of nesting. In 1982, the initiation of egg-laying and the peaks of renestings occurred at the same time in all the colonies where renesting occurred. In that year, very high storm tides destroyed nests in most colonies at one time. On the contrary, in 1985, failures were due to heavy rains and predation, and second nesting waves were again asynchronous.

Synchrony following losses due to predation was less (average standard deviation 4.3 days) than that following losses due to heavy rain and massive flood tides (average standard deviation 3 days, Mann Whitney $U = 3$, $P < 0.01$). Just as with skimmers, predators act as a desynchronizing agent. However, the flood tides in 1979 resulted in less synchrony in some colonies (values of 5 to 7 days), while others were very synchronous (average standard deviation < 1 day). This difference reflects the severity and frequency of flood tides in that year.

Discussion

Common terns in our study colonies nested in small to medium-sized colonies (up to 500 pairs) on salt marsh islands, and in very large colonies on sandy beaches. Marsh colonies were usually a single unit, with no sharp, obvious demarcation between one nesting group and another. The salt marsh colonies allowed us to examine synchrony as a function of colony size, and we found generally no correlation or an inverse relation between colony size and synchrony. Over the years, synchrony tends to be lower in larger colonies, simply because larger colonies attract late arrivals throughout the season leading to a prolongation of nest initiations.

Synchronization and Desynchronization

We believe the lack of a clear relationship between colony size and synchrony is a result of the mobility of common terns, their propensity to shift colony sites during the breeding season, and the arrival of late-nesting pairs. From 1976 to 1990 the number of common tern colonies in Barnegat Bay ranged from 18 to 29. Thus, although colonies are physically discrete, they are not far apart, and are within the normal daily foraging range of the terns. When any colony experiences egg losses due to heavy rains, predators, or storm tides, the terns could reassess their nesting options. Either they can renest in their present colony, move to another seemingly successful colony, or abandon their reproductive efforts that year. We have found that low-level nest loss usually results in renesting in the same colony. However, when there are massive failures, particularly in small colonies, the terns move to another colony.

In addition, there are late nesting terns that simply arrive later than most pairs. Such birds are usually young birds breeding for the first time (Austin 1946; Nisbet and Drury 1972a; DiCostanzo 1980; Ryder 1980; Nisbet et al. 1984), although they could also be birds that had difficulty finding sufficient food resources for egg production, had lost their mates of the previous year, or had lost their territories. Gaston and Nettleship (1981) suggested that some thick-billed murres may "hold back" from egg-laying in order to lay closer to the colony peak.

Nesting Waves in Common Terns

In all three years, there were several waves of nesting in marsh-nesting common terns. Second waves generally involved considerably fewer pairs than the initial wave. This is because not all pairs fail (requiring fewer to renest), and not all failed breeders re-lay. In some cases, terns chicks in one colony suffer lower mortality than those in other colonies. After heavy rains, tern chick mortality was lower than for skimmer chicks, because common terns seem to brood their vulnerable chicks. After a heavy rain, tern chicks are usually dry and fluffy while neighboring skimmer chicks are often wet and shivering or dead.

Conclusion

We found substantial variation in laying chronology, colony size, and synchrony for common terns in large and small marsh colonies. We found interesting differences between the common terns and the associated black skimmers nesting in the same colonies (Burger and Gochfeld 1990a). Terns and skimmers arrive at the same time; terns generally lay earlier, but skimmers have higher synchrony and an earlier peak laying date. Skimmers showed an inverse relationship between number of skimmers and synchrony (smaller colonies were more synchronous), while terns showed no relationship. Terns had much greater variation in number of breeding conspecifics than skimmers, but still did not show increasing synchrony with increasing colony size. Burger (1979a) showed that there is a breakpoint in the colony size curve, above which synchrony no longer increases with colony size. This factor may be operating in the tern colonies. Because birds in one part of a large colony are not in direct communication with birds in a remote part of the colony, the relationship between colony size and synchrony breaks down beyond a few hundred pairs (Veen 1977; Burger 1979a), we suggest that synchrony should be examined in small subsections of tern colonies (Gochfeld 1977). Social facilitation could operate within small subcolonies or neighborhoods, leading to increased breeding synchrony and reduced loss of chicks to predators. Clearly, such factors might have little advantage when the major mortality is due to cold rains or flooding, when, in fact, asynchrony may be associated with higher survival.

FOUR

Habitat Selection and Territoriality

Selection of breeding habitat has profound consequences, since it places an animal in a particular environment for a long period (Partridge 1978). Some habitats have regular and predictable stressors (e.g., daily tidal swings in coastal areas); others, irregular stressors.

An animal species may be found in a particular habitat by chance or by choice (Severinghaus 1982). Description of habitat used by a species does not by itself show selection or preference. Rather, habitat selection implies an active choice by the animal of one set of characteristics over another set which is also available. This choice can be documented by showing that the habitat selected differs consistently and significantly from habitat that is available, but is not used. Thus we distinguish *availability*, space present within the home range of an animal not occupied by a competitor, from *suitability*, space which possesses attributes attractive to and favorable to the species. Habitat selection in breeding birds involves a sequence of choices as follows: 1) selection of a general habitat type (forest, marsh, beach, grassland, etc.), 2) selection of a particular location (e.g., one specific marsh), 3) selection of a territory within that location, and finally 4) selection of a nest site within the territory (Burger 1985b).

Colony site selection is the process whereby a site is chosen for occupancy by a large group of birds. It is difficult to study colony selection since the selection process may occur only once in a lifetime at the time the site is first occupied (Gochfeld 1980b). If the characteristics change dramatically the birds may shift (Burger 1974a; Montevecchi 1978b; Southern and Southern 1980; Gochfeld 1981). In other cases, occupancy continues despite substantial habitat change (e.g., Austin 1932; Southern 1977), and the habitat one studies may differ substantially from what was first chosen. Birds typically return to breed where they have bred successfully in the past, moving to new sites (and repeating the site selection process) only after a failure (Burger 1982a; Greig-Smith 1982). In ring-billed gulls, for example, Southern (1977) found a strong natal site tenacity, and that most of the recaptured gulls that had been banded as adults continued to breed in the same colony. We found a strong tendency of common terns to switch back and forth between the West End and Cedar Beach colonies, and the Austins reported both a strong site tenacity (Austin 1949) and a tendency of birds to move among the Cape Cod, Massachusetts, colonies (Austin and Austin 1956), although birds tended to move as a group, showing group adherence (Austin 1951).

In addition to colony site tenacity, there may be strong nest site tenacity. We have trapped one common tern at the same nest site four years in a row, but the record is a 17-year occupancy of a single nest site (Marples and Marples 1934).

Territory acquisition may be performed by one or both members of the pair early in the breeding season (Burger and Beer 1975). Males often use the territory for displaying to prospective mates. Females may then select the male either on the basis of his displays (as an indicator of his "quality") or on the basis of his territory and its quality (Orians 1969; Lenington 1980). Many species of colonial birds show a high degree of colony and territorial philopatry, returning each year to claim approximately the same space (Southern and Southern 1980). Nest sites are chosen for stability and for protection from the elements (eg., Becker and Erdelen 1982), from predators, and from aggressive neighbors (Burger 1974a, 1977; Dexheimer and Southern 1975; Wiklund 1982).

Higher breeding success for pairs having territories in the center of colonies has been shown for black-headed gulls (Patterson 1965), puffins (Harris 1978), fulmars, (Fisher 1952), gannets (Nelson 1966), double-crested and pelagic cormorants (Siegel-Causey and Hunt 1981), herring gulls (Darling 1938), sandwich terns (Langham 1974), cliff swallows (Brown and Brown 1987), yellow-headed blackbirds (Fautin 1941), red-winged blackbirds (Robertson 1973), and tricolored blackbirds (Orians 1961a,b; Payne 1969). Similarly, lower success rates in peripheral nests have been reported for kittiwakes (Coulson 1968), cattle egrets (Siegfried 1972), adélie penguins (Tenaza, 1971; Spurr 1975), Magellanic penguin (Gochfeld 1980a), and swallows (Emlen, J. T. 1952; Emlen, S. 1971).

Higher breeding success in central territories could be due to a number of factors such as age or experience of the breeding pairs (Crawford 1977; Ryder 1980; Nisbet et al. 1984) as well as lowered predation pressures. In gannet colonies, success may be lower in central nests if young are injured while moving to the edge of the colony to depart from a cliff (Montevecchi and Wells 1984).

Territory size varies among species, but for any species and for each stage of the breeding cycle, environmental constraints such as habitat, colony size, location within the colony (e.g., center versus edge), vegetation pattern, and bird density may influence territory size. In areas of high density, nesting territories must be small, but it is not clear which is cause and effect. Territories may be smallest in preferred habitat (Richards and Morris 1984). The presence of other species may also affect territory size (Orians and Willson 1964; Duffy 1983). The ecological consequences of interspecific territoriality in birds have been discussed by Murray (1971, 1981). Most studies have focused on passerines with Type A territories (Nice 1941) where the territory provides food resources as well as nest sites. In a discussion of interspecific territories in reed warblers in Europe, Murray (1988) notes that time of arrival on the breeding grounds, along with body size, function to permit similar species to establish territories in the same general area. In colonial species, where food is not obtained within the territory, several species may coexist in mixed species colonies. Not only do time of arrival and body size affect territory size and the outcome of interspecific interactions (Burger and Gochfeld 1983), but sheer numbers of individuals and persistence may allow smaller birds to evict larger, earlier-nesting, but less dense species. This has been reported for sandwich terns evicting black-headed gulls (Veen 1977) and elegant terns evicting Heerman's gulls (Barrie 1975).

Common terns often nest with black skimmers (Erwin 1977; Burger and Gochfeld 1990a), and they actively defend their territories against the larger species (see Chapter 5; fig. 5.4). In this chapter, we will discuss the effect of skimmers on territory size and acquisition in common terns. In this special case, where the minimum territory size is determined by the ability to reach one's neighbor without leaving the nest, the actual appearance of the neighbor, whether conspecific, similar-congeneric (e.g., roseate tern), or conspicuously different (e.g., black skimmer), does not determine the nesting pattern. Territorial defense clearly imposes time and energy costs required to obtain and defend the territory and involves potential injury during territorial encounters; while benefits are measured in terms of space for mate attraction, uninterrupted courtship, copulation, incubation, and chick care.

Nest Site Selection

Nest site selection involves choosing a specific site within one's territory for the nest. In many species, this process involves an interaction of the male and female (Tinbergen 1953; Burger 1985b), in others, the female clearly chooses the nest site (Case and Hewitt 1963; Orians 1980). As with the other stages, nest site selection may occur one or more times a year (if nests are destroyed), and may occur every year, or a pair may regularly use previous nest sites. Reuse of previous nest sites has been documented for horned grebe (Ferguson 1981), glaucous-winged gull (Vermeer 1963), herring gull (Brown 1967a), black-tailed gull (Austin and Kuroda 1953), common gull (Onno 1967) and black headed gull (Patterson 1965). We have some evidence from banding that at least some common terns choose the same nest site year after year, but many do not.

It is difficult to assess competition for nest sites because some species almost always nest with other species (Gochfeld 1978; Erwin 1979), and sites that to humans appear suitable may not be adequate for the birds (Burger and Lesser 1978). Nonetheless, competition for nest sites has been shown for swallows (Hoogland and Sherman 1976; Freer 1979), blackbirds (Orians 1961a), alcids (Belopol'skii 1957; Manuwal 1974; Williams 1974; Ashcroft 1979), herons (Meanley 1955; Burger 1978a,b), gulls (Hailman 1964; Ytreberg 1956; Montevecchi 1978b), terns (Morris and Hunter 1976; Burger and Shisler 1978), cormorants (Brun 1974), and penguins (White and Conroy 1975; Trivelpiece and Volkman 1979). Belopol'skii (1957) noted that there is intense competition for nest sites on the cliffs in the Barents Sea. Younger birds are forced to use nest sites close to the cliff edges, and often lose their eggs or young that fall from the nests. This kind of competition also may be common in other marine species that have not been examined. Some gannets remain on their nest sites and defend their territories long after the young have departed (Nelson 1970, 1979), suggesting intense nest site competition with differences in fledging success as a function of nest sites (Montevecchi and Wells 1984).

4.1 GENERAL HABITAT SELECTION IN COMMON TERNS

Common terns nest in a variety of habitats including rocky islands, sandy beaches, freshwater and salt marshes (Austin 1929; Palmer 1941a; Hewitt 1941; Burger and Lesser 1978; Severinghaus 1982; Richards and Morris 1984; Burger and Gochfeld 1988). In salt marshes, they generally nest on a mat, but may nest on the marsh grass itself. To obtain an overview of habitat selection for the coastal common tern population, we examined habitats used by common terns over an extended period of time in coastal New York and New Jersey. For comparative purposes, we examined colonies in New York and New Jersey separately because the birds seem to move within these areas, and few terns nest in the area between northern Barnegat Bay and western Long Island.

New Jersey Salt Marsh and Barrier Island Colonies

In coastal New Jersey, over 100 sites are used by nesting common terns. Based on three years of state-wide censusing (Kane and Farrar 1976, 1977; Galli and Kane 1979), most (89 percent) common tern colonies are located on mat or on *Spartina* in salt marshes, with 4 percent on spoil in salt marshes, and 7 percent on sandy barrier beaches. Almost no tern colonies are on mainland beaches exposed to mammalian predators.

New York Sandy Beach and Salt Marsh Colonies

In New York, common terns nested in four habitat types: interdune beach, spoil islands, rocky islands, and salt marsh. Interdune colonies were located between or adjacent to dunes on the forebeach of barrier islands. We censused the New York colonies from 1976 to 1980, and located 130 colonies (table 4.1). Overall, most colonies (72 percent) were located in salt marshes, but most individuals (65 percent) nested in interdune colonies. This disproportionate distribution is shown in figure 4.1.

Table 4.1. Common tern colony locations by habitat on Long Island. Shown are number of colonies.

Year	*Dune*	*Spoil*	*Salt Marsh*
1976	4	3	7[a]
1977	5	2	11
1978	3	4	23
1979	3	6	28
1980	2	5	24

[a] Salt marshes were undersampled in 1976.

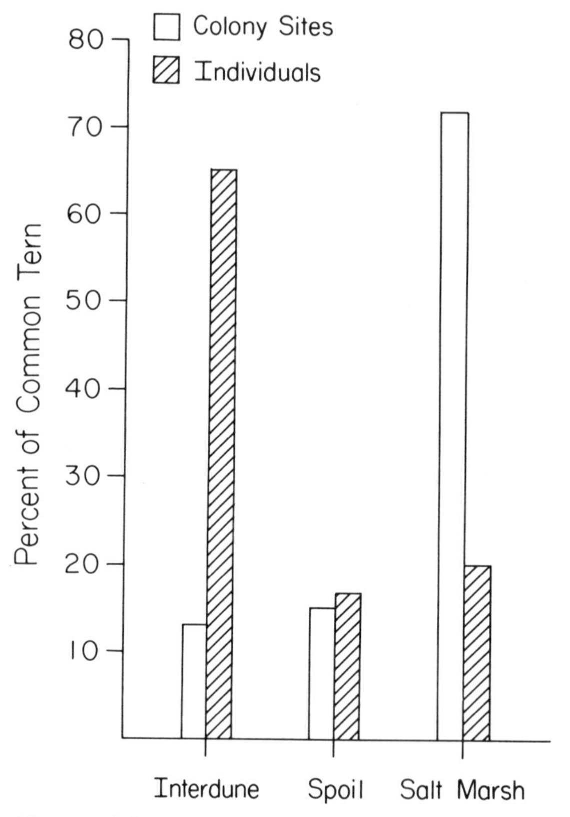

Figure 4.1. Common tern habitat selection on Long Island. The percentage of colonies in each of three habitats are shown with open bars and the percent distribution of nesting individuals is shown by hatched bars. Interdune colonies were consistently larger than those in other habitats except for the Great Gull Island colony.

4.2 COLONY SITE SELECTION IN SALT MARSHES

Colony selection was examined in detail in Barnegat Bay by comparing the colony sites used by common terns with the available habitat. To examine selection, it is necessary to compare the characteristics of the colony sites chosen by the birds with unoccupied sites. From 1976 to 1988, common terns used 35 different colony sites. Some sites were used in most years; others, only once or twice.

Physical Characteristics

Barnegat Bay has 259 islands, and only 2 are completely spoil with no salt marsh vegetation. We defined an island as any piece of land completely surrounded by water, even if the channel was only a few meters wide. Terns selected islands of a limited size range (Yates $X^2 = 30.2$, $df = 6$, $P < 0.001$) relative to the available size, shunning the very small and very large islands (Burger and Lesser 1978). Common terns nested on islands that were closer to land and avoided islands in the middle of the widest areas of the bay.

Of the 259 islands, 28 percent were closer to the mainland and 72 percent were closer to the barrier beach. Likewise, most skimmer (75 percent) and tern (78 percent) colonies were located closer to the barrier beach than to the mainland. Elevation could play a key role in colony site selection because low sites are exposed to frequent inundation by high tides. Vegetation can be used as an index of elevation: *Spartina alterniflora* grows in the lowest areas exposed to daily inundation, *S. patens* grows in higher areas exposed to inundation during spring high tides, and bushes and *Phragmites* usually grow only in areas where flooding is confined to extreme high tides or northeastern storms and hurricanes. Common terns selected to nest on islands with more than 20 percent *Spartina patens*, and many selected higher islands with bushes and *Phragmites* (Yates $X^2 = 41.3$, $df = 7$, $P < 0.001$; table 4.2). Thus, common terns selected to nest on small (<20 ha), but not tiny islands, close to mainland or barrier beaches, and they avoided low islands covered entirely with *Spartina alterniflora*. When we examined the characteristics of unoccupied islands, we found only three with characteristics within the range used by the terns. Unoccupied islands were larger or smaller than occupied ones, and usually nearer to the mainland (Burger and Lesser 1978).

An important feature of the salt marsh habitat is the wrack or mat of dead vegetation which forms stable platforms on the salt marsh grasses. These

Table 4.2. Overall habitat preferences of nesting common terns in salt marshes in Barnegat Bay. Shown are number of islands in each vegetation category with and without nesting common terns (percent = 100% for each column).

	Available	*Terns*
	No Nesting	*Nesting*
All *S. alterniflora*	138 (61%)	1 (3%)
At least 80% *S. alterniflora*	4 (2%)	0 (0%)
All *Spartina:* >20% of each species	14 (6%)	15 (43%)
Over 80% *S. patens*	8 (4%)	2 (6%)
Spartina with <10% bushes	26 (12%)	8 (22%)
Spartina with >10% bushes	14 (6%)	4 (11%)
Spartina with 21–40% bushes and/or *Phragmites*	7 (3%)	0 (0%)
Spartina with >40% bushes and/or *Phragmites*	13 (6%)	5 (15%)
Total	224 (100%)	35 (100%)

platforms are created by the action of wind and tide which carry masses of floating vegetation onto the windward side of islands, depositing much of it in the zone of the highest winter storm tides, areas which often escape flooding in the subsequent breeding season.

In general, common terns avoided nesting on islands with no mats; such islands were usually low and were subject to wash-over by summer high tides. Mat length did not seem to be as critical as mat width, since there was overlap in the length of the mats on islands with and without terns. Although most terns nested on mat, many were forced to nest in the grass itself (fig. 4.2), where they assembled a nest on bent-over grass stems.

Social Factors

Common tern colonies are occupied by a variety of other species as well. Notable among these is the black skimmer whose breeding biology and social behavior has been discussed in a companion book (Burger and Gochfeld 1990a). Since other colonial species are potentially an important part of the habitat (Koskimies 1957), we considered whether common terns might be selecting islands by virtue of their having black skimmer colonies (fig. 4.3). In Barnegat Bay, common terns nested on 35 islands, 24 of which also had

Figure 4.2. Common tern nesting on mat (top) and in *Spartina alterniflora* (bottom) in New Jersey salt marsh colony.

Figure 4.3. Common tern attempting to insinuate itself and carve out a territory among a group of nesting black skimmers.

nesting skimmers in some years. There were no skimmer colonies which did not also have nesting terns. Thus, we concluded that skimmers are strongly influenced by the presence of common terns, but not vice versa. In Barnegat Bay, unlike Long Island, there was no significant correlation between the number of pairs of terns and skimmers nesting in the colony (Kendall *tau* = 0.159, N = 67). Similarly the number of skimmers and the number of terns at colony sites was not correlated across years (*tau* = 0.23).

4.3 COLONY SITE SELECTION ON BEACHES

Long Island also has a series of barrier beaches (fig. 4.4) with back bays, and numerous salt marsh and dredge spoil islands. To compare colony site use in another barrier island–salt marsh complex we compared islands used by nesting common terns from 1975–1978 with those not used. Islands were censused in early June by helicopter, and were censused later in the season by boat.

Figure 4.4. Typical Long Island barrier beach colony (West End II) showing scattered vegetation (top) and common tern nest site (bottom).

For the 25 km area between Jones Inlet and Fire Island Inlet, during the period 1964 to 1985, beach colonies existed at Alder Island, Meadow Island, Short Beach, West End Beach, Meadowbrook, Wantagh, and Cedar Beach. Salt marsh tern colonies were found at Goose Island, South Line Island, North Line Island, Gilgo Island, and Seganus Thatch. The Gilgo and Seganus colonies occupied more than one island.

The Line Island colonies were initiated in 1975 after construction of a sewage outfall pipeline had resulted in deposition of spoil over the marsh. These islands had little natural mat. After terns had been nesting on the narrow spoil zone, the level began to sink and by 1978 it was mostly underwater. Birds began to nest on the nearby salt marsh in widely scattered pairs.

In the vicinity of Jones Inlet (fig. 2.3), the Meadow Island site was occupied for four years, and at one point contained over 1000 pairs. The following spring it was covered with fresh dredge spoil, obliterating the vegetation. Only a few terns and skimmers remained there subsequently. Some moved to the Alder Island site which was occupied for only one year. The Short Beach site was occupied for about three years, and was then reoccupied by about 50 pairs in 1984. Most of these nomadic birds returned to the West End II colony which grew to over 2500 pairs in the mid-1970s. Subsequently, it has declined to zero in 1988, the large concentration of birds relocating about 1.5 km to the east.

The tern population in the vicinity of Fire Island inlet has been more stable. There was substantial movement from island to island in the 1960s, but since 1970 the Cedar Beach site has been occupied continuously and now comprises the bulk of the terns nesting in that region.

Nesting Habitat

Nesting habitat on Long Island salt marshes was very similar to that in New Jersey except that the wrack was primarily dead stems of *Spartina alterniflora* rather than eelgrass. These stems were jumbled together creating a substrate very similar to that of the eelgrass, but it tended to have better drainage and to retain less water after flooding and rainstorms.

On the beach colonies the main substrate was sand mixed with a small amount of shell or fine gravel. The terns avoided the pure fine sand on outer beaches where it was subject to wind-drifting (Burger and Gochfeld 1990b), but tolerated pure fine sand in the more stable interdune areas. Common terns generally avoided completely barren beaches which were favored by least terns. Once the vegetation coverage reached about 10 percent, common terns might nest, although most of the colonies had vegetation coverage in the range of 10 to 25 percent. As areas of the colony became overgrown, common terns moved to less vegetated areas, although in long-lived colonies they appeared to tolerate vegetation that would not have attracted pioneering terns (see also Austin 1932).

More terns nested along the interface between vegetation and open sand than on the more open sand or within the vegetation. This achieved a degree of habitat partitioning, with the black skimmers nesting in the center of open patches of sand (Burger and Gochfeld 1990a) and roseate terns nesting in denser vegetation.

4.4 MOVEMENT BETWEEN HABITATS

In both New York and New Jersey, common terns nest on salt marshes and on barrier beaches. The physiognomy of beach and salt marsh sites clearly differs, and birds may select one or the other. The extent to which adult terns move back and forth among habitats, indicating a flexibility in general habitat preferences, is not known.

In most years, over 500 young terns were banded at West End and Cedar Beach. By trapping adults on nests, it has been possible to determine the origins of many breeding birds and to determine their intercolony movements. The results are illuminating. Extensive banding and trapping results have shown a strong tendency of birds to move back and forth between West End and Cedar Beach.

We also documented a rather free interchange between beach and marsh colonies. Many birds trapped as adults on salt marsh islands had originally been banded as chicks at one of the beach colonies. Repeated trapping year after year showed that some adults bred in salt marsh one year and at a beach colony the next, or—in some cases—birds were trapped in both habitats in a single year, after having failed in an initial nesting attempt and relocating to a new colony.

To consider this in more detail, we benefited from an unexpected experiment of major proportion. In late 1977, the Cedar Beach colony was disrupted by construction of a sewer outfall pipeline which cleared a 70 m swath through the center of the colony (Gochfeld 1981). The 1977 nesting population on this transect was about 525 ± 25 pairs, and knowing in advance of the impending construction, a vigorous trapping program resulted in marking 660 of these birds. No birds nested on the swath in 1978. Efforts to locate these birds included adult trapping programs on adjacent quadrats, on remote quadrats in the same colony, at West End, and on the newly established salt marsh colony at Seganus Thatch 4 km to the northeast.

We trapped 693 adult terns at Cedar Beach in 1978, only 15 of which were birds previously banded on the pipeline swath. By extrapolation to the estimated 5400 ± 200 adults, only about 110 of the 660 banded swath birds would have remained in the colony. Trapping at West End revealed 11 swath adults which extrapolates to an emigration of 133 swath birds, somewhat greater than the number estimated to remain at Cedar Beach. We also trapped 109 terns at the Seganus colony and found 12 swath birds, indicating that at least 30 adults moved from Cedar Beach to the salt marsh (Gochfeld 1981a). Thus the displaced terns moved to both beach and marsh sites.

In New York, we banded extensively from 1964 to 1985, and show data from two years (table 4.3). Clearly, most terns return to the same colony, and many return to the same quadrat. However, some birds did move to salt marsh habitat, indicating that there is plasticity in their habitat choice. Movement of common terns among habitats in Barnegat Bay is more difficult to surmise because there are so many salt marsh colonies. Further, when washouts of common terns occur, they usually either renest in the same colony, or move to another nearby salt marsh colony. However, although we have banded only 200 common tern adults breeding in New Jersey salt marshes, 4 have been located the same year breeding in barrier beach colonies in New York. Similarly, three birds banded in beach colonies in New York have subsequently bred in salt marsh colonies in the same year in New Jersey.

Another unfortunate event provided an unplanned "experiment." In the early 1970s, a large thriving tern colony at Breezy Point, Brooklyn, on extreme western Long Island, was completely eliminated by vandals who set the vegetation on fire during the breeding season. Four years after the abandonment, Breezy Point was recolonized, and by conducting a vigorous program of trapping adults, Post and Gochfeld (1979) documented the origin of the breeding birds. Of known-origin birds, 86 percent had come from the large West End and Cedar Beach colonies, but 21 percent of the remainder were from salt marsh colonies, including one banded as an adult in nearby

Table 4.3. Movement of adult common terns in New York from the colony previously occupied. Shown are number (percent).

	West End	*Cedar Beach*
Remained in previous colony		
Same quadrat[a]	79 (76)	59 (39)
Adjacent quadrat	7 (7)	15 (10)
2–5 quadrats away	5 (5)	22 (15)
Elsewhere	6 (5)	11 (7)
Moved to another sand colony	7 (7)	31 (20)
Moved to salt marsh colony	0	13 (9)
	104 (100)	151 (100)

[a]Quadrats were 20 x 20 m.

Jamaica Bay and one banded as a chick in the Sloop Sedge colony in Barnegat Bay.

These studies indicate that birds raised on beach colonies can breed in salt marsh and vice versa. Moreover, having bred in one habitat does not preclude subsequently breeding in the other. The much more difficult problem is to determine whether these are exceptions to a much stronger tendency to return to breed in one's natal habitat or at least to return to the habitat where one has breed successfully. We have shown experimentally (Burger and Gochfeld 1990c) that common tern chicks under duress will run to the vegetation type in which they were raised, which suggests some degree of habitat imprinting, but additional studies will be required to resolve these questions.

4.5 COMPARISON OF COLONY SITE SELECTION AMONG HABITATS

Throughout the New York Bight from Montauk, New York, to Cape May, New Jersey, there were more common tern colonies in salt marshes, but in New York, at least, more pairs nested on beaches, while most salt marsh colonies held fewer than 100 pairs.

In selecting salt marsh sites for nesting, common terns selected intermediate-sized islands that were closer to the barrier beach than to the mainland, but not so close that predators could easily reach them. Secondly, they selected islands that were above mean high tide (to prevent flooding), but not so high that they contained upland vegetation that could harbor mammals such as rats, fox, or raccoon. Lower elevation islands flood completely during high winter tides which reduces or eliminates rats. In selecting beach or spoil colonies, terns preferred large interdune areas with sparse to moderate vegetation and nested mainly on flat sand sections rather than on the dunes themselves. Overall, the aspects of nesting on beach and marsh were quite similar, sparse vegetation with some birds nesting in grass and others in the open. In both habitats, birds avoided low lying or brushy areas.

4.6 NEST SITE SELECTION IN COMMON TERNS

Salt Marshes

Having selected a salt marsh for a colony site, common terns can then nest in *Spartina alterniflora*, *Spartina patens*, on mat, or among bushes or *Phragmites*. Of the salt marsh islands in Barnegat Bay occupied by terns, over 60 percent of the available space was covered with *Spartina alterniflora*, over 20 percent with *Spartina patens* and the rest with bushes and mat (see table 4.2). Common tern choice of nest site differed significantly from the available habitat (Yates $X^2 = 273$, $df = 3$, $P < 0.001$); they nested primarily on mat and in *Spartina patens*.

Wrack or mat strewn on the marsh varies with respect to three important factors: 1) height above the marsh, 2) color of the mat, and 3) width and length of the mat. Mat height and color vary among mats of appropriate size, or even within one mat, and reflect the age of the mat. Over time, the vegetation becomes bleached, turning from mottled brown and black to white and grey. Similarly, the bottom layers slowly decay, and the mat decreases in height gradually sinking until it is no longer evident.

Terns usually nested on narrow mats, or near the edge of wide mats (table 4.4). This preference could result from competition with skimmers that prefer the centers of wide mats (Burger and Gochfeld 1990a). To study preferences of terns and of black skimmers, we constructed equal-sized mats in 1984 and 1985. By 1985, the mat constructed in 1984 was almost white, contrasting with the new, dark mats left by winter storms. Common terns preferred light mats ($n = 235$ nests) but some did court and nest on dark mats. Over 70 percent of the terns that nested on mats nested on white rather

Table 4.4. Common tern nest placement realtive to the edge of the mat. Given are mean distances (cm) to the edge of mats ± 1 SD. Based on 70–201 nests at Pettit, 41–98 nests at Carvel, and 18–110 nests at West Vol.

	Pettit	*Carvel*[a]	*West Vol*
1979	21 ± 16	18 ± 17	23 ± 12
1983	18 ± 9	25 ± 18	32 ± 19
1984	22 ± 2	14 ± 12	49 ± 51
1985	34 ± 23	36 ± 30	72 ± 61

[a]East Carvel in 1979–1985; West Carvel in 1985.

than dark mats, though only 34 percent of the available mat area was classified as white. Over half the terns nested on mats rather than in the grass, although mats never accounted for more than 20 percent of an island's area. The terns' preference for mats indicates a potential for nest competition between terns and skimmers since the latter nest only on mat (Burger and Gochfeld 1990a).

Terns usually arrive in the area earlier and select the preferred sites on the mats, but skimmers are considerably larger, thus making it easier for them to win in direct aggressive interactions (scc Schoener 1974). Presumably, competition could be reduced by habitat partitioning if each species preferred a different section of the mat. In general, terns nested near the edges of mats (table 4.4), leaving the center for skimmers. In every year on eight islands terns nested closer to the mat edge (outer 25 percent of mat) than did skimmers which generally nested at least a meter or so from the edge (in the inner 50 percent; $P < 0.001$). We confirmed the above observations experimentally, by constructing artificial mats of equal sizes to provide equal nesting opportunities on different islands. In that experiment, terns chose edge sites (outer 25 percent) and skimmers chose central sites (inner 50 percent). The pattern of terns nesting on the edge of mats, and skimmers nesting on the center of mats could result from habitat partitioning or from competition.

We examined this by building large (4 × 12 m) and small (1 × 12 m) mats on three islands (table 4.5). Skimmers nested mainly in the center of these mats and the terns nested all over the large mats, although they still concentrated near the edge. Moreover, on one of the large mats on West Vol, where there were no nesting skimmers, significantly more terns nested near the center. Thus, we infer that the nesting distribution of terns is clearly being

Table 4.5. Location of common tern nests on artificially constructed mats in New Jersey salt marshes. Mean I 1 SD with sample size in parentheses.

	Mean Distance to Edge of Mat (cm)
Experiment I	
Pettit	
Small mats	18.2 ± 14(17)
Large mats	27.8 ± 31(24)
F (ANOVA)[a]	2.70 (P<0.05)
Carvel	
Small mats	10.2 ± 7 (14)
Large mats	19.4 ± 18(19)
F (ANOVA)[a]	3.20(P<0.005)
Experiment II[b]	
West Vol	
Small mats	19.3 ± 16(18)
Large mats	45.4 ± 53(44)
with skimmers	17.9 ± 14(24)
without skimmers	76.5 ± 68(20)
F for terns on large mat	12.0 (P<0.001)

[a] Log transformed data. Compares terns nesting on large with small mats.

[b] In experiment I, large mats were 3 x 10 m. In experiment II, mat was 3 x 12 m., and some sections had terns and skimmers while others had only terns.

affected by the presence of skimmers. Nor is it necessary that the skimmers actually be present when the terns begin nesting, for the terns and skimmers have nested on the same islands for a period of years, and a pattern, observed one year may reflect events transpiring in previous seasons.

We also examined the temporal pattern of territory establishment and nest site selection on the artificial mats constructed on Pettit and East Carvel islands. In the presence of black skimmers, terns clearly occupy the artificial narrow and the naturally occurring narrow mats, before they occupied the wider mats on which the skimmers were nesting (table 4.6).

Therefore, in 1985, we constructed two 3 × 10 m mats on Pettit and Carvel to determine the temporal sequence of nest location in the absence of nesting skimmers (table 4.7). The early-nesting terns nested on the edge in the absence of skimmers, and later-nesting pairs nested in the center. These experiments indicate that there is a complex interaction between the terns' habitat preference and the influence of the skimmers.

Table 4.6. Temporal selection by terns and skimmers of territories on grass, natural narrow mats, and artificial mats in 1984 in Barnegat Bay. Given are number of pairs on territory (and nests).

	Pettit				*East Carvel*			
	Terns			*Skimmers*	*Terns*			*Skimmers*
Date	*Grass*	*Natural Narrow Mats*	*Wide Mats*	*Wide Mats*	*Grass*	*Natural Narrow Mats*	*Wide Mats*	*Wide Mats*
May 4	0	0	0	0	0	0	0	0
May 6	2	6	0	9	12(2)	8	0	0
May 8	6	4	2(0)	10	14(12)	3	0	3
May 10	16(4)	2	2(0)	13(4)	16(14)	2	0	6(4)
May 12	18(8)	2(2)	4(2)	14(6)	18(17)	4(3)	0	12(8)
May 14	24(18)	2(2)	12(10)	14(11)	22(22)	4(3)	0	12(12)
May 16	28(28)	2(2)	20(20)	14(14)	22(22)	4(3)	0	14(14)

Note: Terns also established territories on other, narrow mats not occupied by skimmers.

Table 4.7. Temporal sequence of nest placement of common terns on artificial mats (1985, New Jersey salt marshes). Given are the mean distances in cm (±1 SD) from the edge of the mat of new nests initiated since the last census. Number of new nests in parentheses.

Date	*Pettit*	*East Carvel*
May 16	——— [a]	6.5 ± 5.0 (8)
May 20	6.6 ± 5.2 (4)	19.2 ± 7.9 (6)
May 24	10.3 ± 9.7 (9)	47.9 ± 43.6 (14)
May 28	12.3 ± 7.4 (17)	125.9 ± 106.3 (13)
June 1	84.0 ± 70.3 (10)	——— [a]
June 5	102.3 ± 58.3 (6)	——— [a]

[a] No new nests initiated.

Beach Colonies

Specific features of common tern nest sites and random points have been examined at Cedar Beach (Gochfeld and Burger 1987; Burger and Gochfeld 1988), revealing that common terns select more open areas with less vegetative cover than around random points. However, even on nearly barren areas of the colony, terns usually have some vegetation or other object within their territories where chicks can seek cover. Most nests were placed next to some discontinuity in the sand, if not vegetation, then some stick or other debris. Although almost all roseate tern nests were placed under vegetation, less than 1 percent of common tern nests were under vegetation, and in almost all cases the entire nest and clutch would be readily visible from above.

Discussion

Colony and Nest Site Selection

Common terns nested on two physiognomically diverse habitats: salt marshes and beaches. Their choice of nest site clearly reflects conflicting selection pressures. The dangers of flood tide impose constraints (see Storey 1987a,b) forcing terns to nest on high salt marsh islands with wrack, whereas predation pressures select for nesting on lower islands, devoid of extensive bushes and shrubs. Islands with upland areas can support populations of mammalian predators. Since the terns nest on many islands with skimmers, they face some competition. The skimmers are considerably larger than the terns and can win in overt aggressive interactions.

Our data indicate that there is both resource separation and competition between terns and skimmers. In the absence of skimmers, the terns use wide mats, but still prefer the edges, the interface between mat and *Spartina*. Edge nests are closer to the shade provided by the vertical grass stems.

Banding data document the fact that some terns have nested in both marsh and beach habitats. Sometimes, after an early season failure, birds will move to a colony in a different habitat, suggesting that it is overall physiognomic feature, rather than specific substrate, that influences their nest site choices. Thus both habitats are suitable. Mats, with their open vegetation-expanse, resemble sandy patches of the beach colonies, while nesting in the *Spartina* offers a similar aspect to nesting in the beach grass of sandy colonies. The ability to use both habitats provides alternatives in years when flooding or predation or other disturbance may make one or the other habitat unsuitable.

4.7 TERRITORIALITY AND SPACING

In the discussion that follows, "territory" refers to nesting territory. Accepting the definition of a territory as a space that is actively defended (Nice 1941;

Brown 1964), the best measures of nesting territory size are obtained by direct observation of aggressive interactions. This procedure is time consuming, and of necessity can be applied best to long-term observations of a few nests (Burger 1984c). Therefore, we used the nearest neighbor distance as a surrogate measure to estimate nesting territory size. In a study of herring gulls, Burger (1984c) found a high correlation between territory size determined by aggressive interactions and nearest neighbor distance. The latter measure first gained popularity in phytosociological work by Clark and Evans (1954), and Pielou (1977) discussed the theoretical advantages of this simple technique. We use the nearest neighbor distances for most of our comparisons.

Another common measure of the spatial pattern of nesting is the density or number of nests per unit area. Density and nesting territory size are inversely related, and thus the reciprocal of the density (area per bird) could be used to estimate territory size. However, because of the lack of habitat homogeneity, we found that there are large gaps of unoccupied or unoccupiable space and unless one confines density measures to a small area of uniform habitat, one will get a spurious estimate of the average territory size.

Nearest Neighbor Distance

Relatively few studies report nearest neighbor distances (see Roselaar 1985 for summary). Erwin (1977) reported a mean nearest neighbor distance of 160 cm for a common tern colony. Palmer (1941a) reported that 43 cm was probably the minimum tolerated nearest neighbor distance, which is close to the 44 cm that we measured at the Moriches colony (only one value was as low as 40 cm).

Intercolony Variation

In New Jersey salt marshes, mean nearest neighbor distances ranged from 87 to 514 cm, although most colonies had mean nearest neighbor distances between 100 and 300 cm (tables 4.8 and 4.9). The lowest nearest neighbor distances (smallest territories) occurred on Buster Island in 1976 and on the Lavallette Islands in several years. The colonies on these islands usually contained over 300 pairs, whereas almost all other colonies were considerably smaller. Nests were not placed in the center of territories, since the distance to the closest neighbor in one direction (mean = 214 ± 107 cm) was much less than the two next directions (mean = 310 ± 136 and 487 ± 246, N = 510,

Table 4.8. Comparisons of mean nearest neighbor distances (NND) for common terns as a function of habitat and location for years examined.

Habitat	*Long Island (N.Y.)*	*Barnegat Bay (N.J.)*
Salt marsh islands		
Range of mean NND's (cm)	101–312	87–514[a]
Colony size (pairs)	4–250	2–750
Sandy beach on salt marsh Islands		
Range of mean NND's (cm)	180–310	230–280
Colony size (pairs)	2–225	2–100
Sandy beach on barrier island		
Range of mean NND's (cm)	105–259	80–118[b]
Colony size (pairs)[c]	8–162	2–100

[a] Smaller distances on mat; larger distances in *Spartina*.
[b] Holgate (in different years) and in different subcolonies.
[c] Subcolony.

Table 4.9. Newest neighbor distance (an estimate of relative territory size) of common terns nesting on mat in salt marshes in Barnegat Bay. Given are means ± 1 SD in cm. All nests were measured in these years. For sample sizes, see table 9.7.

Island	*1976*[a]	*1978*
N.W. Lavallette	120 ± 34	110 ± 22
S.W. Lavallette	110 ± 47	118 ± 31
N. Lavallette	126 ± 49	122 ± 36
S. Lavallette	137 ± 52	112 ± 26
West Buster	125 ± 82	391 ± 42
Middle Buster	87 ± 29	271 ± 61
Large Buster	107 ± 45	514 ± 204
East Vol	——[b]	183 ± 62
West Vol	257 ± 117	164 ± 53
Gulf Point	120 ± 62	168 ± 42
East Carvel	170 ± 99	121 ± 56
Pettit	182 ± 67	247 ± 43
Cedar Creek	155 ± 45	228 ± 82
East Ham	——[b]	231 ± 42
West Ham	171 ± 53	167 ± 61

[a] Some values from Burger and Lesser 1976.
[b] No nesting in that season.

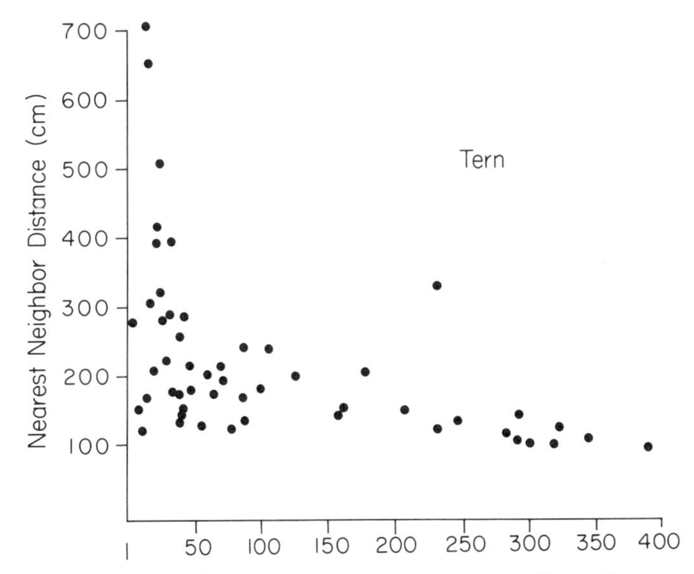

Figure 4.5. Relationship of mean nearest neighbor distance to number of breeding pairs for New Jersey Salt Marsh colonies. (Each dot reprsents one colony-year.)

West End II). Nearest neighbor distance was related to colony size (fig. 4.5). Independent of island area, territories were smaller in colonies with more nesting pairs ($r = -0.87$, $Z = 3.89$, $df = 102$, $P < .001$). Terns nesting in colonies of less than 10 pairs often nested several hundred meters apart; whereas terns in colonies of over 200 pairs usually nested closer than 200 cm, and sometimes nested 50–100 cm apart. The smaller nesting territories are not due solely to the number of birds for the islands had unoccupied areas without increased packing.

In beach colonies, territory size varied greatly because the colonies were large and the habitat diverse. We measured internest distances in several subcolonies at West End and Cedar Beach (table 4.10). Nearest neighbor distances vary greatly even within the same year in the same colony. At West End and Cedar Beach, a tern's average nesting territory size was about 20 sq. m (nearest neighbor distances averaging 250 cm). About 2 percent of nearest neighbor distances were less than 100 cm, and we occasionally found clusters where several adjacent nests were less than 100 cm apart (fig. 4.6, West End II). Terns will on occasion nest much more densely. We have found clusters of nests on Moriches where the mean nearest neighbor distance was 44 cm

Table 4.10. Variations in mean nearest neighbor distance for common terns in densely settled (20 x 20 m) quadrats on the beach colonies. Given are means in cm (±1 SD).

Cedar Beach	*West End*
120±43	100±51
143±58	110±16
160±51	167±84
230±51	193±84
231±120	224±150
220±138	
237±59	
259±142	
300±120	

(range 40–58 cm; fig. 4.6, Moriches). We have not encountered such density at either West End or Cedar Beach, where the maximum density encountered has been 23 nests per 100 sq.m.

Seasonal Changes

Nearest neighbor distances in the salt marsh colonies change seasonally, reflecting the particular constraints of breeding activities (table 4.11). Territories are large initially, decrease during incubation, and increase during the chick phase. Larger spaces are needed in the chick phase to avoid chicks wandering into territories of hostile neighbors or conversely to reduce the risk of adopting a foster chick, and to avoid piracy while fish are transferred to chicks (fig. 4.7; see chapter 6). Two colonies (Log Creek and North Log Creek) show a seasonal pattern of increasing nearest neighbor distance, and these reflect a gradual destruction of nests due to flood tides, and the eventual demise of the colony.

Discussion

Vulnerability of Reproductives

Common terns establish territories that they use for courtship, mate acquisition, copulation, egg-laying, incubation, and chick care. Their spatial requirements thus show daily and seasonally variation, and are often influ-

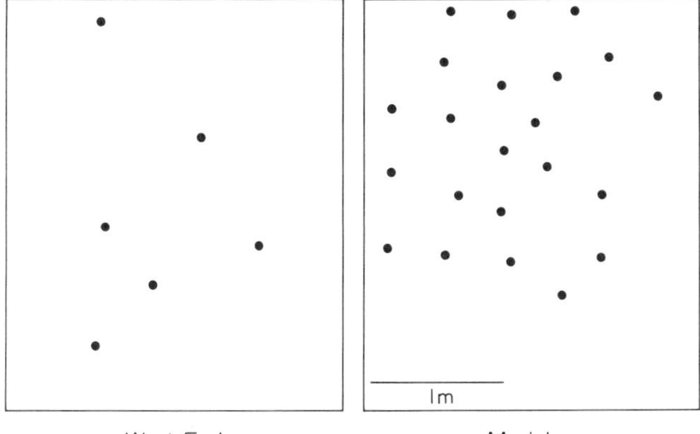

Figure 4.6. Examples of nesting density patterns at two sandy beach colonies in New York. Example on the left shows an unusually dense cluster (mean nnd = 95 cm) at West End; example on the right shows a sample from Moriches illustrating the maximum density we have found (mean nnd = 44 cm) at Moriches. Both patterns are on the same scale, and each dot represents one nest.

Table 4.11. Seasonal changes in territory size (mean nearest neighbor distances in cm ± 1 SD) for salt marsh colonies of common terns in New Jersey (1979).

	N[a]	*Pre-incubation*	*Early Incubation*	*Middle Incubation*	*Early Chick Phase*
Gulf Point	60–110	514±110	245±92	182±32	270±101
Large Buster	32–45	163±42	159±38	98±20	171±42
East Vol	215–300	183±22	111±12	231±26	360±102
West Vol	130–180	230±36	219±18	164±18	220±86
East Carvel	122–210	216±51	151±61	121±12	370±110
West Carvel	31–47	310±101	215±18	165±22	233±53
Pettit	68–90	300±96	251±25	210±18	242±62
Flat Creek	7–13	316±210	346±280	459±310	1010±560
Log Creek	15–18	345±183	482±310	514±322	970±450
Cedar Bonnet	21–26	226±62	216±26	168±46	382±61

[a]Number of nests in the four sample periods (number of active nests varied during the breeding season).

Figure 4.7. Fledgling common tern attempting to eat a large fish on its territory at Cedar Beach. Fish protruding from its beak is vulnerable to piracy.

enced by territorial interactions with neighbors. Life on a very small territory reduces the amount of space one must cover to patrol the borders, while increasing the potential for frequent interactions by virtue of the sheer proximity of neighbors. These aspects of costs and benefits of territoriality will be considered in the next chapter. Many of the functions mentioned do not occur at the same time and exert differing influences on the birds. Courtship and copulation behavior require that the pair be uninterrupted. Thus a fairly large territory is optimal. During incubation, the amount of space required decreases since pairs are required only to protect their eggs, and this they can do mainly by incubating continuously. When the chicks hatch, more space may be required so that chicks have sufficient space to move about and are not molested by neighbors or other intruders.

The above considerations of the functions of territoriality suggest that different aspects of the reproductive effort are vulnerable at different times in the season. Taken altogether these constraints suggest that territory size should vary seasonally and with habitat, since vegetation barriers reduce the space required by isolating chicks from harm. Huxley (1934) first proposed that territories should be able to expand and contract like a rubber disc, with a fixed lower size limit. In common terns, this lowest size during the incubation period is reached at a nearest neighbor distance of 40 cm, a density that we have never seen in the colonies we study, but have found elsewhere, such as at Moriches, Long Island. Our data for common terns generally conform to this pattern. We found terns defending large territories early in the season, smaller ones during incubation, and larger ones again during the chick phase.

Intercolony Variation

There was variation among colonies and years with respect to territory size, as estimated from nearest neighbor distance. Colonies with more nesting pairs were denser, even when space was not limiting. This relationship could result from competition for limited space or from social interaction. It is unlikely that the relationship between nearest neighbor distance and colony size was due to competition for space, since in each colony there was unused suitable habitat and overall density was not different (nests divided by area of colony). This nesting pattern appears to derive from increased social interactions in the presence of more pairs. Pairs that first set up territories are farther apart, but as the season progresses, and as the initial pairs are incubating and therefore less aggressive, new pairs arrive and establish territories among the existing pairs. This phenomenon would be more prevalent in larger colonies where there are more young birds or new recruits (failed breeders from elsewhere) entering the colony.

Interhabitat Variation

There was considerable variability in nesting territory space among habitats (refer to table 4.8). In general, birds nested closest together on salt marsh mat, at intermediate distances on sandy beaches, and farthest apart in the *Spartina*. However, at Holgate, the terns nested at high density on sand, and we found this true as well on casual visits to Moriches and Gardiners Island, Long Island. Small internest distance on salt marsh mat are to be expected, because these areas are limited in space and are highly preferred over *Spartina* as being less likely to flood.

We attribute the small internest distances at Holgate to the propensity of terns to nest in the large skimmer colony, where they are confined to the edge, fringed by dense beach grass. Furthermore, terns in sandy beaches generally nest at distances of 1–2 m apart, while in *Spartina* distances averaged > 3 m, partly because terns were selecting the highest sections where there was extra spoil or decaying vegetation. Moreover, since there was lower intrusion pressure in *Spartina*, terns could more easily defend larger areas with the same effort required to defend a smaller area on a nearby mat.

FIVE

Aggressive Behavior

Aggressive behavior is one aspect of reproductive investment that can be readily observed and measured. Aggressive behavior can range from low cost (in terms of time and energy) behavior such as displays, to high cost behavior such as attacks and fights (MacRoberts and MacRoberts 1980). Fights can result in physical injury and even death (Smith and Hoskings 1955). Presumably time, energy, injury, and predation costs increase with the intensity and duration of aggressive acts.

Aggressive behavior usually occurs when necessary resources are in short supply (competitive behavior), particularly as in the establishment of territories, or in defense of eggs or chicks (antipredator behavior). The resources birds compete for include space and territories, nest materials, mates, and food. Aggression is the major mechanism whereby competitors determine the outcome of disputes over these resources (see Parker and Rubenstein 1981; Hammerstein 1981). Interference competition, the destruction of conspecific eggs and chicks or reduction in conspecific productivity, is an important competitive strategy enhancing the contribution of ones own alleles to future generations (Pierotti 1979). Aggression levels are influenced by the internal hormonal state (Davis 1963; Crooke and Butterfield 1968; Payne and Swanson 1972), and external factors. Typically, aggression levels are higher early in the reproductive cycle. However, the actual relationship between circulating levels of hormones and aggression in nature has been examined infrequently (see Harding and Follett 1979).

The literature on aggression levels in larids usually emphasizes territorial defense (Hunt and Hunt 1976; Ewald et al. 1980; Burger 1983a), with a few important studies of food defense (preventing piracy, Hatch 1970; Hulsman 1976). We also emphasize aggression involving mate guarding and nest defense. Dense aggregations and limited space greatly increase the opportunity and need for aggressive behavior at certain times, and constitute a major cost of being colonial.

In this chapter, we examine variations in aggressive display frequencies and aggression rates as a function of season, type of intruder, and habitat. We also examine copulation interruption, because it usually occurs when the male is distracted by intruders.

5.1 AGGRESSIVE DISPLAYS IN THE COMMON TERN

Tern displays have been described by various authors (Marples and Marples 1934; Palmer 1941a). Moynihan (1955) summarized early descriptive papers and cautioned against interpreting tern displays as homologues of gull displays, the latter having received much more attention from European ethologists. He emphasized the uniqueness, for example, of the fish flight characteristic of many terns, but not gulls or skimmers, and pointed out that it resembles the pursuit flight of the small hooded gulls.

Southern (1938) and Palmer (1941a) provided detailed accounts of postures and displays of the common tern, concentrating more on courtship than on aggressive displays. Some postures are common to both contexts. The "bent" posture with sleeked plumage, termed *Beugestellung* by Southern (1938), is used in both courtship and aggression (see fig. 3.2, top). In the latter context, it alternates with a bill-up display, termed *Reckstellung*, similar to the sky-pointing display characteristic of the territorial encounters of passerines such as icterids (Gochfeld 1975a).

Dynamic displays involve walking toward (fig. 5.1) or flying towards (fig. 5.2) an intruder (stranger or neighbor) that has entered ones territory, and are called the walk to and fly to displays. Palmer (1941a) called attention to the fact that a territorial tern may even attack an intruding stranger that has entered a neighboring territory. Aggression is most often directed toward conspecifics, but in mixed species colonies may be directed toward other species as well (fig. 5.3). Aerial chases occur commonly although actual fights are infrequent. As in the black skimmers, there is occasionally a "flutterup" or "butterfly" pursuit in which the two antagonists rise up vertically on slowly but deeply beating wings. Often the head of the lower bird is pointed up toward the upper bird, and the body may assume nearly a vertical posture with the legs dangling downwards. Similar displays have been reported for murres (Gaston and Nettleship 1981).

In our experience, the upright oblique stance reflects moderate to high level aggression. Moynihan depicted this display (1955; his fig. 28c), and considered it a probable homologue of the oblique display of gulls. There is a similar upright stance used in the courtship "parade" in which the male tern holds its neck extended vertically, with the bill held horizontally.

In view of detailed descriptions available in the literature (Marples and Marples 1934; Southern 1938; Palmer 1941a; Austin 1946) we will not provide detailed description of behavioral postures and displays (see figs. 3.1, 3.2, 3.3), but will examine how common tern displays vary as a function of season, habitat, and location. The most common forms of overt aggression are the walk-to (fig. 5.1, bottom), fly-to (fig. 5.2, top), aerial chase (fig. 5.2., bottom), attack or dive bomb (fig. 5.3, bottom), ground fight, and aerial fight. Walk-to involves walking toward the intruder, usually in an upright posture. Fly-to involves flying toward (but not pursuing) the intruder. Aerial chases involve flying after an intruder, and sometimes result in physical contact or aerial fights. Both members of a pair engage in these defensive behaviors. However, since it is not possible to determine the sex of unmarked common terns visually (they are strikingly monomorphic) and we had only a few known-sex pairs color-marked, we did not examine sex differences in displays in this species.

Seasonal Differences in Displays

During the incubation period, parents must protect their eggs from predators as well as defend a suitable territorial space from neighbors or strangers which intrude on the territory. In the following discussion, we use the term intruder to apply to any bird that enters the territory of a pair of common terns. This can include either neighboring terns, strangers which alight in the territory, or even other species such as black skimmers.

During the chick phase, parents must protect up to three chicks. Small chicks are vulnerable (they can be easily killed), and if they become scattered

Figure 5.1. Aggressive displays: Aggressive upright (top) accompanied by loud calls. Note that carpals (wings) are held tightly against body as opposed to the bent posture (see fig. 3.2). Aggressive upright directed toward a neighboring tern and its chick (bottom), which had sought shelter under a goldenrod at the territorial boundary.

Figure 5.2. Common tern flying upward toward an intruder (top) and chasing an intruder (bottom) (in this case, a roseate tern). Note black skimmer incubating on adjacent territory.

Figure 5.3. Common tern and black skimmer showing mutual aggressive displays. Tern has just flown toward skimmer and landed near the territorial boundary (top). Tern dive-bombing skimmer on adjacent territory (bottom). Territories in this case are about 2 m in diameter, so that fly-to displays are very short.

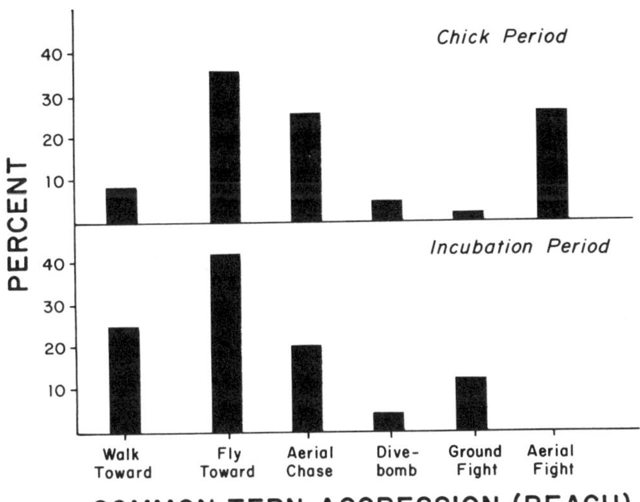

Figure 5.4. Seasonal variation in display frequency for common terns on beach: Use of aggressive displays in common terns in incubation and chick phase. Shown is the percent of total displays in period (N = 1397 interactions).

about the territory they risk attack by neighboring adults. Presumably, the higher the intensity of displays given by defending terns the more quickly an intruding tern would depart. The defending parents might respond differently to intruders as a function of stage in the reproductive cycle, hence the duration of aggressive interactions and the resulting sequence of displays is expected to vary through the season.

We found significant differences in the use of aggressive displays between the incubation period and the chick phase for common terns ($X^2 = 31.0$, $df = 5$, $P < 0.001$, $n = 358$; fig. 5.4). During the incubation period, 65 percent of the aggressive interactions involved walk-to or fly-to displays, whereas during the chick phase these comprised only 43 percent of the interactions. Aerial chases and fights accounted for over 25 percent of the interactions during the chick phase.

Habitat Differences in Displays

Common terns nesting either on salt marsh islands or on sandy beaches can nest in open spots or in vegetated areas. Open spots allow easy movement

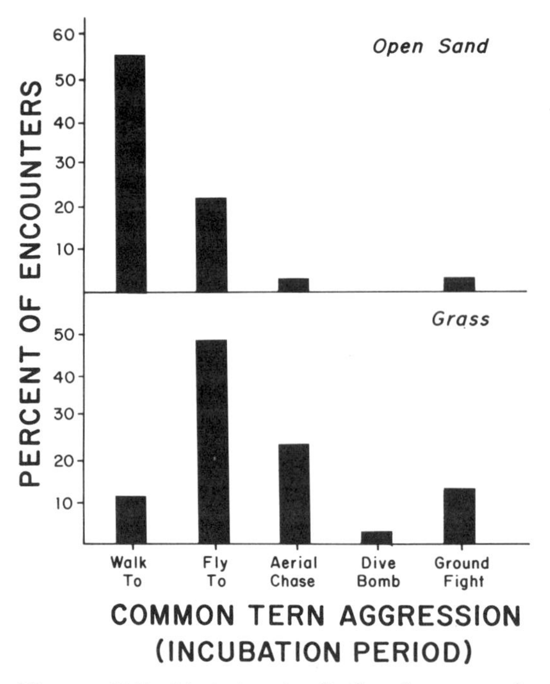

Figure 5.5. Variation in display frequency by habitat (open sand versus grass on beaches) for the common tern during the incubation period (N = 967 interactions)

between defender and intruders (both neighbors and strangers), while birds nesting in vegetation may have vegetative barriers interposed between themselves and intruders. The display types used varied as a function of location in both the sandy beach and salt marsh colonies.

In open areas of sandy beach colonies, over 50 percent of the overt aggression involved walk to displays whereas in the vegetated sections of the same colonies, fly to displays made up almost 50 percent of the displays ($X^2 = 15.1$, $df = 5$, $P < 0.01$, $n = 180$; fig. 5.5). Aerial chases were more common (> 20 percent) in vegetated sections, but relatively rare in open spots. As we watched the terns, it became clear that defenders nesting in vegetated parts of the colony responded quickly to intruders by flying toward them, pursuing them aerially and even attacking them in the air. In open spots, it was possible for the birds to walk quickly toward the intruder, maintaining visual contact.

Salt marsh islands provide an even greater habitat contrast, since the terns

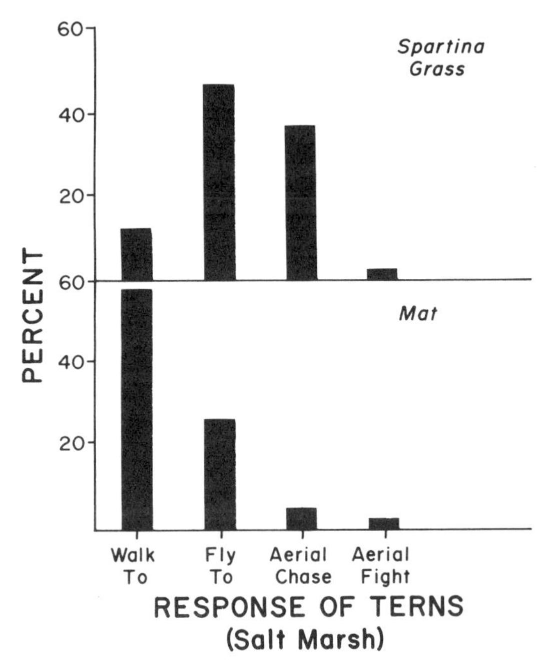

Figure 5.6. Frequency of displays by salt marsh common terns on mat and in *Spartina:* Walk-to and fly-to displays ($N = 511$ interactions).

nest either on mats which are open and provide nearly 100 percent visibility in all directions, or in the *Spartina* vegetation with visual barriers up to 60 cm high. Over 50 percent of the terns nesting on mats perform walk-to displays toward intruders, while 49 percent of those nesting in *Spartina* gave fly-to displays ($X^2 = 47.9$, $df = 3$, $P < 0.001$, $n = 452$; fig. 5.6). Similarly, the terns nesting in *Spartina* performed more (40 percent) aerial chases and fights than those nesting on mats (17 percent). Thus, a similar pattern of habitat influence on choice of displays emerges for both the salt marsh and beach colonies.

Discussion

Dense nesting aggregations, whatever their benefit or selective advantage, impose upon the nesting birds increased opportunity for aggressive interactions with their neighbors. When space for nesting is at a premium, the frequency of intruders seeking to carve out territories will increase. Moynihan

(1960) emphasized that even in densely colonial species, aggressive displays were well ritualized and important.

Common terns used different proportions of displays and overt aggressive behavior as a function of stage in the reproductive cycle. The terns also performed a higher proportion of high intensity displays (ground and aerial fights) during the chick stage compared to incubation. The relative increase in overt aggression in the chick stage compared to incubation reflects the change in vulnerability of the reproductive unit, and perhaps increased parental investment (Andersson et al. 1980). Chicks are more vulnerable than eggs because they no longer remain in one place, beneath the parent. Chicks can succumb to both territorial clashes, as they wander to the periphery of the territory, or to predation attempts as they move about in the open. Eggs however, are usually covered (> 90 percent of time), and because they still benefit from the crypsis of their coloration when not covered.

Burger (1981a, 1983a, 1984) found that chasing in herring gulls also increased markedly in the early chick phase. The greater the duration of the aggressive interactions, the greater the exposure of chicks to predation, heat or cold, a cost referred to as aggressive neglect by Hutchinson and MacArthur (1959) and Ripley (1961). Furthermore, the high intensity displays (chases and fights) could result in injury (although this is a rare event in common terns).

Common terns nest on sand in open spots, vegetated spots, and on open mats or in the salt marsh grasses. In open spaces, common terns employ walk-to displays, while in vegetated areas they fly toward intruders. This reflects, no doubt, the ease of walking over open ground or flat mat toward an opponent that is completely visible. In comparing the displays of terns nesting in grass on sandy beaches with those nesting in the *Spartina* of salt marshes, we found that the latter perform more aerial chases and virtually no ground fights. This difference reflects the density and pattern of the vegetation. In sandy beaches, the vegetation is usually clumped with clear open sandy zones intermingled with the grass (terns avoid dense stands of grass); this contrasts with the homogeneously dense stands of *Spartina* where there are simply no places to land and fight. Although we have not witnessed tern mortality from fights, we have seen adults that have perished from becoming entangled with *Spartina* stems.

5.2 LEVELS OF AGGRESSION

Aggressive interactions are a major cost of territoriality in most colonially-nesting species, although aggression is also a response to other resource-based competition and to predators. In this section, we examine how levels of aggression vary by season (1981 to 1983), time of day and habitat. We used the frequency of aggressive displays per tern per minute of observation as our measure of aggression level.

Seasonal Variations in Aggression Levels

Common terns nesting on mats in salt marshes had increased aggression levels towards conspecifics (both neighbors and strangers) in the preincubation phase, and again during the hatching period (fig. 5.7). During incubation, aggression levels were very low, but increased at hatching. As the chicks grew older, aggression levels decreased. Common tern aggression directed at skimmers peaked during the egg-laying period of skimmers, and reached a secondary peak during the hatching period of the skimmers (fig. 5.7). Thus, the pattern of aggression varied as a function of the species of intruder.

For common terns nesting on an open sandy beach, aggression levels were also highest in early May during the preincubation period, and again in late June when chicks were over two weeks old (fig. 5.8). Thus aggression levels for common terns differed in the two habitats in two ways: 1) levels were generally higher in salt marshes than in sandy beaches, and 2) aggression peaked during hatching in salt marshes but peaked when chicks were over two weeks old in the sandy beaches where such half-grown chicks were more likely to wander through neighboring territories.

Habitat Variations in Aggression Level

As mentioned above, there were major differences in aggression levels between birds nesting in salt marshes, and those nesting in open sandy beaches. However, aggression also varied within each major habitat type. During the preincubation and egg-laying phases in salt marsh colonies, birds nesting on mats engaged in significantly more aggression than those nesting in *Spartina* (mean = 2.4 ± 1.2 per hr/pair versus 0.4 ± 0.2 hr/pair, Mann Whitney $U = 0$, $P<0.001$). For birds nesting in *Spartina* aggression decreased after egg-laying, and remained relatively low throughout the reproductive cycle.

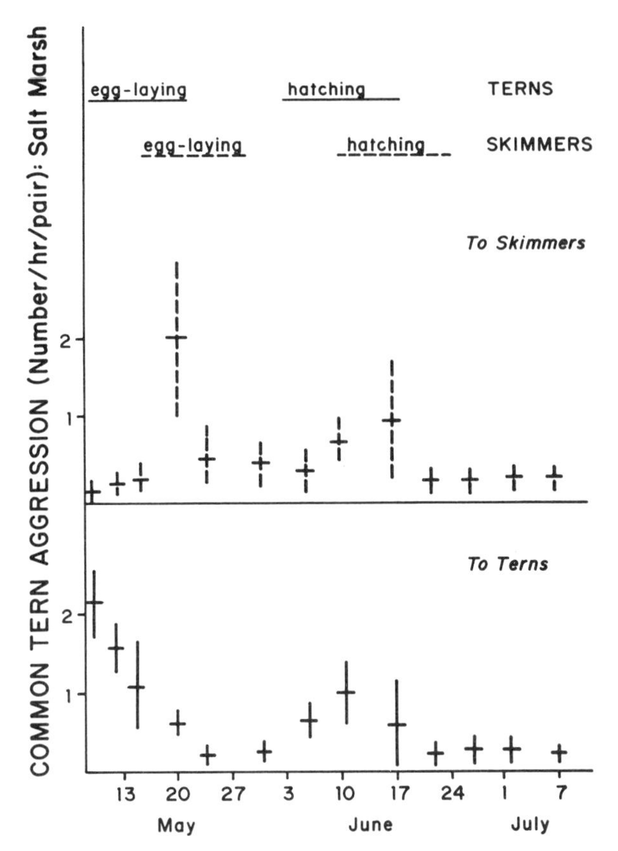

Figure 5.7. Variations in common tern aggression levels by season in salt marshes. Shown are means ± 1 SD (N = 1110 interactions).

Aggression levels also varied by location in the sandy beach colonies where we examined aggression in the edge, intermediate, and central area of the colony (refer to fig. 5.9 for map of locations). Although the overall pattern of aggression was similar, the levels varied (fig. 5.10). Terns nesting in the center had higher levels of aggression during peak times (preincubation, chick stage) than terns nesting in intermediate or edge locations. Also, in early May, aggression levels decreased first in the center, and last in the edge locations (fig. 5.10). In addition, aggression levels in the chick stage (after June 12) were highest in the center, and lowest for the birds nesting on the edge. Common tern aggression levels also varied by time of day (refer to chapter 3).

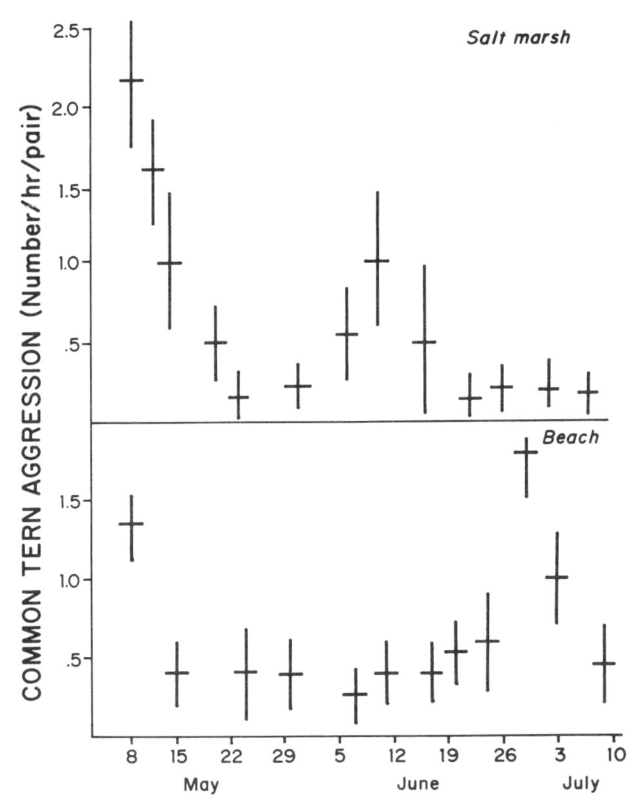

Figure 5.8. Variation in conspecific aggression levels in common terns on salt marsh (top) and beach (bottom) colonies. Shown are means ± 1 SD (N = 2410 interactions)

Discussion

Aggressive behavior is an obvious part of reproductive costs for most colonial species. The relatively dense nesting of birds in colonies results in frequent opportunities for conflicts with neighbors. In gulls and terns, the importance of aggression in colony dynamics was recognized early (Tinbergen 1953, 1959) and considerable attention has been devoted to examinations of aggression rates.

Seasonal levels of aggression vary during the breeding cycle in gulls, terns, and probably most species. Territorial aggression rates are highest during the pre-incubation phase in southern black-backed gull (Fordham 1972a,b),

BEACH COLONY

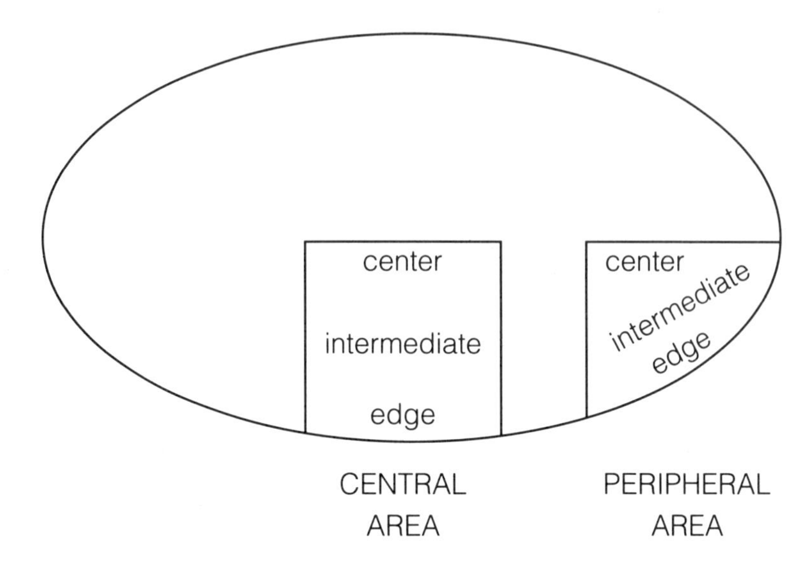

SALT MARSH COLONY

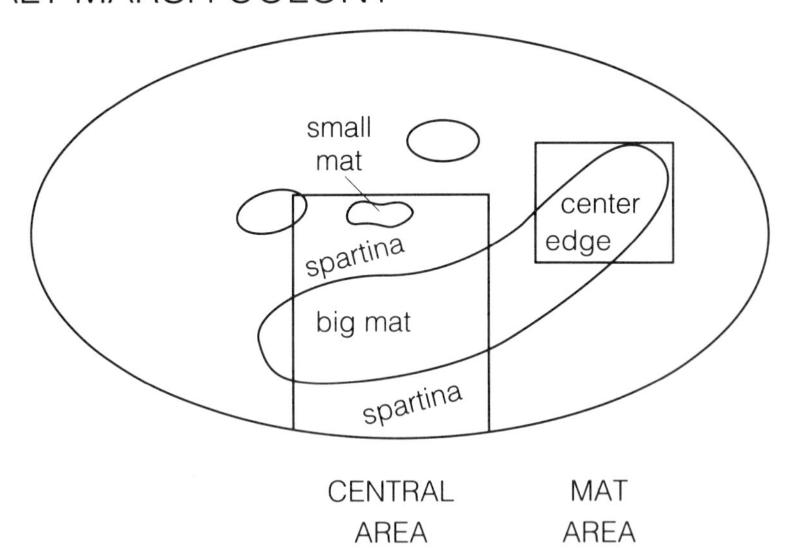

Figure 5.9. Schematic map of Cedar Beach and salt marsh colonies showing middle and peripheral study areas and center, intermediate, and edge sections.

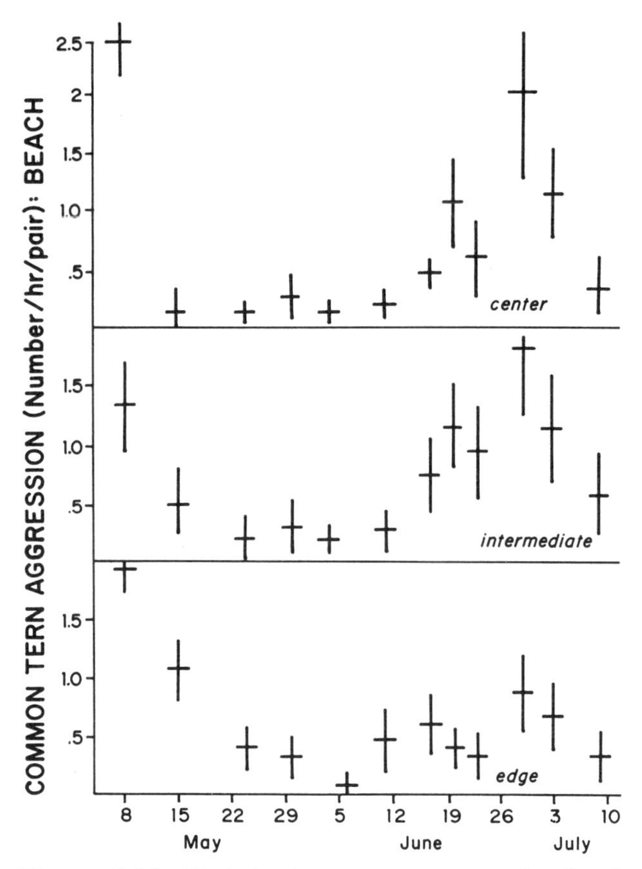

Figure 5.10. Variation in tern aggression in beach colonies: center versus edge. Shown are means ± 1 SD ($N = 2410$ interactions).

lesser black-backed gull (MacRoberts and MacRoberts 1972), black-headed gull (Hutson 1977), laughing gull (Burger and Beer 1975), arctic tern (Lemmetyinen 1971) and herring gull (Burger 1981b). For other species (eg., great black-backed gull), aggression increases seasonally (Verbeek 1979), sometimes with a peak at hatching (Burger 1983a). Recently, Dulude et al. (1987) reported that territorial aggression was highest in the post-hatch period because of the threat of piracy. In other species, rates of aggression peak at different times in the reproductive cycle in different colonies. For example, in common terns the highest aggression rates have been reported both early in the reproductive cycle (Lemmetyinen 1971) and at hatching (Veen 1977).

In most of the above studies, aggressive behavior was recorded without regard to the cause of the aggression. As mentioned at the beginning of this chapter, aggression results from defense of several resources or in defense of eggs or chicks from predation. However, not all aggression even among conspecifics is attributable to territorial defense, for it can relate to defense of mates, or of chicks or eggs from cannibalism. Thus, it is not surprising that authors have found differences in the timing of the peak of aggression. We suggest that this timing can be used to generate hypotheses concerning the function of the aggression for each species in each colony. Levels or rates of aggression can thus serve in the functional analysis of the selective factors underlying aggression.

In our study, seasonal aggression levels varied as a function of species of intruder, habitat, and location. In all situations, aggression levels were high in the preincubation and egg-laying periods. We suggest that the function of aggression during this period is either related to territorial defense or mate competition. Similarly, the levels of aggression directed at heterospecifics were high during the egg-laying period but not nearly as high as towards conspecifics. This interspecific aggression is most likely directed at territorial defense and at protecting eggs.

In common terns, aggression toward conspecifics showed a secondary peak at hatching in salt marsh-nesting birds, and when chicks were a week to three weeks old on sandy beaches. We believe these differences relate to behavior of chicks and territory size differences. Aggression near hatching involves protecting chicks from predation attempts or from becoming victims in territorial clashes. For terns, conspecifics do not usually eat chicks. However, territorial clashes can result in injury (or death) to chicks that wander. Older chicks that learn the boundaries of territories tend to favor specific shelters, and do not leave readily unless disturbed. This suggests that for terns nesting in salt marshes, territorial clashes (at least on mats) are an important cause of injury or mortality to chicks. Indeed, terns nesting on mats nest quite close together (see chapter 4) and territorial clashes are frequent.

In sandy beaches, common tern intraspecific aggression peaked when chicks were one to three weeks old, suggesting that either they succumb or are injured in territorial clashes or are otherwise vulnerable to attack when they wander onto neighboring territories. As also mentioned above, terns attempt to pirate food when parents transfer food to their young. We suggest that the increased levels of aggression with older chicks are a direct result of piracy attempts by neighboring chicks or adults cruising over the colony looking for potential food. This hypothesis can be tested by examining piracy or adoption behavior (see chapter 7).

For common terns, aggression directed toward skimmers peaked during the egg-laying period in all habitats. To some extent, this represents territorial defense since skimmers were intruding to establish their own territories. Common tern aggression directed toward skimmers also showed a lower secondary peak during hatching. At hatching, and until the chicks are about a week old, they are very vulnerable to skimmer attacks. Skimmers occasionally eat young tern chicks. The hypotheses discussed above suggest that aggression levels reflect competition for space and mates, protection of chicks from territorial clashes or predation attempts, and protection of food from piracy. These alternatives can be distinguished only by examining the behavior of intruders which is discussed below.

Other Factors Affecting Levels of Aggression

Levels of aggression are also influenced by many other factors such as sex of the defender and intruder (Tinbergen 1953; Vermeer 1963; Pierotti 1981), age of the intruder (Burger and Beer 1975), whether the intruder is a neighbor or a stranger (Patterson 1965), vegetation barriers (Burger 1977), intrusion pressure (Ewald et al. 1980), nesting density (Butler and Trivelpiece 1981), time of day (Delius 1970), and habitat (Burger 1981b). Clearly, birds nesting in colonies are required to expend varying amounts of time and energy in territorial defense depending upon their specific nesting conditions.

We found that levels of aggression in common terns varied as a function of habitat. Common terns showed generally higher intraspecific aggression levels in salt marshes than on sandy beaches. We believe this difference relates to space limitations on the mat areas compared to sandy beaches. Usually, not all the terns nesting in a given colony can fit onto the available mat, and we infer that they are forced to nest in *Spartina*. Mats are generally higher quality habitat than *Spartina* because their higher elevation renders them less vulnerable to flooding. By contrast, in the sandy beach colonies we studied, the preferred habitat was sand with moderate vegetation. It was the prevalent type, and extended beyond the boundaries of the colony. The peripheral areas may not have been preferred, but nonetheless afforded suitable habitat. Thus, it appeared that competition for preferred habitat was more stringent in the marsh than in beach colonies. In every year, the beach colonies had abundant open space with few or no nesting terns.

Aggression also varied as a function of the species of intruder. Terns were generally more aggressive toward conspecifics than toward heterospecifics. Also, terns mate-guard only against conspecifics. Tern aggression in defense of the food brought back to mates or to chicks was directed only against other terns.

Aggression levels also varied as a function of location in the colony. Terns in the center of the colony were more aggressive than those in intermediate or edge locations. However, during the egg-laying phase, aggression was higher on the edge than in intermediate locations; and during the chick phase, aggression was higher in intermediate than edge locations. These differences reflect the resource being sought. In the egg-laying phase, intruders were seeking territories or mates. Having failed to insert themselves in the center, they usually moved to the edge to establish a territory. However, during the chick stage, intruders were often attempting to steal fish and there were more fish being brought back (per unit area) in the center and intermediate areas compared to the edge.

5.3 INTRUSION PRESSURE

Levels of aggression are determined in part by the resident bird's motivational state, itself influenced by the cost or value of the resource, and by the number of intruders (hereafter referred to as the intrusion pressure). Intrusion pressure refers to the pressure exerted on a territory holder by the appearance or behavior of conspecifics entering the defender's territory. Intruders are both neighbors (birds sharing territorial boundaries) and strangers. Territorial birds can respond to an intruder, or they can ignore it. Aggression directed toward neighbors can be increased or decreased depending on the internal motivation of the resident. Birds respond differently to a neighbor in a specific location depending on its own internal hormonal state, activities (whether it is incubating or not), reproductive stage (nest, eggs, young chick, older chick), and other factors.

Species of Intruder

In both salt marshes and sand-beach colonies, terns and skimmers are exposed to intruders of both species. In 1982 and 1984 we recorded the species of all defenders and intruders in the center of the West End beach colony,

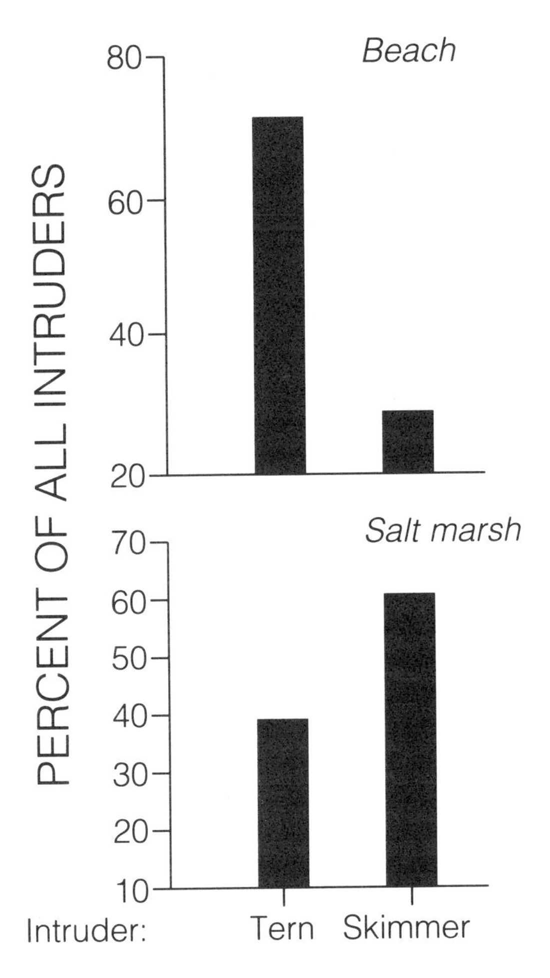

Figure 5.11. Intruders in common tern territories in beach (top) and salt marsh (bottom) (*N* = 1908 interactions.

and on the mat in a salt marsh colony (Pettit Island). We selected study plots with an approximately equal number of nesting pairs of terns and skimmers. In sandy beaches, terns had significantly more conspecific intruders than skimmer intruders (fig. 5.11). Terns in salt marshes had more skimmer intruders than tern intruders ($X^2 = 23.0$, $df = 1$, $p < 0.001$).

Types of Intruders

Several years of observations on common terns made it clear that intruders (strangers) can be seeking territories, seeking mates, or attempting to steal fish from other terns. In general, territory-seeking intruders land in an open space and look around quickly. Usually, these intruders are immediately chased, and either they leave entirely or continue to return. If the intruder is persistent, it may succeed in habituating or fatiguing the resident defender, resulting in its eventual acceptance. In this manner, a tern succeeds in carving out a territory among already established pairs. Early in the season, such territory seeking birds are usually single birds (presumably males), but one also sees pairs of terns attempting to establish territories. We have observed these pairs court and copulate on the edge of the colony, and then fly into the colony apparently in quest of a territory.

Once paired, common terns engage in courtship feeding, a behavior which has been studied in several species of terns (Nisbet 1973a; Moller 1981; Morris 1986) and gulls (Niebuhr 1981). Most males are extremely active in bringing back fish to present to their females (Nisbet 1973a; Smith 1980; Kilham 1981; Morris 1986; Wiggins and Morris 1988). This behavior pattern continues from the early territory acquisition stage through the egg-laying period and at a reduced rate during incubation. In some years, late-nesting male common terns may have lower courtship feeding rates than those nesting earlier (Morris 1986). In common terns where we had marked individuals of known sex, we observed males carrying fish land on occupied territories and present fish to females which were not their mates. They would either land next to a female, or land and walk toward a female whose mate was away. Such males are usually chased by the resident female, either immediately or occasionally after the female had eaten the fish. Very infrequently, early in the season, a female will join the new male, apparently abandoning her old mate (see below). Mate-seeking males with fish continue to appear in the colony until the early chick stage.

Pirates are present in the colony throughout the reproductive stage, stealing fish when males return to courtship feed their mates, and later stealing from parents bringing fish to chicks or from the chicks themselves. A small number of terns may specialize in piracy, usually flying low over the colony in search of other terns carrying fish. These pirates usually attempt to steal the fish from flying terns, but some individuals, both adults and nearly fledged young, watch terns bringing back fish for mates or chicks and try to steal the fish during transfer. Few adults, but many chicks engaged in this type of piracy. Some young terns become proficient in stealing fish from chicks on a neighboring territory. Piracy is discussed more fully in chapter 7.

Methods for Examining Intruder Pressure

We examined intruder pressure from strangers and neighbors for common terns nesting in salt marshes (1982) and sandy beaches (1981, 1983). Observations were usually conducted form 0600–1800 three or four days a week from blinds situated in the colony. We followed the technique of having two people walk to the blind. One remained in the blind while the other walked away. The procedure was reversed to retrieve the observer. This ruse results in birds returning to their nests more quickly than if the observer walked to the blind alone and didn't emerge. We defined an intruder that landed with a fish as a mate-competitor, one landing without a fish as a space competitor, and one attempting to steal fish as a pirate.

Mate Competition

We present the following evidence for intruders with fish being mate competitors: From our observations, it became clear that terns which were carrying fish when intruding on a territory attempted to present these fish to females. These intruders would circle periodically, land again, and repeat the process of presenting the fish. They specifically approached the resident bird, and the resident bird was almost always alone. In a study area at West End, where most of the terns had been marked, we identified as females the recipients of a courtship feeding attempt. In most pairs, we could distinguish the males and females by the bands they wore. In all cases where we could distinguish, the fish-carrying intruder approached the resident female while her mate was away.

By contrast, tern intruders without fish did not land near nor approach resident birds, but landed equidistant between two established pairs. Often there were two intruders. We examined the landing location for 980 common tern intruders, relative to female terns standing on territory (West End beach colony, 1981 and 1982). These were females who were also being courtship fed by their presumptive mates. Intruders with fish usually landed within 1 m, whereas intruders without fish landed farther away ($X^2 = 567$, $df = 1$, $P < 0.001$; fig. 5.12).

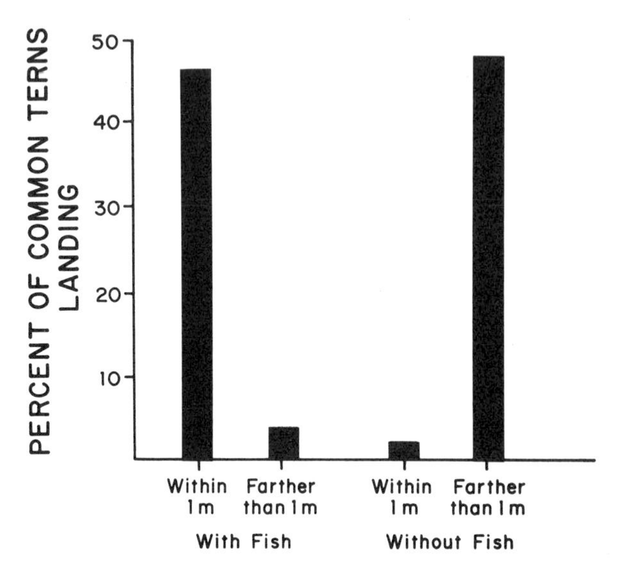

Figure 5.12. Landing distance of common tern intruders with and without fish. Shown are the distances they land from a tern on territory (N = 1452 interactions).

We also examined the types of displays given to tern intruders with and without fish (table 5.1). Intruders with fish were responded to with higher intensity aggressive displays such as aerial chases and fights than intruders without fish ($X^2 = 273$, $df = 8$, $P < 0.001$). Intruder pairs that landed without fish also were responded to more intensely than were single intruders without fish. Similarly, defending terns responded immediately to single intruders with fish and to intruder pairs with fish, but delayed somewhat before responding to single intruders without fish (table 5.1, ANOVA, $F = 49.3$, $df = 2{,}124$, $p < 0.001$).

Tern intruders with fish were sometimes observed to walk toward females on territory, but those without fish never did. Females sometime ate these fish (3 percent of 182 cases, May 10–25, 1983) and then chased the male. Early in the season, a small number of females ate the fish, did not chase the males (1.5 percent of 262 cases, May 5–15, 1981, 1982), and may have subsequently mated with these males.

Taken together, these data indicate that common tern intruders with fish land closer to potential mates, and are responded to more quickly than single

Table 5.1. Percent of aggressive displays of common terns directed toward different types of intruders (West End, 1982).

	With Fish	*Without Fish*	
Display Type	*Singles*	*Singles*	*Pairs*
Number of events	108	200	94
No response	4	11	0
Walk-to	4	41	14
Fly-to	18	48	74
Aerial chase	56	0	12
Fight	18	0	0
Mean Time to Respond (sec)[a]	0.61	3.86	1.12
±1 S.D.	1.1	1.2	1.2

[a]To intruder landing within territory.

intruders without fish. This difference is not due solely to distance from the intruder because the intruder pairs were also responded to quickly but they landed far from defending terns.

Frequency of Different Types of Common Tern Intruders

During the preincubation to mid-incubation period most of the intruders were seeking space for territories in both habitats (fig. 5.13). However, in salt marsh colonies, only 18 percent appeared to be seeking mates compared with 38 percent in sandy beach colonies ($X^2 = 69.7$, $df = 1$, $P < 0.001$). In salt marsh habitats, most territory-seeking intruders were single birds (86 percent), although some pairs did attempt to insinuate themselves between already established pairs in both mat and *Spartina* sites (fig. 5.14).

Intruders in salt marsh habitats were generally expelled (95 percent) once initial territory acquisition had occurred in early May. In beach habitats, however, fewer intruders were expelled. Territory-seeking intruders in sandy beach areas were expelled 85 percent of the time, whereas mate-seeking intruders (those carrying fish) were only expelled 79 percent of the time. This slight difference is statistically significant ($X^2 = 5.52$, $df = 1$, $P < 0.02$; fig. 5.15). Overall, only 81 percent of intruders in beach colonies were expelled. The greater proportion of intruders chased in salt marsh compared to beach colonies probably reflects the greater average densities relative to the total available habitat.

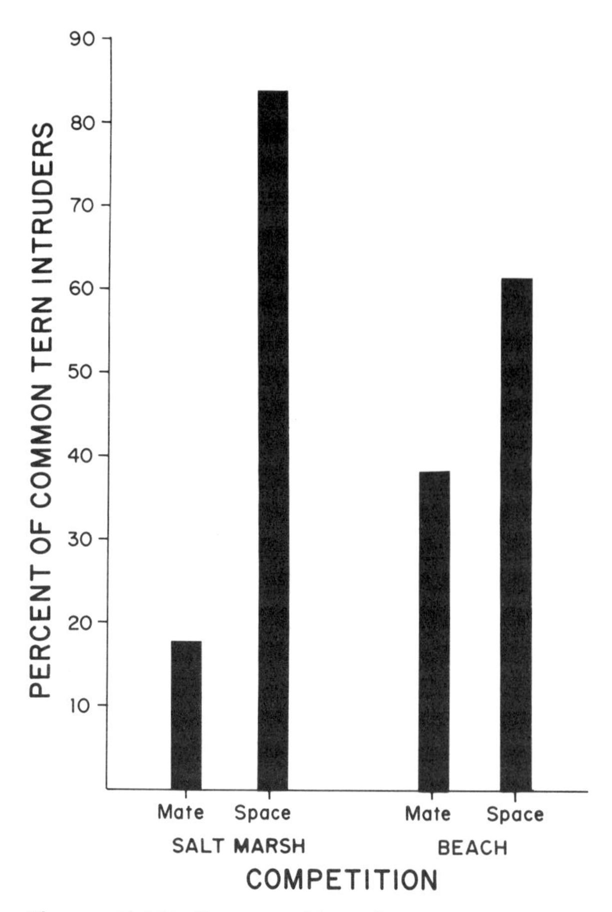

Figure 5.13. Percent of intruders on common tern territories that were engaged in mate or space competition ($N = 1452$ interactions).

In the above discussion, we examined intruder pressure early in the season from the viewpoint of territory defenders standing on territory. However, food pirates generally attempt piracy aerially, chasing terns carrying fish. Thus, we also computed the relative composition of intruders summed over the entire season (May 8–July 20) for central sections of salt marsh and beach colonies (table 5.2). The absolute numbers should not be compared, since the number of pairs differed, but the percent of intruder types indicates the relative costs for these colonially nesting terns. There were significant differences in the types of intruders present in salt marshes and sandy beaches

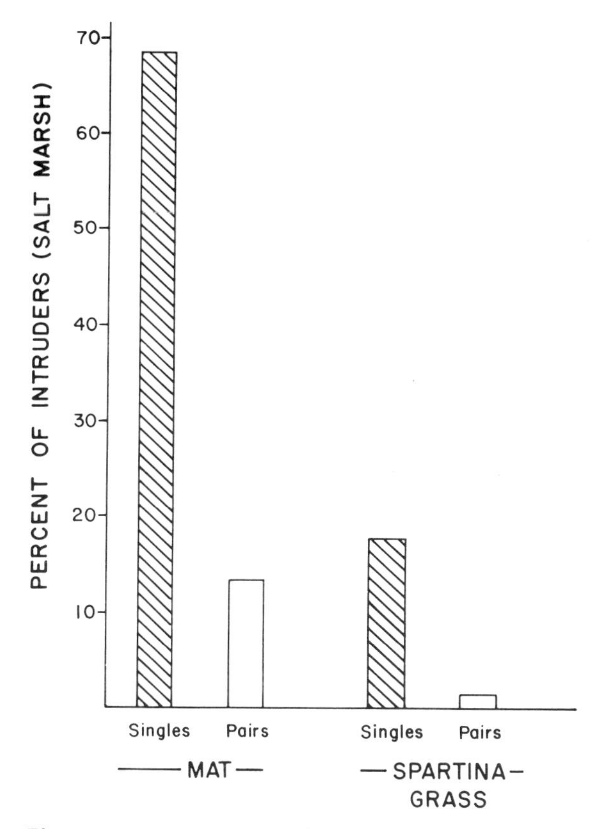

Figure 5.14. Percent of all intruders on salt marsh colonies that were single birds or pairs landing on mats or in *Spartina* (N = 511 interactions).

($X^2 = 173$, $df = 3$, $p < 0.001$; table 5.2). Food pirates were an important cost to beach-nesting terns, accounting for 30 percent of the intruders overall, mostly during the last weeks of the chick phase. This corresponds to a decline in the number of small fish in the inlets (Safina and Burger 1985).

Seasonal Differences in Intruder Types in Common Terns

The seasonal variations in intruder types (neighbors versus strangers) was examined in detail in a sandy beach colony (West End) by recording all the intruders in a central section of the colony with 25 nesting pairs. By using

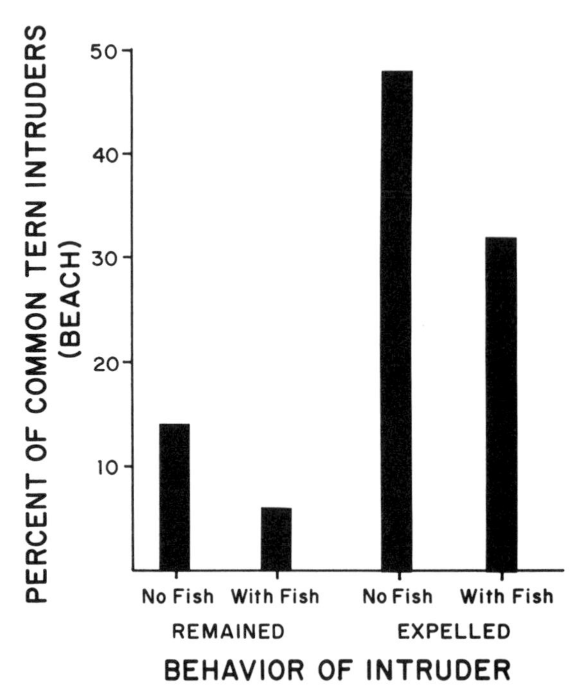

Figure 5.15. Percent of all common tern intruders (with and without fish) that were driven away or remained on territories (N = 139 interactions). The four bars sum to 100 percent.

this method, we could also count aerial as well as ground pirates. Otherwise, aerial piracy attempts are difficult to assign to particular pairs.

Mate-seeking intruders (those with fish) peaked in mid-May during the peak of egg-laying in both marsh and beach colonies (figs. 5.16, 5.17). Thereafter, the frequency of mate competitors remained low with a slight secondary peak in early June. Territory-seeking intruders peaked early in May just as egg-laying began and decreased thereafter, although there continued to be terns seeking territories throughout the reproductive cycle. Pirates (both ground and aerial) were in low numbers throughout the reproductive cycle except during the most active courtship feeding period (mid-May) and during the chick stage (after mid-June). During the chick stage pirates accounted for most of the intruders (fig. 5.16).

In salt marsh colonies, mate and space competition showed similar patterns, but piracy attempts were less frequent (fig. 5.17). Similarly, mate and

Table 5.2. Identity of intruders in common tern territories. Shown are the percent of conspecific intruders for the reproductive season of salt marsh and beach nesting terns.

Intruder Type	*Salt Marsh*[a] N %	*Beach*[b] N %
Mate competitor	96 (19%)	359 (26%)
Space competitor	385 (75%)	612 (44%)
Egg or chick predator	2 (0.3%)	0 (0%)
Food pirate	28 (6%)[c]	426 (30%)[c]

[a] Mat habitat. Initially there were 46 nesting pairs, but only 38 pairs fledged young (1982).

[b] Central area of beach. Initially there were 65 nesting pairs, but only 49 fledged young (1983).

[c] Includes piracy attempts of birds flying over the area as well as those piracy attempts on the ground (not included in previous analyses).

space competition were restricted to a shorter period of time due to more synchronous breeding (see below). Salt marsh colonies had a slight increase in intruders at hatching (fig. 5.17).

Habitat and Location Differences for Common Tern Intruder Types

Types of intruders differed significantly in salt marshes ($X^2 = 4.68$, $df = 1$, $P < 0.05$) and sandy beaches ($X^2 = 13.0$, $df = 1$, $P < 0.001$; fig. 5.18). In salt marshes, over 85 percent of the intruders landed on mats, and only 14 percent landed in the *Spartina*. On mats, 21 percent of the intruders were mate competitors compared with 36 percent in the *Spartina*. Further, the presence of intruding mate competitors varied by location and date in the beach habitat (fig. 5.19). Thus, we examine intrusion pressure in the two habitats in detail below.

Salt Marsh Habitats

We examined two parts of Pettit Island from May 7 to July 5, 1984. The middle area included three habitat zones: Large Mat, Small Mat (defined as having a width of less than 1 m), and *Spartina*. We also looked at a separate mat area, which was further subdivided into central and edge (outer 0.5 m section of the mat zones). On mats, significantly more intruders landed in the center zone than the edge (Goodness of fit, $X^2 = 14.2$, $df = 1$, $p < 0.001$;

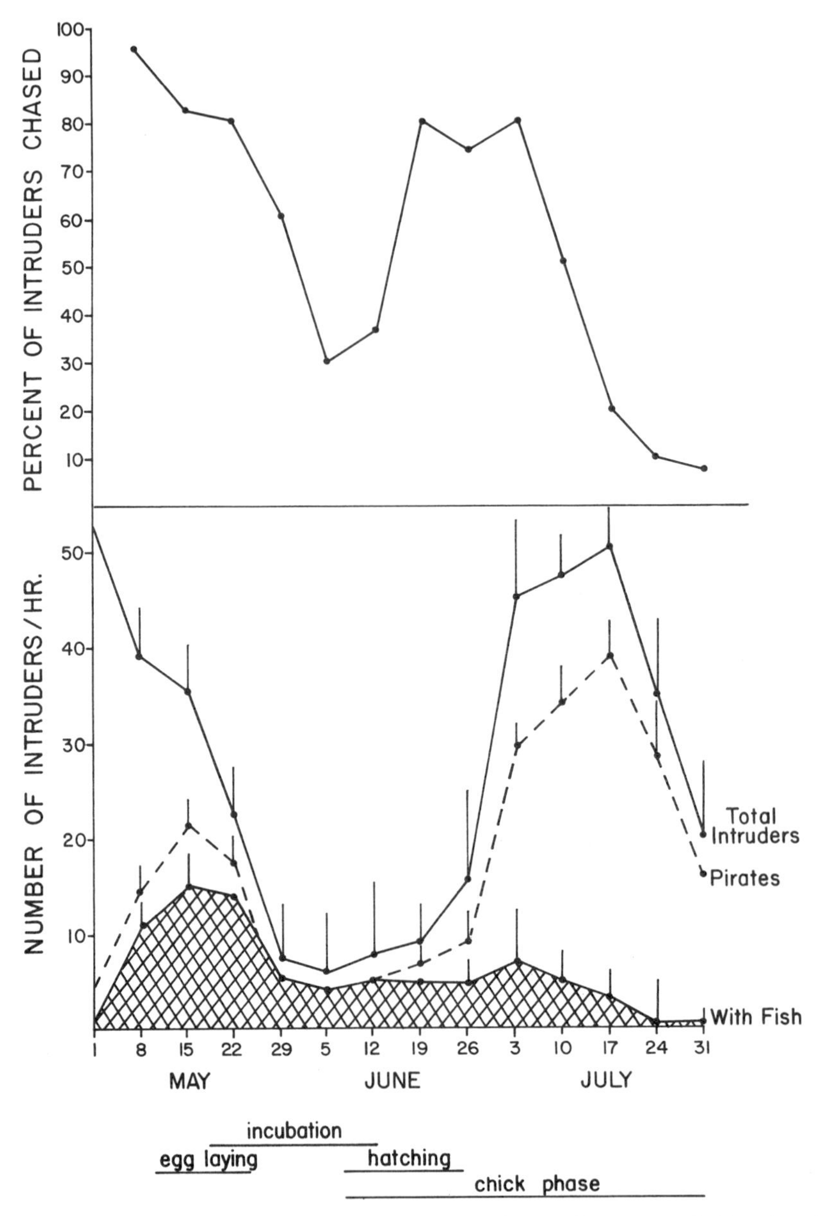

Figure 5.16. Seasonal pattern of number of intruders (N = 511) in beach colonies and percent of intruders (with or without fish) that were chased by territory owners. (Vertical bars are ± 1 SD).

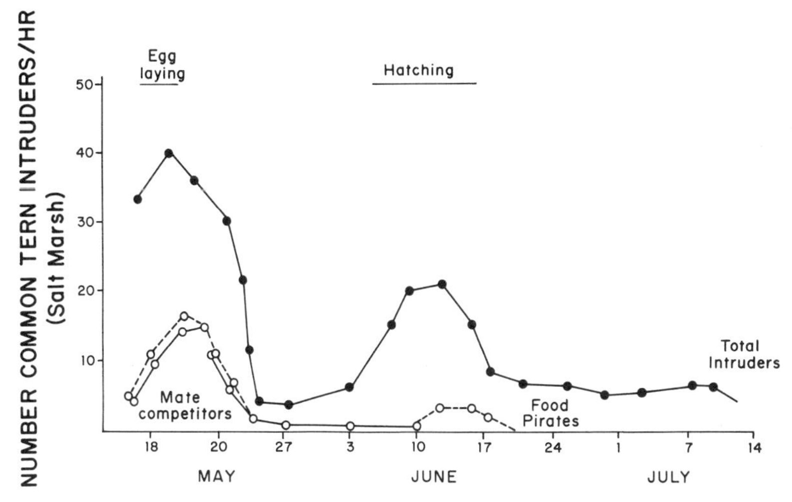

Figure 5.17. Seasonal pattern of number of intruders (N = 1397) in marsh colonies and percent of intruders (with or without fish) that were chased by territory owners.

fig. 5.20). In the middle area, most intruders landed on the large mats, rather than on the small mats or *Spartina* ($X^2 = 147.4$, $df = 2$, $p < 0.001$). There were significant differences in intruder rate by date and location ($X^2 = 13.1$, $df = 4$, $p < 0.005$; fig. 5.21). A higher proportion of intruders landed on large mats early in the season, compared to later. By the end of May, in the middle plots fewer than 50 percent of the intruders landed on the large mats and about 25 percent landed on small mats and 25 percent landed in *Spartina*.

Within mat locations, similar differences occurred in intruder pressure ($X^2 = 29.8$, $df = 3$, $p < 0.001$; fig. 5.22). Early in May, almost all intruders landed on the center of mats. By late May, almost all intruders were landing on the edge of mats, and in early June, intruder pressure was nearly equal on the center and edges of the mat.

The mean rate of mate competition per nesting female was higher on large mats compared to small mats (fig. 5.23). Food delivery rates to mates, however, were not significantly different (fig. 5.23). Thus, equivalent food was brought to females standing at their nest sites in all three salt marsh habitat types, but birds seeking mates more frequently landed on mats.

Common terns nesting on mats in salt marshes were also exposed to

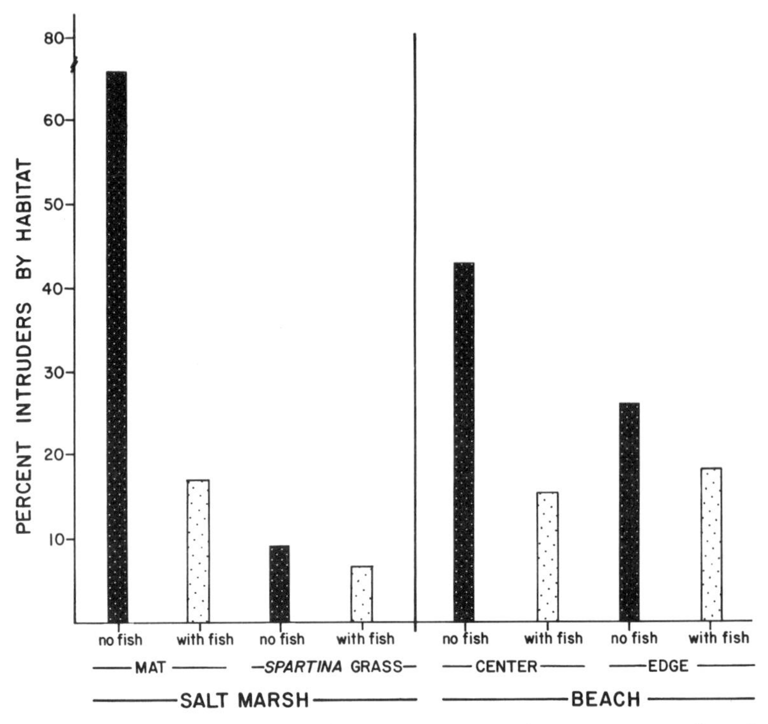

Figure 5.18. Percent intruders in common tern territories, with and without fish, in different habitats and center versus edge locations (N = 1908 interactions).

another type of intruder, not heretofore mentioned. As dusk falls single birds and pairs without territories come to roost on the mat areas. Roosting birds normally do not choose to roost on *Spartina*. Figure 5.24 shows a composite graph for four nights on one sample plot on Pettit Island.

Also, at dusk mates of incubating birds return to the colony. If their mates are on the mat, these returnees land next to them. But if the nests are in *Spartina*, they may land temporarily at the nest site, but then move to the nearby mats where they spend the night (fig. 5.24). Only a small proportion remain on *Spartina* territories. Thus, in the *Spartina*, the number of birds remains relatively constant as night falls. However, on the mats, the incubating birds are joined by their mates and by roosting birds without nests. In the late afternoon, there were 3 or 4 nests with both birds present, about 20 nests with only a single bird present, and no nonincubating pairs on mats. By

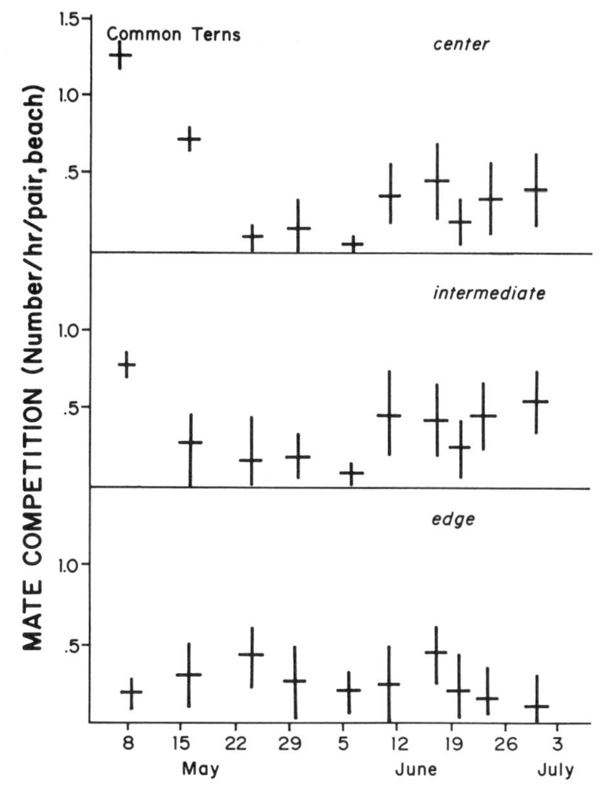

Figure 5.19. Mate competition (number of unmated birds landing with fish) as a function of location in beach colonies ($N = 359$ interactions).

dark, the 20 singles had been joined by their mates, and an additional 15 mates of birds nesting in *Spartina* had settled, as well as single birds apparently not nesting (fig. 5.24).

During this transition period, birds with territories on the mats may respond to the first few evening intruders, but because of the sheer numbers they soon give up and cease or downgrade their aggression. By dark, the birds have fallen silent and there are few subsequent squabbles if the colony is not disturbed.

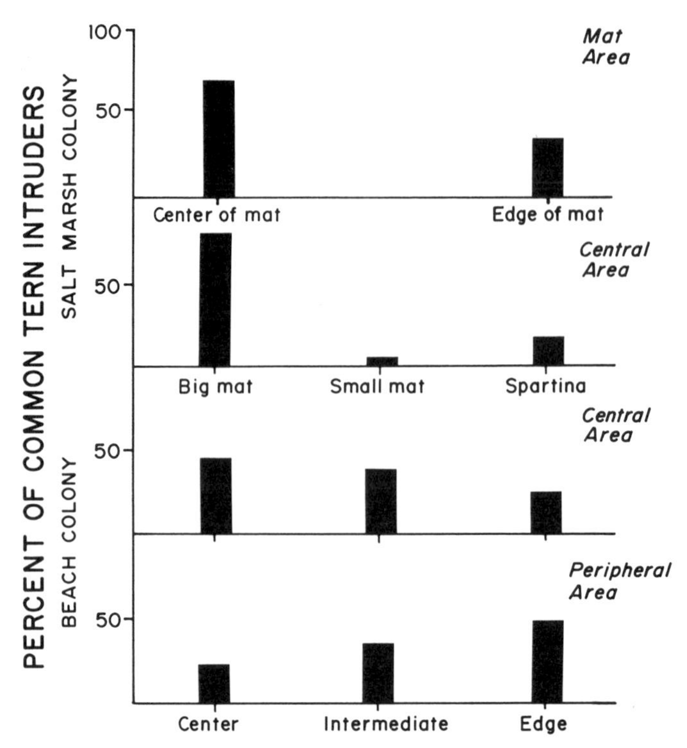

Figure 5.20. Frequency of common tern intruders on salt marsh mats and sandy beach colonies by type of area.

Beach Habitats

In beach colonies, we examined middle and peripheral areas, each of which we divided again into three zones, center, intermediate, and edge. The Cedar Beach colony is 900 m long, and the areas we chose (fig. 5.9) represent major subdivisions of the colony. In the middle study area, 48 percent of the intruders landed in the center whereas in the peripheral site almost 50 percent of the intruders landed on the edge and less than 25 percent landed in the center ($X^2 = 63.5$, $df = 2$, $p < 0.001$; fig. 5.25). We also examined the seasonal effect of intruders as a function of habitat in the middle study area (fig. 5.26). Intruder pressure was heaviest in the center in early May, and heaviest in intermediate and edge locations in early June.

In the middle area, 24 percent of the intruders in the center zone were territory-seeking birds whereas at the edge, territory seekers accounted for 37

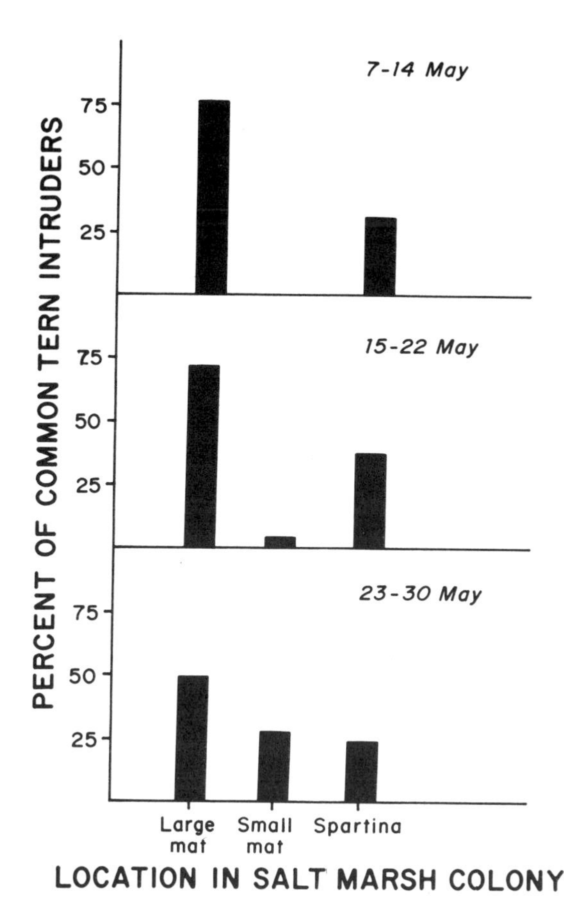

Figure 5.21. Frequency of common tern intruders ($N = 315$ intruders) in center of salt mash colonies by date and location.

percent of the intruders. For birds nesting in the intermediate zone, only 30 percent of the intruders were mate competitors.

More of the mate competitors entering the middle area of the tern colony landed toward the edge rather than in the center ($X^2 = 13.7$, $df = 2$, $P < 0.002$), and a higher proportion of them in the edge zone were repelled ($X^2 = 13.9$. $df = 2$, $P < 0.002$; fig. 5.27).

The pattern of intrusion pressure by mate competitors in the West End sand beach colony varied significantly by location and date. These data represent intruder pressure for particular pairs, thus aerial pirates that cruised over the colony or attacked flying terns with fish were not counted. Peaks in

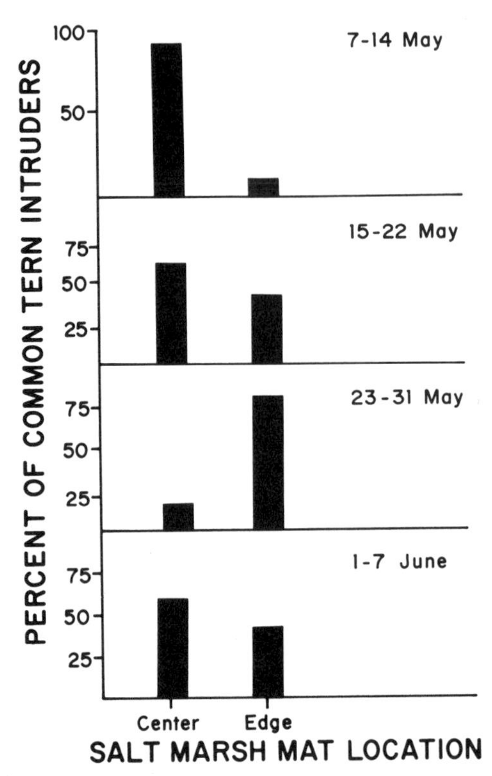

Figures 5.22. Frequency of common tern intruders ($N = 511$) in center and edge of salt marsh mats.

intruder pressure due to initial territorial establishment and territorial competition were earlier in the middle and later in the peripheral areas. Territory-seeking intruders tried to insinuate themselves first into middle areas, and later into edge or peripheral sections of the colony. Intruder pressure during the chick stage (after early June) was similar in all areas, although it decreased more rapidly toward the periphery.

Discussion

Species of Intruders

The species and frequencies of intruders varied by habitat for common terns. In both habitats, we studied terns that had skimmers nesting nearby. In beach

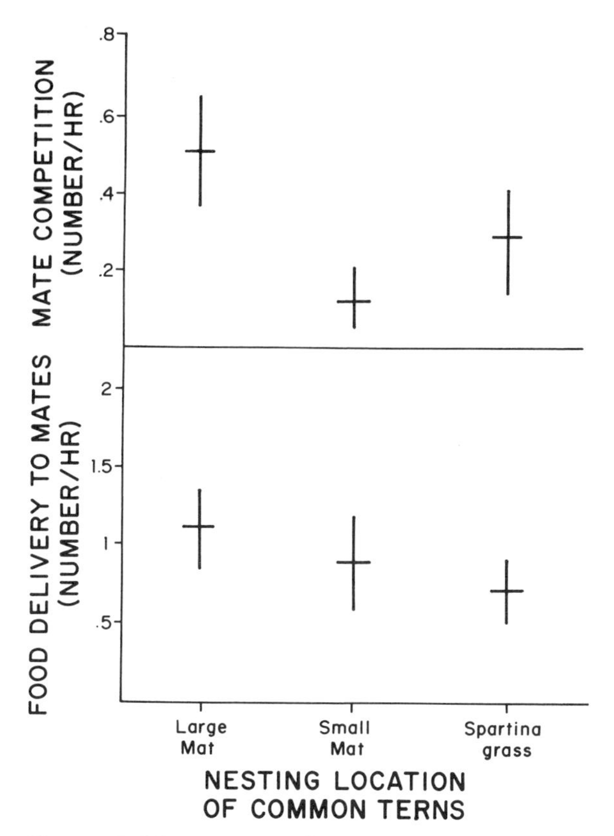

Figure 5.23. Intensity of mate competition and food delivery rates for terns nesting on large and small mats and in grass.

colonies, terns were exposed to more tern intruders, while in salt marshes terns were exposed to more skimmer intruders. This difference derives from the relative composition of nesting assemblages. In salt marshes, skimmers often dominate the subcolonies on mats and there may be equal numbers of terns and skimmers, whereas beach colony subgroups usually have many more terns than skimmers. Second, where skimmers nested with terns in salt marshes, skimmers used the central parts of the mat and terns either preferred or were forced to use the edges. Skimmer nests often formed a compact clump while the terns were spaced out along the edge of the mat. Thus, in salt marshes, terns often had more skimmer neighbors than did terns in beach colonies.

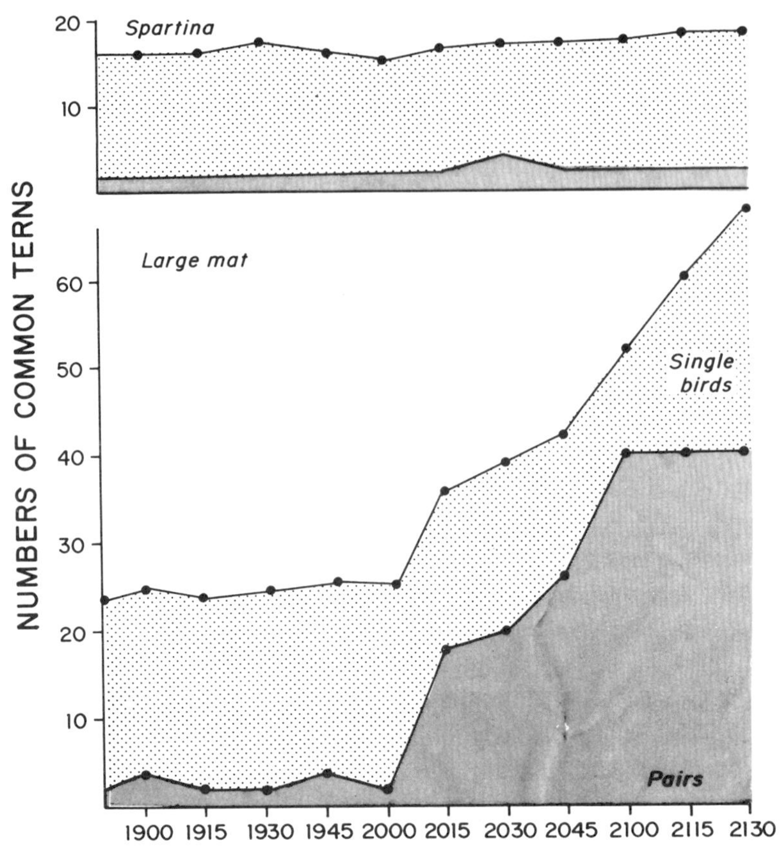

Figure 5.24. Nocturnal roosting of common terns on Pettit Island. Shows the number of birds arriving at each study plot by time of day. *Spartina* plot had 16 nests and mat plot had 24 nests.

Types of Intruders

Ewald et al. (1980) were the first to emphasize the importance of intrusion pressure in the examination of territoriality and aggression. They found that as territory size increased, time spent per act of aggression increased, supporting Tullock's (1979) assumption that as territory size increases, the cost of defense increases. This is not a simple linear relation, however, since Ewald et al. (1980) showed that intrusion pressure was greater on smaller than on larger territories, and Butler and Trivelpiece (1981) showed that great black-

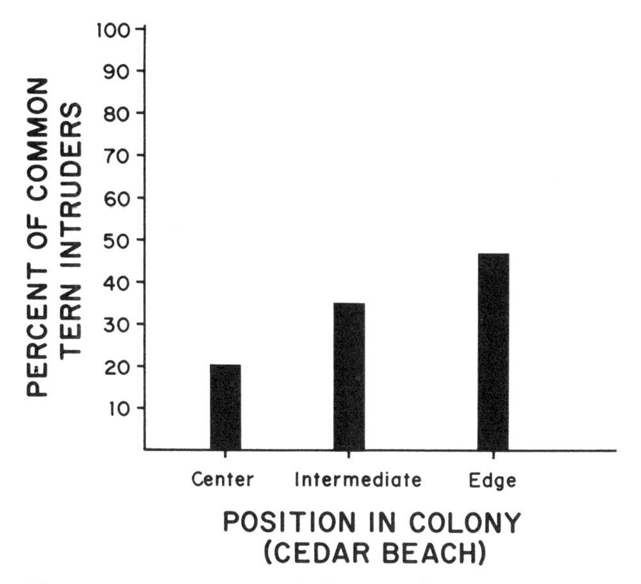

Figure 5.25. Percent of all common tern intruders (N = 1397 intruders) by location in beach colonies.

backed gulls in high density areas (with small territories) engaged in significantly more bouts of agonistic interactions.

Ewald et al. (1980) discussed two types of intruders: territory seekers that usurp space, and neighbors (holding adjacent territories) that might injure or kill chicks in territorial clashes. Burger (1981c) distinguished and quantified aggression directed at neighbors and non-neighbors in herring gulls, noting also the intrusion of cannibals. Cannibalism is well-known in the large white-headed gulls (Parsons 1971). McNicholl (1979) described space-seeking intruders in forster's tern.

Some intruders are seeking mates, and the existence of mate competition has been noted for several species (MacRoberts 1973; Gladstone 1979; Birkhead 1982). Similarly, piracy of food from parents returning with provisions for their young has been recorded in several species of gulls and terns (see review in Burger and Gochfeld 1981a). However, piracy and mate competition have usually been discussed separately, and not in the context of the total intruder pressure that territory defenders in nesting colonies face.

Thus, several papers (reviewed in the appropriate sections below) have noted the occurrence of competition for space, mates, and food (brought

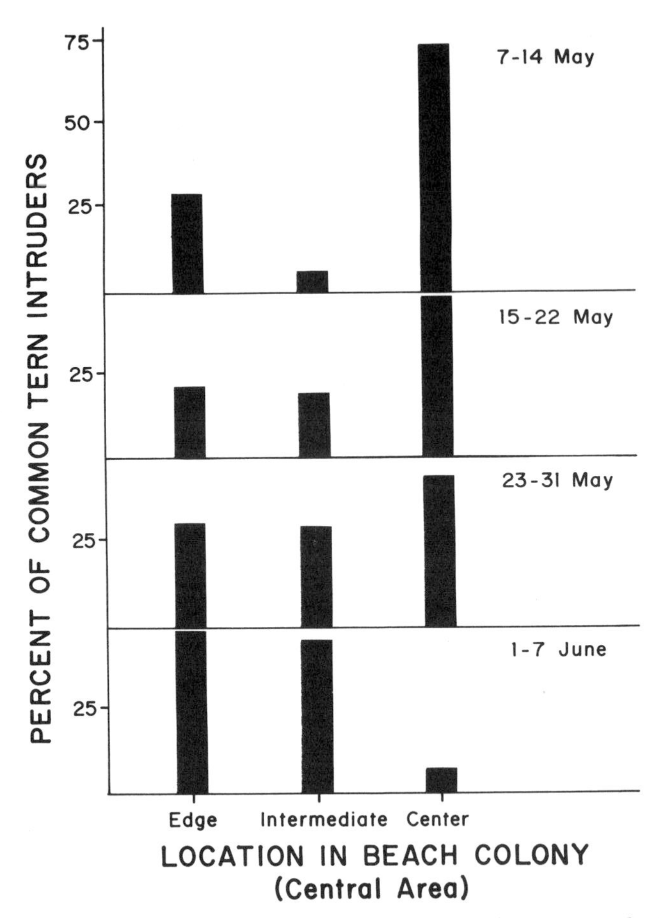

Figure 5.26. Seasonal variations in intruder pressure by habitat in beach colonies (N = 1397 intruders). Total intrusion sums to 100 percent for each week.

back to feed chicks), but have not examined the relationship between them in any single colonial species. Since competition is an important aspect of the cost of coloniality, we examined the overall relationship of these costs in common terns.

During the initial phase of the breeding cycle in common terns, space competitors accounted for 82 percent of the intruders in salt marshes, but only 62 percent in beaches, with mate competition accounting for most of

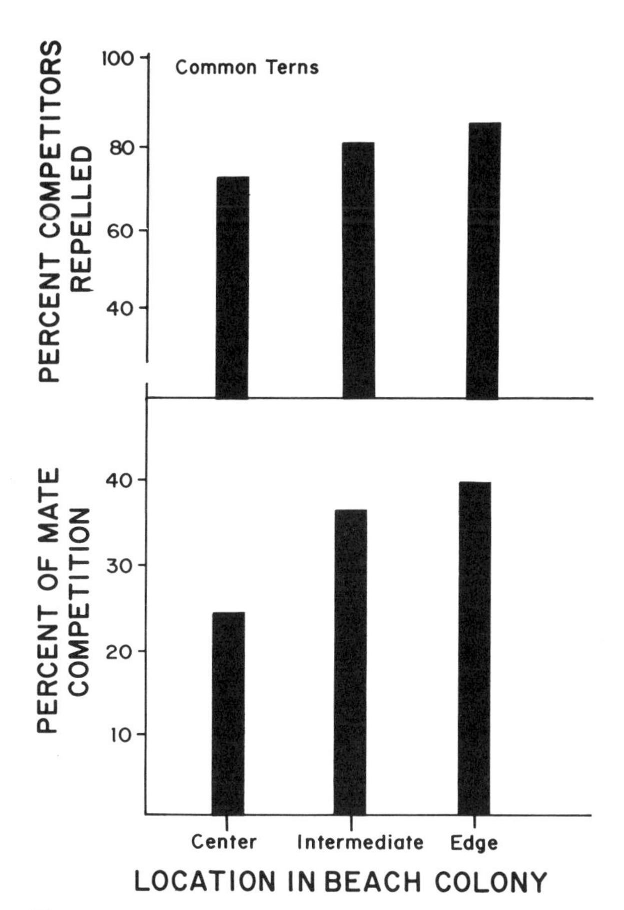

Figure 5.27. Mate competition in common terns (competitors sum to 100 percent) and percent competitors expelled as a function of location in beach colonies ($N = 359$).

the remainder. A small proportion due to food piracy mainly involved aerial chases and could not be attributed to the costs for a specific pair.

The relatively high cost of space defense in salt marshes reflects the fact that high-quality nesting space in salt marshes is more limited than in beach colonies. Common terns nesting in salt marshes prefer to nest on mats because they are less vulnerable to flood tides (see chapter 8). On most salt marsh islands, mats are relatively long and narrow, and only occur at the highest tide mark from the previous winter. When the mats become fully occupied, terns seeking nest sites must nest in suboptimal habitats in the adjacent *Spartina*. These locations are lower in elevation, and require the terns to construct elaborate nests. Thus, new territory-seeking terns (or pairs) attempt to obtain territories on mats before settling for *Spartina* sites.

In sandy beach colonies, terns also compete—usually for central locations or those with a particular nest density. However, these sites may be less limiting since the relative amount of center area in beach colonies is greater than the amount of mat on salt marshes. Relatively, mate competition was greater in sandy beach colonies than in salt marshes. We believe this difference relates to differences in the timing of the breeding synchrony between salt marshes and sandy beaches (see below).

We then compared total intruder types for the entire season, using a sample area rather than observing intruders at individual nests. This analysis identified pirate intruders as a more important component of intruder pressure in sandy beach colonies than in salt marsh colonies. This difference may relate to differences in food availability (relative success of foraging parents or greater distance to feeding grounds) and to the more synchronous nature of the chick feeding stage in salt marshes.

Seasonal and Habitat Differences in Intruders

Common terns nesting in salt marshes and sandy beaches experienced seasonal and habitat differences in intrusion pressure. Overall, mat areas in salt marshes and middle areas in sandy beaches had higher intrusion pressures than *Spartina* or peripheral areas. In salt marshes, intruding terns preferred the center of large mats. In beach colonies, there were more intruders in the center of middle areas, but on the edge of peripheral areas.

These differences generally reflect space or mat competition. Presumably intruder pressure for space is highest in preferred places where birds wish to establish territories, and mat competitors are most prevalent where their chances of success are greatest. In both salt marshes and sandy beaches, space competition accounted for over 60 percent of the intrusions. Thus, habitat differences in intrusion pressure reflect largely territorial constraints (see "territory competition in common terns" below).

There were also differences in intrusion pressure by habitat and date. In salt marshes (where over 80 percent of intruders are seeking space), intruder pressure was highest on the center of large mats early in the season, but this tendency decreased throughout May reflecting a decrease in available space on mats. The central parts of large mats eventually are packed with territorial birds and no new birds can insert themselves between existing pairs. Intruders then attempt to land on smaller mats, or in *Spartina*. The seasonal shifts in salt marsh common tern intruders largely reflect competition for space and will be discussed below.

In middle areas of beach colonies, intruder pressure is highest in the center zone in early May, and it decreases steadily thereafter. Again, this difference reflects the temporal decreases in available space in preferred locations. Center zones eventually become occupied as late-arriving birds settle to nest, and eventually there is no more room for space competitors to land. Mat competitors continue to land in middle areas, however, but since nesting tends to be earlier in the preferred habitat, the potential for mat acquisition and promiscuous matings decreases, and late in the season mate competitors land almost exclusively in intermediate and edge zones. Through this process terns in the colonies we studied maintained nearest neighbor distances on the order of 120–200 cm. Late arrivals did not succeed in fitting into such crowded quarters but we and others (Nisbet pers. comm.) have seen much denser nestings (e.g., 40 cm) for common terns. Such areas are often very synchronous. It remains a puzzle why some terns tolerate such dense nesting while others permit no intruders to raise the density. In the following sections we will discuss territory competition and mate competition in common terns.

5.4 TERRITORY COMPETITION

Intruders competing for space were relatively more numerous in salt marshes than on beaches, and overall aggression levels during the territory acquisition phase were correspondingly higher in salt marshes. Taken together these data indicate that competition for space is greater in salt marshes than in sandy beaches. As mentioned above, this difference relates both to constraints imposed by synchronous nesting and to competition for suitable space. Quality of specific sites will be discussed below.

The reproductive activities of common terns nesting in salt marshes were clearly more synchronous than terns nesting on sandy beaches. This resulted in higher intrusion rates and higher aggression rates during the territory acquisition stage in salt marsh nesting common terns. However, the advantage of fairly synchronous territorial competition is that the high intensity aggression period is shorter, and terns nesting in salt marshes have lower space intrusion rates during incubation, resulting in lower aggression rates during this time (late May to early June).

In contrast, on beach colonies, intruders competing for space continue to arrive throughout the reproductive cycle, because late-arriving birds are continually attempting to establish territories. Birds attempting to establish territories later in the breeding season may be younger birds (Ryder 1980; Massey and Atwood 1981), birds that recently arrived, birds that have lost mates, eggs, or chicks (Burger 1982a), or birds that are attempting second clutches (Wiggins et al. 1984). Recently Wiggins et al. (1984) reported a low frequency of common terns laying second clutches while they were still feeding chicks from the first brood.

Most common terns will re-lay or re-nest after a failed breeding attempt. In large colonies, nest failure of some pairs still leaves a large enough nucleus of breeding pairs so that failed breeders will re-nest in the same colony. In smaller colonies, nest failures may substantially reduce the number of breeding pairs, perhaps resulting in insufficient social stimulation for failed breeders to renest. This results in emigration, with birds that have failed searching for other suitable colonies.

In our study, the sandy beach colonies were usually very large (over 1000 pairs), while the large salt marsh colonies were usually less than 600 pairs (except for the Pettit colony in 1990 which had over 1000 pairs). Large-scale renesting was characteristic of beach colonies, but less frequent in the salt marsh. It is possible that the lack of large scale renesting (and some shifts in territory location) may be due to lack of adequate social stimulation (chapter 3).

Within sandy beach and salt marsh colonies, intruder pressure for space varied. In salt marshes, over 85 percent of the intruders competing for space landed on mats. This is not only due to the higher value of mats as nesting sites, but to the ability to land conveniently. Mats are open surfaces that allow a tern to land unobstructed by vegetation, without risk from protruding stems of *Spartina*, or becoming entangled in vegetation. The lack of unobstructed space in *Spartina* also presents a problem because there is less maneuvering room for displaying to potential mates, for courtship feeding mates, for prospective mates to land, for mates to land before assuming incubation duties, and for chicks to wander about.

Spartina has an advantage as a nest site, particularly as the season progresses. *Spartina* grows. By June, a bird displaying to potential mates becomes less and less visible from the air. Presumably, space competitors select sites that allow easy transmission of their courtship display, and this results in more intruders seeking space on mats rather than in *Spartina*.

There are also major differences in the frequency of territorial intruders in beach colonies. Central areas have more intruders seeking space than edge areas. Many authors working with colonial seabirds have stressed the importance of nesting in the center. Generally, center-nesting birds have higher reproductive success because they are less susceptible to mammalian predators that enter from the edge of the colony (Patterson 1965; Nelson 1966; Coulson 1966; Spurr 1974; Harris 1978). There is an opportunity for circularity here since older, more experienced birds tend to nest in the center; hence, higher productivity can be due to experience as well as location.

However, predation rates from aerial predators are sometimes higher in central areas because predators may prefer to hunt where nesting density is higher (Burger and Lesser 1978). Dry-land beach colonies have greater exposure to mammalian predators that enter from the edge (including humans) than do marsh islands. Thus, in these colonies, edge areas are suboptimal; and fewer intruders attempt to establish territories there.

Presumably, the relatively higher costs of territorial defense by central nesting terns are balanced by the benefits of nesting in the preferred habitats and locations. Since the territorial costs of nesting on salt marsh mats exceed those of nesting in *Spartina*, and those costs of nesting in central areas exceed those of edge areas on beaches, the high cost locations must result in higher reproductive success for their choice to be adaptive evolutionarily, other things being equal. And indeed in chapter 9, we will show that terns nesting on mats in salt marshes or in central areas of beach colonies do have higher reproductive success than those in nearby, suboptimal areas.

Copulation Interruptions

An additional cost of nesting in colonies is the interruption of copulations by intruders. These interruptions occur when the copulating male dismounts to chase an intruder, or when the female becomes distracted and no longer presents her cloaca. The cost of interrupted copulations presumably is slight for solitary nesting species since competition for space and mates is reduced, and conspecifics seldom land nearby.

During the period immediately preceding egg-laying, and during egg-laying itself, common terns copulate several times a day. Multiple copula-

Table 5.3. Copulation interruptions in common terns at West End Beach.

	1982		1983		1984		All Years	
	Total	*%*[a]	*Total*	*%*[a]	*Total*	*%*[a]	*Total*	*%*[a]
Pre–egg-laying	66	5%	90	7%	72	8%	228	6%[b]
Egg-laying period	349	8%	125	8%	252	14%	726	11%[c]

[a] Percent of total copulations interrupted.
[b] Contingency table $X^2 = 19.4$, $df = 1$, $P < 0.001$.
[c] Contingency table $X^2 = 0.04$, not significant.

Table 5.4. Copulation interruptions of common terns as a function of habitat.

	Total Copulations		*Percent Interrupted*
Sandy beach[a]	954		10%
Mat in salt marsh[b]	93		28%
Spartina in salt marsh[b]	152		2%
Interhabitat Difference X^2		43.2[c]	
P <		0.001	

[a] West End, 1982–1984.
[b] Carvel and Pettit Islands, 1983–1984.
[c] The X^2 value is based on 2 × 3 contingency table.

tions may be essential for pair-maintenance, for stimulation of males to courtship feed the females, for high fertility, or to minimize sperm competition. In any case, interruptions may result in wasted time, energy, and sperm and necessitate further copulation attempts later on.

For common terns, 6 percent of the copulations in the pre egg-laying period were interrupted whereas 11 percent were interrupted during egg-laying (table 5.3). The frequency of interruptions of copulations varied by year, and for terns it was three times as high in the egg-laying period in 1983 as in 1982. Copulation interruptions also varied by habitat (table 5.4). Common terns nesting on mat in salt marshes were interrupted more than terns nesting in sandy beaches, and terns nesting in *Spartina* were interrupted far less than the above two groups.

5.5 MATE COMPETITION

All sexually reproducing animals, whether they breed solitarily or in social groups, have the task of finding a mate. Breeding in colonies enhances the opportunity for finding mates by increasing the exposure of an unmated individual to quantities of potential mates. The opportunities for finding mates presumably are greater if animals seeking mates spend some part of their time in groups where other unpaired individuals congregate. Such vertebrate assemblages could include foraging flocks, roosting groups, courting groups or leks, as well as breeding colonies.

In some species females may choose males on the basis of the quality of their territories (i.e., richness of food, Verner 1964; Verner and Engelson 1970; Lenington 1980). However, where males provide parental care, females should consider the males' willingness and ability to provide such care (Orians 1969; Trivers 1972; Nisbet 1973a). Even where territory quality does influence female choice, other factors such as experience or body size are also correlated with female choice (Yasukawa 1981). More recently authors have tried to isolate the features of the male that may influence or correlate with mate choice. In birds, mate choice relates to song repertoire (Searcy 1984), plumage color (Klint 1980; Roskaft and Jarvi 1983), male courtship activity (Kruijt et al. 1982), mate attentiveness (Wishart 1983), nest quantity (Garson 1980), nest quality (Collias and Victoria 1978), male fat reserves (Petrie 1983), handicap traits (Zahavi 1975), and arrival time on the breeding ground (Alatalo 1984). Indeed, in an experimental study of pigeons, Burley (1981) established that females use multiple criteria in mate choice—including plumage color or pattern, age, and previous reproductive experience.

In a colonially nesting, monogamous species such as terns, females could use a variety of characteristics such as territory size, territory location, arrival time at the colony, or specific male traits for mate choice. Generally, mate selection has not been examined in detail, although the assumption is that females can use courtship activity and fish provisioning (Nisbet 1973a). Since males courtship feed the females prior to egg-laying, females could assess their potential willingness and ability to provide care for the young (Nisbet 1973a; Wiggins and Morris 1986).

In our view, advantages of colonial nesting include increased potential mate access, a particular place to go in search of mates, and opportunities to encounter mates throughout each reproductive cycle. Whenever a bird arrives at the breeding colony, whether early in the season or late, its potential for finding a mate is enhanced by the presence of many birds at the colony.

However, the presence of unpaired birds seeking mates is a disadvantage to birds that are already paired, in that birds may seek to win paired birds away from their current mates. The courtship or promiscuous behavior of unpaired birds imposes several costs on paired birds: 1) males must mate-guard their females to prevent having her mate with another male (referred to as extra-pair copulation, kleptogamy, or cuckoldry (Gowaty 1982)) 2) males must prevent mate loss, and 3) females must fend off the suitor (if she wishes).

Mate Guarding

Mate guarding, a tactic aimed at increasing the male's confidence of paternity (Mumme et al. 1983), has been recorded for herring gulls (Morris and Bidochka 1982), magpie (Birkhead 1979, 1982), captive ring dove (Lumpkin et al. 1982), bank swallow (Hoogland and Sherman 1976), mountain bluebird (Power and Doner 1980), starling (Power et al. 1981), acorn woodpecker (Mumme et al. 1983), little blue heron (Werschkul 1982a), and brown-headed cowbird (Gochfeld 1977b). Thus, mate guarding can occur in both colonial and solitary-nesting species. Promiscuous or extra-pair copulation has been reported for a variety of species (see Gladstone 1979; Werschkul 1982a,b). Presumably, mate guarding lowers the risk of males raising offspring not their own (Trivers 1972).

Mate guarding and prevention of promiscuity are behaviors that have been recognized, and their prevalence and costs discussed. Yet there are few data that quantify these costs for individual males. This is due in part to the difficulty in actually witnessing such extra-pair copulations, and also to uncertainty as to the frequency with which extra-pair copulations result in fertilization. Birkhead (1979) did examine intruder pressure in magpies, and found that the number of fly-in intruders (potential males seeking promiscuous matings) was high during egg-laying but was almost nonexistent before and after that period.

Mate Competition in Common Terns

In common tern colonies mate competition occurs before and during egg-laying, and paired males respond accordingly. The mate competitors are most probably unpaired birds seeking mates, but later in the season such unsuccessful birds or even paired birds may attempt to mate promiscuously with a paired female, or lure away the female from her mate. We documented the presence of mate competition in common terns nesting both in salt marshes and sandy beaches, although the relative abundance varied.

In all habitats, the presence of mate competitors was highest during the egg-laying period, although it continued into June (in salt marshes) and into July (on sandy beaches). The presence of these mate competitors was costly to both members of a pair, because both males and females chased away mate competitors that landed with fish, and they did so immediately. Mate competitors were chased more quickly than territory competitors. The response time may be so rapid because promiscuous copulation or forced extra-pair copulation (sometimes referred to as rape) can occur very rapidly, whereas territory establishment is a time consuming process. The cost of delaying a chase is greater if the intruder is seeking matings than when it is seeking space. Clearly, it is adaptive for males to chase away other males, both because it would prevent promiscuous matings and reduce the probability of their female deserting them for another male.

Females have four options when a mate competitor lands: she can 1) ignore the intruder, 2) chase him away, 3) copulate with him, or 4) join the new male and desert her mate. We found that females generally chased mate competitors, and they did so immediately. Infrequently, they ignored them, stole their fish, and then chased them away, and—rarely—they copulated with them. Females may chase mate competitors because: 1) their individual distance has been violated (males usually land within a meter of them), 2) they are preventing them from usurping space for a territory if they later obtain a mate, or 3) they are assuring their mate of their fidelity. It seems unlikely that it is merely a matter of individual distance, since in the late evening female terns let roosting birds land equally close without chasing them.

It is possible that by chasing intruding males, females are actually preventing them from usurping space. A male with a fish may be using it to disguise his intention of seeking space by appearing nonthreatening. However, we discount this since males with fish usually land close to females and not equidistant from other territorial birds. Presumably, by landing equidistant from established pairs, an intruder would lessen his chances of being chased. We favor the hypothesis that females are assuring their mates of their fidelity. In a crowded tern colony, birds are flying in all directions. During much of the preincubation stage, females are alone on the territory and males, engaged in courtship feeding, go off to forage. After returning with a fish and feeding it to the female, males may depart immediately. Thus, males are repeatedly going out and returning to the colony with fish. Further, the direction of their return varies as they forage in different locations. Just before males land at the nest site, they give an identification call, and the female becomes attentive and greets them. If a male approaches and lands from behind without calling, he may be initially attacked by his mate. Thus, males can be nearby without the female being aware of their presence.

Mate guarding has probably evolved to increase the certainty of paternity, and presumably males will be less likely to invest parental care if they observe or suspect their mates of copulating with another male (Erickson and Zenone 1976), although this remains to be studied. Females have already invested time and energy in the pair-bonding process with one male, and disruption of the current pair bond in favor of the new male may delay breeding and the date of egg-laying, with adverse results. Testing these hypotheses in the field will provide a substantial challenge to future investigators.

In colonial birds, there is clear evidence that birds that lay earlier in the season usually have higher reproductive success (Fisher 1971; Milne 1974; Parsons 1975, Spaans and Spaans 1975). Thus, unless her current mate is not delivering adequate food during the courtship period, it is to the female's advantage to stay paired with the male on the formerly established territory. The new male would either have to expel the old male, or establish a new territory elsewhere. We suggest that just as females may be assessing the male's quality by his ability to defend a territory and courtship feed, males may be assessing the female's ability to repel intruding males.

Variations in Mate Competitors by Habitat

As mentioned above, mate competitors account for a larger proportion of the intruders in sandy beach colonies compared to salt marsh colonies. This is true both in an absolute sense (number per pair per hour) and in a relative sense. Overall, the mean aggression rate in salt marshes and beaches was similar. The clustering in time of mate competitors in sandy beaches is attributable to 1) lack of breeding synchrony (potential mates are available longer), 2) attractiveness of large colonies to younger birds that may breed later, and 3) existence of larger area with lower average density.

The beach colonies examined (West End, Cedar Beach) were large (from

1000 to 6000 pairs) while the salt marsh colonies generally had fewer breeding pairs (usually, fewer than 500 pairs). Larger colonies may attract more younger birds than smaller colonies (Gochfeld 1980b). Further failed breeders in search of others at the same reproductive stage (for social facilitation, Darling 1938) may seek out larger, less synchronous colonies. Post and Gochfeld (1979) documented that birds that failed at small salt marsh colonies were trapped later in the same breeding season in the large beach colonies.

Finally, on the mat sections of salt marsh colonies, terns nest very close together, whereas density is generally lower in most sections of sandy beach colonies. Thus, there is more room for intruders of any type to land in beach colonies, and fewer terns would chase them.

Within the beach colonies, there was proportionately more mate competition in edge compared to central locations despite the greater "desirability" of the center. Edge areas are generally less dense, allowing more room for intruders to land without being attacked. Further, in salt marsh colonies proportionately more mate competitors landed in *Spartina*, whereas more territorial intruders landed on mats. These findings coroborate the suggestion above that mate competitors often land in less dense areas.

Thus we see that several patterns begin to emerge with respect to aggressive behavior. Areas preferred by birds are selected early in the season. The intent of the intruder (territory quest, mate competition or piracy) is as important an influence as density and the natural responsiveness of territorial birds in determining the response of the territory holder and in affecting the outcome of the encounter.

SIX

Predation, Vigilance, and Antipredator Behavior

Predation is one of the primary selective forces during reproduction (Krebs 1973; Gochfeld 1985; Wittenberger and Hunt 1985). Predators can specialize on eggs, young, or adults. Solitary nesting species rely mainly on cryptic coloration, on being widely spaced, and/or having distraction displays (Lack 1968; Tinbergen 1959, 1963, 1967; Gochfeld 1984). Colonial species rely on antipredatory behavior of the group (Cullen 1960; Krunk 1964; Bianki 1967; Slobodkin 1968; Alexander 1974; Wittenberger and Hunt 1985; Conover 1987), camouflaging coloration of nests and eggs (Patterson 1965; Tinbergen et al. 1967), or nesting in vegetation (Houde 1983), or in inaccessible places (Cullen 1957). Within colonies, nesting dispersion is a compromise between nesting far enough apart so that cryptic eggs and young are not detected by predators (Tinbergen 1956; Krebs 1973), and nesting close enough for effective group interactions (Lack 1954; Crook 1964; Pulliam 1973).

Most seabirds have minimized predation by nesting on remote or inaccessible places such as offshore islands, rocky cliffs, or trees. Although such habitat choices often minimize exposure to mammalian predators, avian predators still pose a threat for many species (Cullen 1957; Kruuk 1964; Muller-Schwarze 1973). Further, the large concentrations of nesting birds are attractants, and in some cases may serve to recruit new predators to the colony as the season progresses (Gochfeld 1980b).

Coloniality may arise from a passive retreat to a predator-free location (Alexander 1974), or may itself represent an effective adaptation to thwart predators. Large numbers of individuals may confuse a predator, making it difficult to single out a prey item, as Hamilton (1971) suggested for bird flocks and Milinski (1977a,b) suggested for fish schools. A milling flock of mobbing terns can confuse a predator, reduce its vigilance, and actually increase its risk of being injured or eaten itself (Milinski 1984). The rapidly growing literature on vigilance in groups (first emphasized by Lack 1954) provides ample support for the benefits of living in groups. Positive and active antipredator behavior in the form of mobbing constitutes an additional factor. Lack (1954) recognized the adaptive significance of mobbing for the individual within the flock. In this chapter we review the main predators on common terns and examine the factors affecting predation, including vigilance and antipredator behavior.

6.1 PREDATORS IN COMMON TERN COLONIES

Common terns in salt marsh and sandy beach colonies face both avian and mammalian predators with some regularity. Some predators are native species whereas others are relatively recent introductions or human commensals, which the terns have encountered only in the past two or three centuries. Some of the native species have only recently spread onto the barrier beaches and may still be uncommon in salt marshes.

Mammalian Predators

A wide variety of mammals have been implicated as preying on terns, and inland colonies face predators such as coyotes and river otters (Verbeek and Morgan 1978) which do not occur in the coastal colonies we studied.

Red Fox: Foxes can have a major impact on a seabird colony (Kruuk 1964; Patton and Southern 1978; Southern et al. 1985). Sargeant et al. (1984) recently provided a detailed account of fox predation on waterfowl. We have not seen red fox at either Cedar Beach or West End, but they are present on salt marshes in both New York and New Jersey (e.g., Captree, Clam, Mordecai, Hester, and Sandy Islands). The population is sparse. Foxes take adult terns and cache them, thus removing more than they may eat in a single visit. Common terns and hearing gulls have deserted islands when foxes have appeared.

Raccoon: We have had only one documented episode of raccoon predation in our tern colonies which occurred at Holgate. However, in 1990 we found the first evidence that raccoons were present on the New York barrier beach. A dead adult raccoon was found at Cedar Beach, and raccoon tracks were found repeatedly in the colony. Although we suspected predation on both common and roseate tern eggs and chicks, we could not ascertain what these animals were eating. Marshall (1942) reported nocturnal raccoon predation on common terns as a major factor leading to reproductive failure.

Striped Skunk: Less numerous than raccoons on Long Island, and apparently rare or absent from the barrier beach, skunks could certainly become significant predators on tern eggs.

Mink: Mink are rare on Long Island (Connor 1971) and in New Jersey (van Gelder 1984). On one salt marsh island, Thorofare, we found the characteristic devastation wrought by a mink (Burger 1974a). The mink killed all young terns in three days, leaving the slightly devoured carcasses in piles containing 3 to 12 bodies.

Long-Tailed Weasel: Weasels are widespread and common in New York and New Jersey, but are scarce on the coastal plain and barrier beach (Connor 1971; Van Gelder 1984). Weasels have been reported to devastate colonies in Massachusetts (Austin 1929). On occasion, we have found chicks with their brains eaten in a manner indicative of a weasel kill, but have never seen weasels.

Grey Squirrel: Connor (1971) lists this otherwise ubiquitous mammal as rare on Long Island barrier beaches, and Van Gelder (1984) doesn't list it for this habitat in New Jersey. We have not seen squirrels on salt marshes. However, during the course of our study, squirrels colonized the Long Island barrier beach. The first squirrels were seen at West End in 1975 and at Cedar Beach in 1977. At both colonies, certain individuals (one only at Cedar Beach) developed a preference for tern eggs. They leave behind eggs broken in half, with a characteristic finely jagged margin caused by their teeth (fig. 6.1).

Figure 6.1. Predators leave characteristic signs. This common tern egg was chewed open by a grey squirrel. Grey squirrels leave a finely jagged border on the shell.

Domestic Dog: Feral dogs or pet dogs running free are a problem in bird colonies. We have seen dogs running back and forth snapping up eggs and chicks. The characteristic clouds of terns circling noisily overhead without diving, quickly call attention to any dog that wanders into a colony. The birds do not attack the dog but circle about 3 to 5 m above it, as if aware that the dog might leap up and capture a mobbing bird as wild canids often do.

Domestic Cat: Feral cats are moderately common on the Long Island barrier beaches. A cat lived in the West End tern colony in 1983, and was regularly seen hunting at dusk, until it was killed by an automobile as it carried a half-grown tern to its den. In 1989 and 1990, cats were seen at Cedar Beach and killed many tern chicks, although at least one of the cats was killed by a car. At the Holgate barrier beach colony, a single cat killed at least 20 adult terns, leaving their wings and backs strewn about the colony. At Holgate, cats from nearby houses occasionally enter the colony and kill many adults and chicks. Panov and Zikova (1987) reported severe cat predation in Russian gull colonies, and predation by feral cats appears to be a very widespread problem in seabird colonies (Veitch 1985).

Norway Rat: Rats are noted predators on many species of seabirds (Kepler 1967; Atkinson 1985), and rats are generally the most significant nonhuman mammalian predator on terns. Rats maintain a large population and can wipe out entire sections of a tern colony. Austin (1948) detailed the impact of rats on Cape Cod colonies and Gochfeld (1976) described the occurrence and predation of rats at Cedar Beach. Rats thrive only in areas where bushes, dense grass, other tangled vegetation or debris such as driftwood create shelter. Although the Austins (1933, 1948) undertook vigorous rat control programs, few such programs have been conducted in the New York Bight colonies. Fish and Wildlife personnel conducted an effective control program at the Holgate Division of the Brigantine (now Edwin B. Forsythe) Refuge. Prior to this program, many tern nests were destroyed by rats, whereas after the control program the colony thrived. Few of the salt marsh islands sustain a rat population. However, rats severely reduced the large tern colony on Lane's Island, New York, in 1974 and 1975, with the greatest impact on the roseate terns nesting in the denser cover.

Avian Predators

Avian predators have easy access to all of the colonies we studied. The main groups of avian predators are raptors, herons and gulls. The former are a threat to adults and chicks, the latter a threat to chicks and eggs. Grackles, crows, and shorebirds such as oystercatchers and turnstones, prey on tern or skimmer eggs.

Gulls: Gulls prey on a variety of birds including ducks (Dwernychuk and Boag 1972; Joyner 1974), terns (Parsons 1971), shearwaters (Corkhill 1973), murres (Johnson 1938), gulls (Burger 1979c), cormorants (Kury and Gochfeld 1975), murres (Gaston and Nettleship 1981), and puffins (Nettleship 1972). Gulls are also important as pirates (Nettleship 1972). At West End and Cedar Beach, a few gulls perch on utility poles adjacent to the colony waiting for the opportunity to capture vulnerable half-grown chicks. These few individuals represent specialists, whereas most of the thousands of gulls nesting at nearby colonies do not become tern predators. A few of these gulls attempt to nest on the edge of the tern colonies (fig 6.2). In salt marshes, herring gulls often start to breed adjacent to tern colonies, and the terns eventually desert the site. When there are only a few pairs of gulls, the terns may persist but suffer high egg predation. Terns often respond by moving into the *Spartina* where their eggs are less visible to the gulls, but where they are more vulnerable to flooding. Elimination of these pioneer gull nests is essential in preventing gulls from establishing a colony which usually causes terns to abandon the site. Expanding gull populations have had a major impact on tern colonies in New England and the Great Lakes (Nisbet 1973b; Shugart and Scharf 1983; McKearnon and Cuthbert 1989).

Black-Crowned Night Heron: Some individuals specialize on young terns, stalking the colonies at night and taking mainly young birds (Hunter and Morris 1976). We have occasionally heard their calls after dusk, and in 1990 night heron tracks were seen at Cedar Beach. The episode of nocturnal predation on newly hatched chicks at West End in 1971 may have been attributable to night herons. Collins (1970) reported one eating a roseate tern chick on Great Gull Island, New York. We have seen no evidence of night heron predation in our colonies.

Marsh Hawk or Northern Harrier: Two pairs of harriers breed within 5 km of Cedar Beach, and from 1981 to 1990 they repeatedly entered the western part of the colony to capture tern chicks. A harrier appeared over the dune, where it was greeted by a rising mass of terns. It flew directly to the central area of the colony, sometimes deviating slightly in response to the mobbing terns. The hawk then dove to the ground seizing a chick, usually one running for nearby vegetation, and left the colony, escorted by as many as 500 adult terns. We saw such forays between three and four times per day. We also saw two adult terns captured. Harriers breed near salt marsh tern colonies, and we saw evidence of their predation frequently.

Peregrine Falcon: Peregrines were deliberately introduced in 1970 to the New Jersey salt marshes by hacking young birds. A few breeding pairs have become established, and they prey on a variety of birds including terns. They take both adult and young common and least terns, and we once watched a peregrine snatch an adult common tern out of a flock that was mobbing it.

Figure 6.2. Herring gull that have established territories on the edge of the Cedar Beach colony. Such pioneering pairs usually include at least one subadult bird, and often fail. If successful, they may form the nucleus of a growing gull colony which eventually displaces the terns.

Ruddy Turnstone: Turnstones are well known as tern egg predators (Crossin and Huber 1970; Parkes et al. 1971; Loftin and Sutton 1979; Brearey and Hilden 1985; Farrow et al. 1986; Morris and Wiggins 1986). From 1983 through 1990, up to six turnstones visited Cedar Beach and fed on tern and skimmer eggs. In 1985, a single turnstone on Egg Island, New Jersey, destroyed almost all tern eggs before it was removed. In 1988, a single turnstone on Marshelder destroyed eggs in about fifteen tern nests before migrating from the island. Brearey and Hilden (1985) reported that turnstones may destroy up to 67 to 100 percent of nests of common gull and sandwich tern.

American Oystercatcher: Oystercatchers have undergone a remarkable explosion in the past thirty years (cf. Post and Raynor 1964), and are now a common and conspicuous salt marsh nesting species. In Barnegat Bay, oystercatchers show an affinity for tern colonies (see below). On Long Island, oystercatchers nest mainly on salt marsh islands (Lauro 1986; Lauro and Burger 1989), and there have only been two oystercatcher nesting attempts in a barrier beach tern colony. Some tern colonies have suffered extensive egg predation by oystercatchers, however, the oystercatchers nest comfortably in the midst of terns and as a rule they are not mobbed. Veen (1977) reports Eurasian oystercatchers preying on chicks as well as eggs of terns.

Short-Eared Owl: Over a 20-year period we have only 5 sightings in barrier beach colonies and 6 in salt marsh colonies. Owl pellets and feathers of this species have been seen in both West End and Cedar Beach colonies during periods when a spate of nocturnal killing of adult terns occurred. Short-eared Owl predation on terns was described nearly a century ago by Mackay (1898).

Great Horned Owl: We have never seen or heard this owl on the barrier beach in summer. However, they do nest on the mainland within 10 km of the tern colonies, and could easily make this feeding flight. In New England, Great Horned Owls are very important predators on nesting terns (Floyd 1928; Austin 1940; Nisbet 1975).

Owl predation is characterized by decapitated and partially eaten bodies of adults and large young. On the barrier beach, there have been two occasions when the vast majority of common tern chicks disappeared between 2 and 7 days of age. With no evidence of diurnal predation, we waited through the night, but to no avail. Either owls or night herons could have been the culprit. In two salt marsh colonies we have found evidence of horned owl predation, judging from decapitated adults and young amid several great-horned owl feathers.

Fish Crow: Crows of various species are important predators on colonial waterbirds (e.g., Croze 1970; Burger and Hahn 1977). Although it is not uncommon on the barrier beach, this crow is very seldom observed at either the West End or Cedar Beach tern colonies, perhaps because the terns vigorously mob crows. Fish crows have been severe chick predators at the Sore Thumb (Town of Babylon) Least Tern colony only 5 km from Cedar Beach, taking virtually all young in 1984 and 1985. In salt marsh colonies we have seen surprisingly little evidence of fish crow predation, although in one colony where human disturbance was frequent, crows ate 40 percent of the tern eggs.

Blue Jay: This potential egg predator is not a common breeding species on the barrier beach and does not generally get to salt marsh islands during the breeding season. However, blue jays occasionally have reached the Great Gull Island common tern colony where they were mobbed vigorously (Cooper et al. 1970). However, they were not observed eating tern eggs.

Common Grackle: Grackles have increased dramatically in both New York and New Jersey in the past twenty years. Accordingly, the frequency of grackle predation has increased, and we have seen grackles peck and eat tern eggs at West End beach and in several salt marsh colonies.

Boat-Tailed Grackle: This species is gradually spreading northward in New Jersey and now nests on several of the tern islands which have the necessary bushes. The first Long Island breeding was reported in 1981 (Gochfeld and Burger 1981a). The population size is still low, and although the grackles breed on some of the same salt marsh islands with terns, we have documented only one case of predation on tern eggs.

Red-Winged Blackbird: We have seen blackbirds destroy tern eggs in both beach and marsh colonies, and this has been reported at Great Gull Island as well (Pessino 1968). They typically fly in while a nest is not attended, and eat the eggs close to the nest. When blackbirds forage on the ground in a colony, they are not generally mobbed, although a tern nesting nearby may swoop at and chase a blackbird.

Reptiles: We have not found any snakes on the barrier beach or salt marsh islands, although garter snake predation on terns has been documented in New England (Floyd 1929; Lazell and Nisbet 1972), and snakes are important predators on colonial swallows (Plummer 1977). Diamond-back Terrapins, while not predatory, are regularly recorded in beach tern colonies when the females come ashore to lay eggs. They are vigorously mobbed by terns that form dense flocks diving low over the turtles as they move through the colony (fig. 6.3). It is possible that females could damage eggs as they crawl through the colony in quest of a nest site, but they would be only a minor cause of damage.

Ants: Ants have been implicated as significant predators in certain tern colonies (MacKay 1895; Jones 1906; Austin 1929). They enter pipped eggs and kill hatchlings and young chicks. This has been reported for spoonbills (Ramsey 1968), killdeer (Jackson and Jackson 1985), as well as for terns. The two ant species recorded at the beach colonies are *Tetramorium caespitum*—the pavement ant—which has extreme population fluctuations and is mainly a scavenger, and the much more common *Lasius neoniger*, which is a major threat to hatching and newly hatched chicks. Elsewhere a nocturnal species, *Solenopsis molesta* is suspected to cause mortality of young tern chicks (Spendelow, pers. comm.).

6.2 FREQUENCY OF PREDATION

Overall Occurrence

We examined predator occurrence in the New York and New Jersey colonies in all years and documented the relative frequency of intrusion during a year of intense sampling (1982). Table 6.1 summarizes the occurrence of predators in the marsh and beach colonies based on the total of colony-years shown. The table indicates the percent of colonies in which the predator was seen at least once in a particular year, and whether it was actively preying on terns. Therefore it reflects potential predation.

Several conclusions are clear: 1) barrier beach colonies suffer higher

Figure 6.3. (top left) Dense group of common terns mobbing a diamondback terrapin. (bottom left) Another photo shows proximity of terns to turtle. One bird has landed to peck at the turtle's head. Such groups usually number fewer than 50 birds and seldom reach 100. By contrast harriers (top right) elicit mobbing by hundreds of terns which escort the hawk out of the colony (bottom right).

presence of mammals than salt marsh islands, 2) non-native mammalian predators are more common in both habitats than native mammals, 3) avian predators are equally common in both habitat types, and 4) predators are more abundant on barrier beach than salt marsh colonies.

Predation Attempts

Table 6.1 also indicates whether we have observed specific predators eating eggs, chicks, or adults. Clearly, there is a wide variety of predators that will prey on eggs and chicks, and few that prey on adults. In 1982, we observed predator presence for 420 hours at West End Beach in a 30 × 100 m area. Although 22 avian species entered the colony, only the 11 listed in table 6.1 were potential predators that elicited a response by the terns. Other species such as green heron, glossy ibis, red knot, least tern, mourning dove, horned lark, and Eastern meadowlark were ignored. Species such as killdeer, piping plover, and willet regularly nested in the tern colonies. They were not predators and were usually ignored by the terns.

Overall, the number of intruders observed in this section of the colony varied from 1 or 2 per day to 58 in early June when chicks were hatching (fig 6.4). Intrusion rates were higher from mid-May to mid-June, when eggs were available, compared to July when large chicks were present. These observations show that there are a variety of avian species that enter the colonies regularly. No mammalian species were observed during the day, although a female cat frequented the West End colony and killed tern chicks to feed to its kittens. Shorebirds and passerines usually landed and walked about in the colony whereas gulls and hawks seldom did.

Turnstones in Colonies

Turnstones were present in 11.4 percent of the sand colony-years, and only 0.9 percent of the marsh colony-years. Yet in most years when they were present, they were few in numbers, entered infrequently, remained in the colony for only a short period of time, and had little effect. However, three instances deserve special mention. In 1984, 6 turnstones were present in the Cedar Beach colony May 27–29. They remained in the colony all day moving from one part of the colony to another, eating eggs. Another group of 4 turnstones were in the same colony for five days in 1985. Turnstones

Table 6.1. Occurrence of predators in common tern colonies. Presence based on entry or fly-overs at least once per year.

	Percent of Years Present			*Observed to Eat*		
Species	*Salt Marshes*	*West End & Cedar Beach*	*Holgate*	*Eggs*	*Chicks*	*Adults*
Colony-years[a]	282	39	13			
Mammalian	%	%	%			
Red fox	1	0	23	x	x	x
Raccoon	0	0	23	x		
Mink	1	0	0		x	
Grey squirrel	0	20	0	x		
Dog	1	29	31	x	x	
Cat	0	6	38		x	x
Norway rat[b]	4	62	31	x		
Avian						
Herring gull	97	47	61	x	x	
Laughing gull	43	15	38	x	x	
Owls[c]	1	23	15		x	x
Harrier	0	26	15		x	x
Peregrine	2		23		x	x
Ruddy turnstone	1	11	15	x		
American oystercatcher	36	10	38	x		
Fish crow	26	29	38	x		
Common grackle	7	35	15	x		
Boat-tailed grackle	23	0	0	x		
Red-winged blackbird	9	44	46	x		
Other						
Diamond-back terrapin	3	27	15			
Ants	0	100	15	x	x	

[a]Total number of colony-years or occupancies during the period 1976–1988 for salt marshes and Holgate and for 1969–1988 at West End and 1970–1988 at Cedar Beach.

[b]Most rats noted in salt marsh rats were found on Moredcai, a salt marsh island with a narrow strip of sand.

[c]Great-horned and short-eared owls were both potential predators, and their effects could not be distinguished easily.

worked in groups, one approaching the tern nest and eliciting a chase, allowing others to reach the nest and peck the eggs.

On May 28, 1985 one injured turnstone that was unable to fly arrived on a salt marsh island (Egg), and remained until it was removed on June 8. There was no other easy source of food on the island but bird eggs, and the turnstone made use of this food source. When we removed the turnstone, it

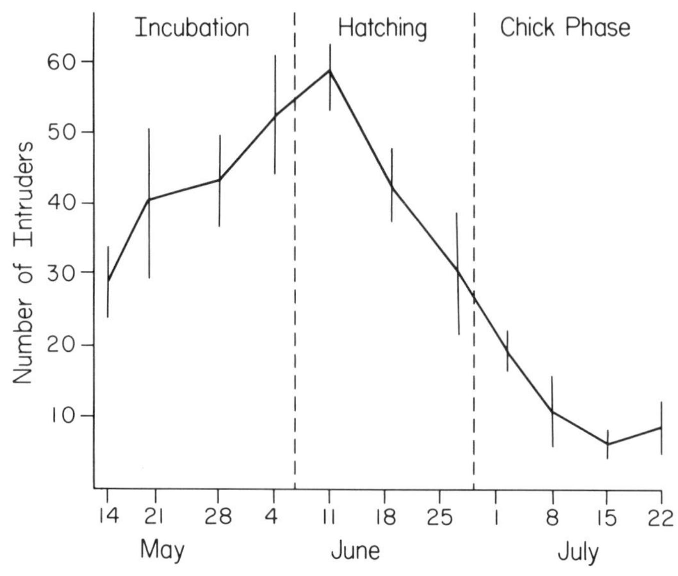

Figure 6.4. Frequency per day of potential predators intruding in tern colonies (after Burger and Gochfeld 1990a; mean ± 1 SD)

had already suffered severe damage to one eye—apparently caused by the bills of attacking terns. These three instances indicate that although turnstone visits are usually infrequent and brief, they can be long and continuous, involving deliberate rather than opportunistic predation. Turnstones have previously been reported as egg predators on terns (Crossin and Huber 1970; Parkes et al. 1971; Loftin and Sutton 1979; Brearey and Hilden 1985; Morris and Wiggins 1986).

Severity of Predation Threat

The severity of the potential predation threat is a function of the kind, frequency, and predation method of the predators, the evasive and defensive potential of the victim, and the quality and quantity of defensive behavior of nearby conspecifics. The latter two points will be discussed below. The former point is exceedingly important as it delineates the nature of the predator problem any animal faces. Predators are limited in their choice of potential prey by such factors as their own size, prey size, and the evasive behavior of the prey. The predators that occur in the tern colonies we examined fall into three main categories: egg predators, egg and chick predators, and chick and adult predators.

Eggs remain in the nest and can be eaten by being pecked open; therefore, they are particularly vulnerable to a large number of predators. Shorebirds (turnstone, oystercatcher) and passerines (grackles, blackbirds) can peck and eat unattended eggs where they find them. Other predators such as gulls and crows can either eat them at the nest site, or they can carry them away in their bills. These types of predators present a threat mainly to the eggs, not to the adults themselves. Thus, without risk, adults can come very close to such predators in their attempts to thwart them.

Egg predators were present in all tern colonies, although their relative frequency varied by habitat, colony, and year. Egg predators are also limited in that eggs are only available for a few weeks during incubation. By being highly synchronous, terns could reduce egg predation by predator swamping (Nisbet 1975; Gochfeld 1980b), although this is not effective if predators are continually recruited to the colony.

Egg and chick predators include species such as night heron and gulls that may be opportunistic predators, or that may chose to specialize on eggs and chicks (Tinbergen 1953; Veen 1977). Whereas egg predators are limited to finding uncovered eggs, chick predators can often prey on chicks from broods that are scattered or unattended. Further, predators that prey on eggs and chicks cause nest failures and induce re-laying, thereby increasing the period when prey are available.

Chick predators pose a severe threat to fitness, because they remove the reproductive unit when it may not be possible to invest in a replacement in the current year, and they remove offspring that parents have invested in for several weeks or in some species even months (Stonehouse 1956; Serventy et al. 1971; Harris 1973). Terns that lose their clutches early in incubation will relay. Terns that lose chicks early in the season may renest if other birds in the colony are still incubating. Predators that eat adults and large chicks are the most serious threat, because they eliminate the opportunity for further reproductive effort. Such predators include hawks, owls, foxes, dogs, and cats. Since these predators can kill adults, the adults are generally more wary and terns may abandon the colony when faced with such predators.

Larids generally lack effective nocturnal predator defense mechanisms (Southern and Southern 1979; Southern et al. 1985). Their only response is fleeing from the predator (Kruuk 1964), leaving eggs and chicks exposed (Emlen et al. 1966). If they flee soon enough they can save themselves, but not their offspring. Nocturnal predators, therefore, can take a heavy toll on eggs and chicks. Moreover they may induce nocturnal desertion (Marshall 1942; Nisbet 1979), thereby prolonging incubation. Owls and foxes, for example, frequently kill adult terns (Kadlec 1971; Burger 1974a; Patton and Southern 1977; Southern and Southern 1979). In a nine-year study of two species of gulls nesting in a habitat with foxes, Southern et al. (1985) found that both species suffered total or nearly total reproductive failure during all but one year. The number of breeding gulls declined, but nonetheless some continued to breed at that site.

Predator Threat as a Function of Habitat

We found some differences in the predation threat in the two main habitats. Avian predators have equal access to salt marsh and sandy beach colonies, and factors other than acessibility govern their impact. There were more mammalian predators and equal numbers of avian predators on sandy beaches compared to salt marsh colonies. Thus, it is clear that the predator threat may be more severe in sandy beach colonies. The threat could be counterbalanced by predator swamping, whereby the large numbers of eggs and young available exceed the predator's capacity or by active colony defense.

Mammalian predators are generally absent from the salt marsh islands. Many of the islands are in the middle of the bay, and are difficult to reach by swimming. Where marsh islands are close to mainland or barrier beaches predators such as cats, rats, and foxes occasionally swim there. In 1985, rats moved onto Mordecai Island, separated from the barrier beach by only 100 m. This caused the complete failure of the colony. In some winters, Barnegat Bay freezes over and foxes walk across the ice to salt marsh islands. However, without a good food source, they usually return to the mainland or barrier beaches.

6.3 ANTIPREDATOR BEHAVIOR

Coloniality may increase the threat of predation, because colonies of birds are conspicuous and offer a potential predator promise of prolonged reward. Even within colonies, some species rely on the cryptic coloration of their eggs and nests (Patterson 1965; Tinbergen et al. 1967). More importantly, colonial species benefit from the enhanced opportunity for early warning and for active antipredator defense (Cullen 1960; Kruuk 1964; Hamilton 1971). In the following section, we discuss the antipredator behavior of common terns. We examine vigilance behavior and parental response to predation and present experimental results examining the effect of predator type, season, prey density, and colony size on antipredator behavior of common terns.

When an intruder approaches a common tern colony, certain individuals spot the intruder and fly up and toward it uttering loud "kip kip kip" calls. These birds may begin to mob the intruder. In response to these alarm calls, young chicks crouch motionless in their nest while older chicks run for cover.

Potential Threats

Many colonial species breed in inaccessible places to minimize the threat of mammalian predators (Tinbergen 1953; Kruuk 1964; Robinson 1985). This usually means nesting in trees, on cliffs, or on islands. Nonetheless, many colonies of birds are located on mainland areas where mammalian predators pose a threat. Even on salt marshes in New Jersey, severe winters accompanied by a freeze-up of the bay allow mammals access to islands. Much important early research on predators at seabird colonies concerned mammalian predators (Kruuk 1964). In many cases, however, birds are the primary predators at seabird colonies. Croze (1970) proved an excellent study of carrion crow predation at a gull colony.

At the very least, defense requires time and energy that could be devoted to other activities (Trivers 1972; Buitron 1983). Defensive behavior also can result in death of the adult (Myers 1978). The threat to most colonial birds can be divided into threats to adults and threats to their eggs and chicks. The response to these threats varies. Birds should be more aggressive in defense of eggs or chicks when they are not directly threatened. When they themselves are threatened, they will flee, even when the cost to eggs or chicks is high. Thus, there should be selection for a level of defense that is a compromise between the survival of the current brood versus the survival of the adult and production of future broods (Trivers 1972, 1974). When the threat to young is high and threat to adults is moderate, adults should mob intruders, but from a safe height (moderate avoidance distance). As the threat to the adult declines, mobbing should grade into overt attack.

Theories on parental investment predict that parents should increase the time and energy spent and the risks incurred in defense of offspring as their young approach fledging (Buitron 1983). The older the offspring, the greater the cost of replacing it in the same season (Dawkins and Carlisle 1976). Others have suggested that defense should relate to egg or chick vulnerability (Andersson et al. 1980), rather than cumulative investment. Chicks are differentially vulnerable depending on their age and circumstances (Burger 1984c,d). There are other costs to mobbing and attacking; young may be left exposed and vulnerable. Furthermore, different predators pose different threats and presumably parents respond accordingly. An alternative prediction, therefore, is that parental response is greatest when and where their offspring are most vulnerable.

Passive Avoidance

Birds nesting in colonies can passively avoid predation just by being a member of a large group of potential prey (Hamilton 1971). "Predator swamping" occurs when predators can take only a limited number of prey each day, and being in a large group thereby reduces any one individual's chance of becoming a victim. Nisbet (1975) showed that when a predator takes the same number of prey each day, it is an advantage to nest during the peak of nesting (Nisbet and Welton 1984).

Darling (1938) invoked this relationship as evidence for the importance of reproductive synchrony. The swamping effect has even been suggested for the nesting pattern of solitary species such as yellow warblers (Clark and Robertson 1978). With heavy predation, birds nesting either early or late in the season would suffer proportionally higher rates of predation, and field data on common terns exposed to horned owl predation supports this prediction (Nisbet and Welton 1984).

The spatial dispersion of nests is also an important factor influencing their detectability by predators. In a classic experiment, Tinbergen et al. (1967) showed that increasing nest density increased future predation rates on the cryptic eggs of the black-headed gull. Within colonies, dense nesting is effective in deterring some avian predators, while sparse nesting deters others (Burger and Lesser 1978). In some cases, individual birds can avoid predation by nesting in the center of the group (Hamilton 1971), for nest predation is often lower for central nesting birds (Coulson 1968; Veen 1977; Siegel-Causey and Hunt 1981).

Active Avoidance

Vigilance and Early Warning

Birds nesting in colonies derive advantages from other colony members through increased vigilance, early warning, and antipredator behavior. Animals must be vigilant to detect predators, as well as to search for mates, food, nest sites and other requirements. Factors affecting the time devoted to vigilance include group size (Powell 1974; Siegfried and Underhill 1975; Lazarus 1979a,b; Grieg-Smith 1981a,b), parental care constraints (Lazarus and Inglis 1978), habitat (Underwood 1982; Metcalfe 1984), distance from cover (Barnard 1980a,b), and brightness of plumage (Lendrem 1983). In most studies, the time an individual bird devoted to vigilance decreased with increasing group size.

Studies of vigilance have generally focused on foraging groups of birds (Powell 1974; Abramson 1979; Lazarus 1979; Bertram 1980; Curaçao 1982), on herds of ungulates (Underwood 1982; Alades 1985), or on schools of fish (Magurran et al. 1985). Interest in vigilance as a sociobiological phenomenon can be summarized briefly as follows. The time which an individual devotes to vigilance is time unavailable for feeding or other activities. Animals in groups can "share" the responsibility of vigilance, and each individual will be able to devote a smaller proportion of time as its share.

Clearly, as colony size increases, animals breeding in colonies should be able to decrease their vigilance (Ricklefs 1980; Feekes 1981). But such benefit depends on what else they must do with their time. Vigilance behavior is important in colonies of meerkats (a mongoose, Moran 1984), and in prairie dogs (Hoogland 1979a,b) where decreased need for individual alertness is an important benefit of coloniality. Similarly, decreased individual alertness and defense could also be a benefit of avian colonies although this aspect of coloniality is poorly documented.

An individual's time devoted to vigilance should decrease with increasing group size until some optimal group size is reached beyond which no decrease in individual vigilance can occur. Similarly, the amount of time that at least one member of a group is vigilant will increase with group size until vigilance for the group reaches 100 percent. Further, birds may coordinate vigilance to reduce the amount of time when no individual is vigilant, thus avoiding a lapse when they might fail to detect a predator (Wickler 1985). We examine vigilance behavior and describe the costs associated with vigilance in terns nesting in salt marshes and beaches.

Figure 6.5. Human intruders, including the researchers, also elicited mobbing by hundreds of terns.

Active Defense

In addition to early warning, another advantage of coloniality is group defense. Any bird can always distract, threaten, or attack and pursue a predator. In colonies, many birds perform such behaviors together, often serving as effective deterrents to predators (Kruuk 1964; Lemmetyinen 1971; Zubakin et al. 1984). Mobbing is the most common form of group defense, and can involve several birds or the entire colony dive-bombing and attacking a predator, often escorting it from the colony. Mobbing is readily elicited by human intruders (figs. 6.5, 6.6), and in colonies where birds are accustomed to human intrusion, they nearly constantly dive and strike as one moves through the colony. Occasionally one can actually snatch a mobbing tern out of the air (fig. 6.6), in a manner reminiscent of wild canids, such as coyotes, which we have watched leap upward attempting to seize mobbing magpies. Work by Curio and colleagues (e.g., Curio 1967, 1975, 1978; Vieth et al. 1980) and by Altman (1956), Andrew (1961), Shalter (1978a,b), and many others have employed mobbing as a paradigm to explore the innate and learned components of behavior, to investigate cultural transmission of enemy recognition, and to investigate the select features of mobbing itself.

Figure 6.6. In colonies such as Cedar Beach, where terns were habituated to humans, the terns approached closely (left), frequently striking the intruder. It was occasionally possible to snatch a mobbing tern out of the air (right), much as canine predators attempt to do.

Mobbing behavior is very widespread among birds (Shalter 1978a) and is certainly not restricted to colonial species. It occurs in a variety of solitary-nesting species including, for example, American robins (Shedd 1983), flycatchers (Shalter 1978b), chickadees (Shedd 1983), crows (Brown 1985), and lapwings (Elliot 1985). Mobbing is characteristic of several species of terns (McNicholl 1973; Fuchs 1977b; Lemmetyinen 1971; Veen 1977) and gulls (Conover and Miller 1979; Boshoff 1980). Even breeding chickens will mob predators (Curtis 1986). By contrast, the Pelecaniiformes do not mob predators (e.g., Siegel-Causey and Hunt 1983; our personal observations).

Mobbing intensity varies as a function of species of predator, stage in the reproductive cycle (Lemmetyinen 1971), and the potential threat. Several studies have documented the effectiveness of mobbing in deterring predators (Horn 1968; Lemmetyinen 1971; Craig 1974; Veen 1977; Bildstein 1982; Zubakin and Avdanim 1982; Kaverkina 1984; Montevecchi 1977, 1978a; Shields 1984), and in some cases mobbing effectiveness increases with the number of birds mobbing (Hoogland and Sherman 1976; Shields 1984; Robinson 1985; Elliot 1985).

Birds initiating mobbing actively recruit neighbors with characteristic vocalizations, thereby alerting others and enhancing the effect (Curio 1978; Frankenberg 1981). This occurs virtually instantaneously in a tern colony. Rarely is a mobbing bird captured by the bird or animal being mobbed (Cade 1967; Myers 1978; Denison 1979; England 1986). Some species which do not actively mob or attack predators appear to benefit from the aggressive behavior of other species with which they nest (Koskimies 1957; Neuchterlein 1981; Burger and Gochfeld 1990a). Furthermore, even within mobbing groups, some birds (usually the one whose nest is most threatened) may be more aggressive in mobbing than other birds, and these active mobbers incur the greatest risk (Shields 1984). Being attracted to the site of mobbing also allows birds to learn about predators from watching the interactions of conspecifics with predators (Conover 1987).

6.4 VIGILANCE AND EARLY WARNING

For optimal vigilance in colonies, some birds should be facing in each direction. On windless days ($n = 18$), common terns in one of our study plots had between 23 and 30 percent of the birds facing in each of the four compass quadrants. As wind velocity increases, more birds tend to face into the wind, although some still face in most directions until wind speed reaches about 15 km (Gochfeld 1978e). At higher wind speeds, all birds face into the wind.

Upflights: "Alarms," "Dreads," "Panics"

A characteristic phenomenon in gull and tern colonies is the so-called "panic" or "dread" (Marples and Marples 1934), a synchronous upflight where an entire section of the colony or the whole colony, flies up noisily, lapses into a profound momentary silence, swoops down toward the ground out over the water, then circles back to return to the colony.

Marples and Marples (1934) report "dreads" lasting usually less than one-half minute, although the ones we have timed may degenerate into circling flights that keep the birds off their nest for as long as two minutes. Occasionally, these are actually triggered by the presence of a predator on the ground or overhead, in which case the circling birds may end up mobbing the predator. In other cases, a nonthreatening stimulus appears to initiate the upflight. We have seen upflights following the appearance of a distant jet airplane or a child's balloon, and King (1977) implicated a hare and the hissing sound of a local wind phenomenon. In most cases, particularly early in the season, the behavior appears spontaneous with no discernible or threatening stimulus. Marples and Marples (1934) were impressed by the seeming spontaneity of these flights. The very terms "panic" or "dread" suggest that fear is the motivating or causal factor, which is by no means certain. Hence the term "upflight" is preferable, and we use it to refer to the massive flights without apparent stimulus. Becker (1984) reported "upflights" in response to herring gull predation, but these appear to represent concerted mobbing by large numbers of terns rather than a "dread" phenomenon.

Frequent disturbances during incubation can greatly prolong the incubation period (Austin 1933; Nisbet and Cohen 1975) or even lead to embryonic mortality. Leaving chicks exposed invites predation. Whether the departure is due to an alarm flight or to attack a predator, the outcome is an interruption of incubation and a change in egg temperature. Ripley (1961) and Emlen et al. (1966) have discussed aggressive neglect as a significant cost attached to parental antipredator aggressive behavior.

We examined the upflight or "panic" phenomenon to determine how much of a cost it might represent, and how it varied seasonally and by habitat. The number of upflights is highest prior to egg-laying, and remains high through the egg-laying period, declining thereafter (fig. 6.7). In sunny weather, there were 25 percent fewer upflights than on cool, cloudy days. Moreover, on sunny days terns stayed off their eggs for less time (mean = 25 ± 18 sec, $n = 121$ events) than on cloudy days (mean = 81 ± 69 sec, $n = 80$ events). This is likely to reflect the vulnerability of eggs to overheating, particularly in sandy habitats (Austin 1933), a topic which invites further research.

An upflight can involve an isolated neighborhood of a few pairs, an entire subcolony, or the whole colony. Spectacular upflights have occurred in the Cedar Beach colony, for example, where virtually the entire 6000-pair colony flew up, apparently in response to a high-flying jet. More often only part of such a large colony would respond. In small colonies, it is not unusual for the entire colony to fly up, particularly if a predator is actually present. However, in only about 10 percent of cases could we identify an actual predator, and the terns' behavior in other cases suggested that no predator was present.

As the season progresses, a decreasing number of birds join in these

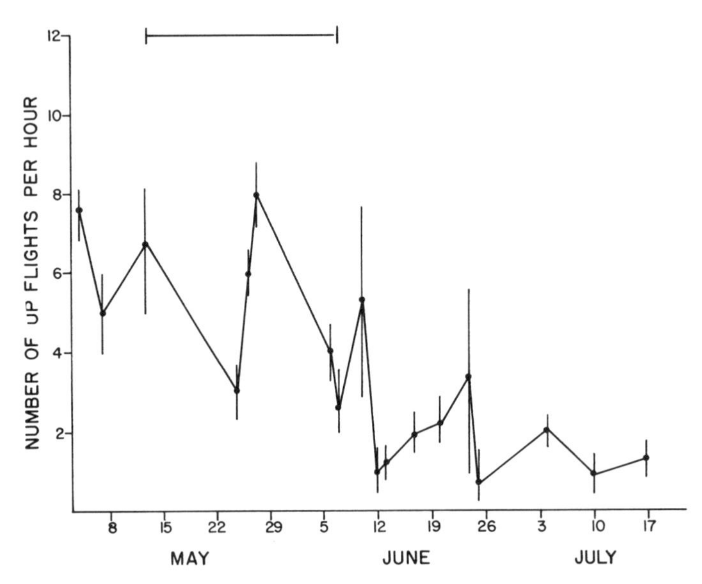

Figure 6.7. Seasonal rate of upflights at Cedar Beach tern colony (1984). Averages are per hour (± 1 SD). Horizontal bar spans peak egg-laying period.

disturbances. This was noted in both marsh and beach colonies, and was more marked in the latter (fig. 6.8). We examined in detail a colony of about 500 common tern pairs at Pettit Island, as well as a section of the West End II colony with about 500 pairs.

Benefits of Vigilance

Numerous studies including our own have shown that for members of a group, the time devoted to vigilance can be decreased compared with solitary individuals. One testable prediction from this hypothesis is that the birds nesting in dense parts of the colony with many neighbors to share the vigilance duties should devote less time to vigilance than those in less dense sections.

In 1985, we examined vigilance in dense (> 7 pairs within a 5-m circle of the focal bird) and sparse (< 5 pairs within 5 m) sections of one salt marsh and two parts of a sandy beach colony (fig. 6.9). Unlike most studies of vigilance where the converse of vigilance is time spent feeding, we watched

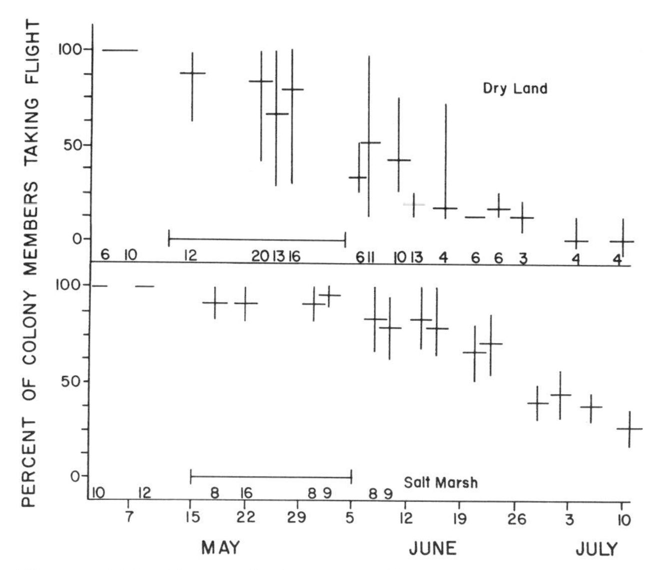

Figure 6.8. Seasonal change in the percent of colony members to engage in upflights for terns nesting on salt marsh and sandy beaches. Horizontal line equals incubation period (no chicks present). Shown are mean ± 1 SD.

incubating birds who only needed to rest or sleep. Lendrem (1983) has provided an examination of the tradeoff between resting and vigilance.

We recorded how much time the focal common tern slept (head retracted and eyes closed) in one-minute samples. Each bird was sampled twice with one minute intervening between the two samples. We then selected at random another focal tern at least 10 m away and repeated the counts. We watched 10 terns in the high-density area and 10 terns in the low-density area at each of the three colonies, West End, Cedar Beach, and Pettit.

To check the data for consistency we compared the first samples with the second samples, and found no differences between these two across colonies or densities (ANOVA). For all samples combined, the mean time devoted to sleeping was significantly higher in the high density (mean = 54.7 ± 7.2 sec) than in low density areas (mean = 27.7 ± 1.4 sec). We examined the effect of colony and density on the time devoted to sleep using a two-way ANOVA with repeated measures (Cody and Smith 1985). We found significant effects of colony ($F = 31.9$, df = 2, $P < 0.001$), density ($F = 199.6$, $df = 1$, $p < 0.0001$), and the colony x density interaction ($F = 20.9$, $df = 2$, $P < 0.001$).

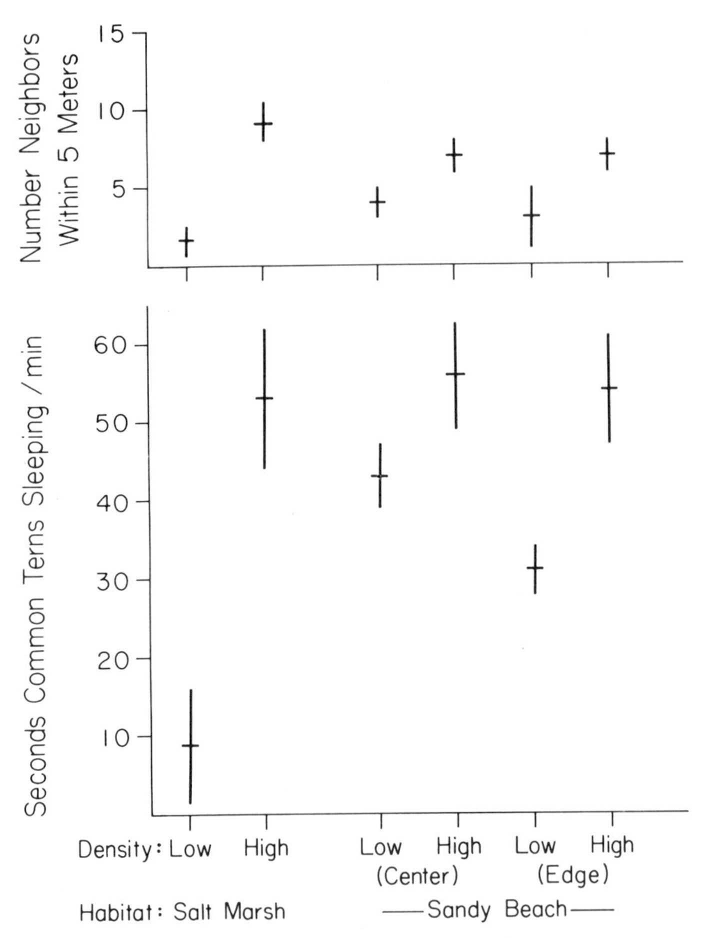

Figure 6.9. Time devoted to sleep by incubating common terns in high- and low-density sections of two beach colonies and one marsh colony. Shown in top graph are the measures of density for each study site. Shown are mean ± 1 SD.

By comparing the *F* values, we infer that density had the greater effect. Clearly, birds nesting in high-density areas can devote more time to sleeping than those in low-density areas.

These observations were made during incubation, when the birds were settled and were familiar with their neighbors. It seems likely that earlier in the season, when birds in high-density areas may have a higher rate of territorial encounters, a different pattern would emerge. The interaction effect was interesting. In the high-density areas, terns in all colonies spent the greater part of the time sleeping. In low-density areas, there were striking differences between colonies, with birds at Pettit spending much less time sleeping than those in the low-density parts of the beach colonies (fig. 6.9).

Many of the common tern colonies also had black skimmers nesting in them. This suggests that the skimmers may derive some benefit from nesting with the terns, in the form of information parasitism regarding approaching predators, and defensive parasitism as well (Koskimies 1957; Erwin 1979; Young and Titman 1986). In both sandy beach and salt marsh colonies, significantly more of the upflights (about 60 percent) involved only terns than involved both species or skimmers alone (fig. 6.10). Although the skimmers did not actually fly up at all times, they became alert. They awoke, used head-up postures, and turned their heads from side to side, sometimes appearing nervous and ready for flight. They waited to see what the threat was before they flew up. We conclude that skimmers monitor the behavior of neighboring terns, responding only to some of their warnings and relying on their tern-response to initiate their own (Burger and Gochfeld 1990a).

Discussion

Alone or in groups individual birds devote time to vigilance. A pair of birds can maximize vigilance by alternating their periods of alertness. If both members are alert, they should monitor different directions. We demonstrated that the latter is true of skimmer pairs (Burger and Gochfeld 1990a), while in common terns, usually only one member of the pair is present at the nest site during the day. Halkin (1983) showed that nesting skimmers tuck their bills toward the outside of the group, thereby increasing their vigilance.

Presumably the seasonal decline in upflights represents a decrease in fear or a decrease in what might be termed "ambivalence" regarding the colony itself (Austin 1944). A similar seasonal decline in upflights (referred to as "dreads") was reported by Morris and Wiggins (1986). By contrast, Becker (1984) reported an increase in upflights as the season progressed, but this was in response to herring gull intrusions. In the salt marsh colonies, the proportion of terns participating in these upflights is much higher—usually the entire colony—compared to beach colonies. This reflects in part a difference

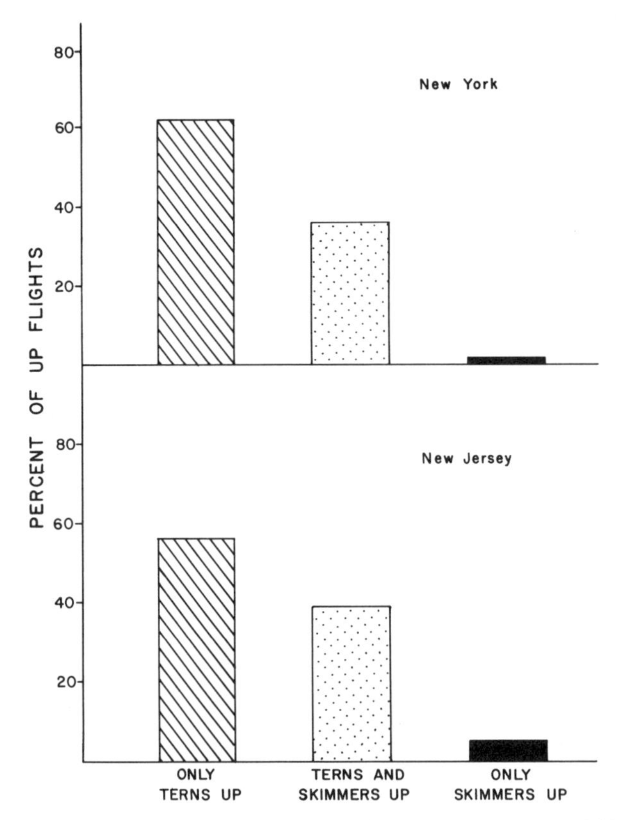

Figure 6.10. Percent of upflights or panics ($N = 200$) engaged in by only terns, by both species, or by skimmer alone, for beach (top) and marsh (bottom) colonies (1984).

in colony size. The salt marsh colony we examined in detail, Pettit, is small and nearly circular, and it is possible for birds at any point on the island to observe upflights occurring anywhere else in the colony. On the other hand, the large beach colonies are spread out and birds may be 500+ m away from the focal point where a disturbance is initiated.

Although we examined only a small subsection of West End and Cedar Beach we found that these sections, delineated by habitat, behaved as social units. Even for birds nesting at the edge of the subsection, where the edge was contiguous to another group of birds rather than water, their neighbors were quite likely to remain sitting rather than join in the upflight. Thus, we suggest that the difference is at least partly due to some social cohesiveness within a neighborhood, rather than simply the size of the area being surveyed.

6.5 NATURAL ANTIPREDATOR BEHAVIOR

Antipredator behavior includes early warning of the presence of a predator and defensive behavior by colony members, including mobbing, attacks, and strikes. Terns fly over, dive, and occasionally strike a predator in the colony, thereby interfering with the hunt, and encouraging the predator to leave (Curio 1975, 1978).

Mobbing has been reported to be effective in deterring avian predators in several gulls and terns: Sandwich tern (Veen 1977), common tern (Bullough 1942; Collins 1970; Lemmetyinen 1971; Veen 1977; Erwin 1979), Arctic tern (Lemmetyinen 1971), black-headed gull (Kruuk 1964; Patterson 1965; Fuchs 1977), kittiwake (Andersson 1976; Montevecchi 1979), kelp gull (Boshoff 1980), laughing gull (Montevecchi 1977, 1979), and herring gull (Schoen and Morris 1984). However, Becker (1984) reported that herring gull predation proceeded despite mobbing by terns. Mobbing is also effective against a few mammals such as seals (Barker and Hand 1981), stoat and hedgehog (Kruuk 1964), and against humans (Veen 1977). Sometimes, birds will even mob feathers (Kilham 1982). Nonetheless, colonial nesters do not always mob predators, or their mobbing may be ineffective.

In general, gulls and terns do not mob predators at night, but instead they temporarily abandon the colony (Marshall 1942; Southern and Southern 1979; Nisbet and Welton 1984) or they may permanently abandon if disturbance or predation is intense. It seems that gulls and terns mob predators when they pose a threat to their eggs and chicks, but not when predators pose a direct threat to themselves (Kruuk 1964). This hypothesis will be tested in this section.

Although authors have noted the effectiveness of mobbing in deterring diurnal predators, the factors affecting mobbing behavior in colonies have seldom been investigated in detail (Barash 1976). Several authors have noted seasonal effects; mobbing is greatest at hatching (see Veen 1977; Kruuk 1964; spurr 1974). However, Montevecchi (1979) found no seasonal difference in defense in kittiwakes. Lemmetyinen (1971) found no differences in the rate of mobbing in colonial and solitary-nesting terns, but he did not examine the effect of colony size per se.

If one advantage of nesting in colonies is increased benefits from antipredator behavior such as mobbing, it should be possible to demonstrate increased benefits and decreased costs for birds nesting in large rather than small colonies. We examine the effects of seasonality and colony size on mobbing behavior of terns below.

Types of Responses to Predators

There is a range of responses to predators from silent departure from the nest and colony to active, overt attack on the predator.

Nest Departure: At the approach of a potential predator terns fly up from their nests, sometimes silently, but usually accompanied by an alarm call. Departing terns either fly very high, circling above the nest, fly out to greet the intruder, swoop low over the nest or intruder, or fly directly away from the nest, out over the water. In extreme cases terns may depart from the colony as a group.

Mobbing and Attacking: Mobbing and attacking grade into one another. Mobbing usually involves a number of terns flying around, hovering over, and swooping at an intruder. Attack involves dive-bombing and, finally, striking the intruder. Terns may mob predators singly or as an entire colony, all milling around the predator. The distance a mobbing bird will approach the intruder (approach distance) varies from zero (physical strike) to several meters, depending on the threat the predator poses to the adult terns themselves. Mobbing can involve merely flying in a dense milling flock above the predator, although one usually thinks of the dive-bombing attacks as more characteristic.

Physical attack

An attacking tern flies rapidly and directly toward the intruder's head, and attempts to strike it with its bill or occasionally its feet. Then just as rapidly the tern flies back upward in an arc after its attack. Birds often fly off upwind about 10 to 20 m and then fly nearly parallel to the ground at a height of about 2 m. The flat trajectory approach is usually not accompanied by a strike. Birds that are striking usually fly in a steep U-shaped course, starting off about 5 m above the intruder, and plunging headfirst towards it, then halting at the strike point, and rising upward again on rapidly beating wings. If there is a wind, the tern faces into the wind and allows itself to be blown backward to its starting point and repeats the attack. The interval between such strikes averages 2.7 sec.

An individual tern whose nest appears threatened may also engage in dive-bombing. For instance, if a skimmer approaches close to a tern nest or moves through its territory, the tern may fly up and engage it in the U-shaped dive-bombing approach, usually without actually striking the skimmer. The skimmer's response is to raise its head and bill just at the moment the tern reaches the bottom of its trajectory. In such cases, usually the only attacker is the tern whose nest is threatened. Figure 6.11 shows similar defensive behavior by a herring gull which raises its bill to ward off an attacking common tern.

Common terns frequently strike human and other intruders. This behavior tends to be colony-specific, being very frequent in the large beach colonies and less frequent in some of the smaller salt marsh colonies. We believe the difference is due to the frequency of human intrusion and colony size, rather than to habitat. Some potential predators may be mobbed but never attacked. Generally if the predator is a real threat to the adult (e.g., a canid), the terns will neither swoop down nor dive-bomb it, but instead mill about in a noisy umbrella about 5 to 8 m overhead. In colonies where terns are habituated to humans, they dive and strike repeatedly. With practice, we have learned to catch by hand an attacking tern out of the air (fig. 6.6), and there is no doubt that a fox could do so as well.

Terrestrial Attack: On occasion when terns are confronted by a neighbor attacking its chicks or a predator at the nest, it may stand next to the nest and attack a predator by lunging. In some cases, the pecks are accompanied by vigorous wing flapping. In the tern colonies we studied, this behavior was infrequent.

Distraction Displays: Many ground nesting birds give displays which appear to lure a predator away from the nest (Gochfeld 1984). Black skimmers perform such displays regularly (Blus and Stafford 1980; Burger and Gochfeld 1990a), several species of gulls do so occasionally (Burger 1974b; Gochfeld 1984), and we have noted incipient distraction displays in least terns (Gochfeld 1974). But, in general, most species of gulls and terns do not perform distraction behavior, and we have not noted any such behavior in common terns.

Figure 6.11. Herring gull being dive-bombed by common tern in antipredator experiment. Gull simply raises head and bill to keep tern from striking it.

Table 6.2. Antipredator responses of common terns to different species of predators, based on observations in salt marsh and beach colonies. + indicates behavior was observed at least once.

Predator	*Chase*	*Mob*	*Attack*	*Strike*
Harrier	+	+	+	
Oystercatcher	+		+	
Ruddy turnstone	+		+	+
Laughing gulls	+	+	+	
Herring gull	+	+	+	+
Great black-backed gull	+	+		
Black skimmer	+		+	
Common tern	+		+	+
Common and fish crows	+	+	+	
Red-winged blackbird	+	+	+	+
Common grackle	+			+
Red fox		+		
Grey squirrel	+	+	+	
Domestic dog		+		
Domestic cat		+		
Human		+	+	+
Terrapin	+	+	+	+

Response to Specific Predators

Over the years, we studied the responses of terns to a wide variety of predators. Table 6.2 summarizes our data on the responses of terns to predators, and some of the interactions are described below. From table 6.2, it is clear that terns will regularly chase most avian predators; they mob the larger avian predators (except crow and great black-backed gull). Except for herring gulls they actually strike few species. Common terns chase and attack squirrels, which are egg predators only, and rabbits as well as terrapins which pose no threat. Clearly, the response of a mobbing tern varies as a function of the species of predator, behavior of the predator (height above the colony, hunting), their own reproductive stage (eggs, hatching eggs, or young), and proximity to their own nest.

Figure 6.12 shows data collected at both salt marsh and beach colonies. The mean number of terns that responded to each predator type varied, with more terns mobbing herring gulls, turnstones, and red-winged blackbirds, all

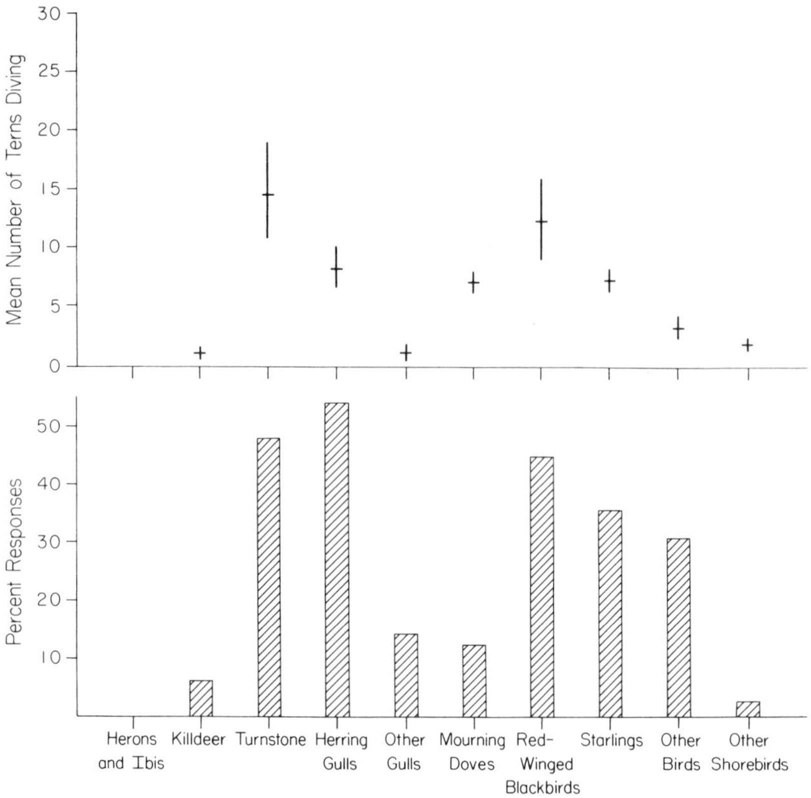

Figure 6.12. Mean (± 1 SD) number of common terns diving at intruders of different species (top), and percent of intrusions responded to by common terns faced by different predators (N = 284, bottom). (Averaged for all colonies).

of which are frequently egg predators. Over the season, their response rate was never more than 55 percent for any given species.

Seasonality, or reproductive stage, affected both the frequency of responses and the mean number of terns engaging in each mobbing incident (fig. 6.13). More terns responded to all intruders combined in early May and from mid-June to mid-July than at other times. Early May is the egg-laying period when terns may be less attentive to their incomplete clutches. In late June and early July, there are many small chicks vulnerable to predation.

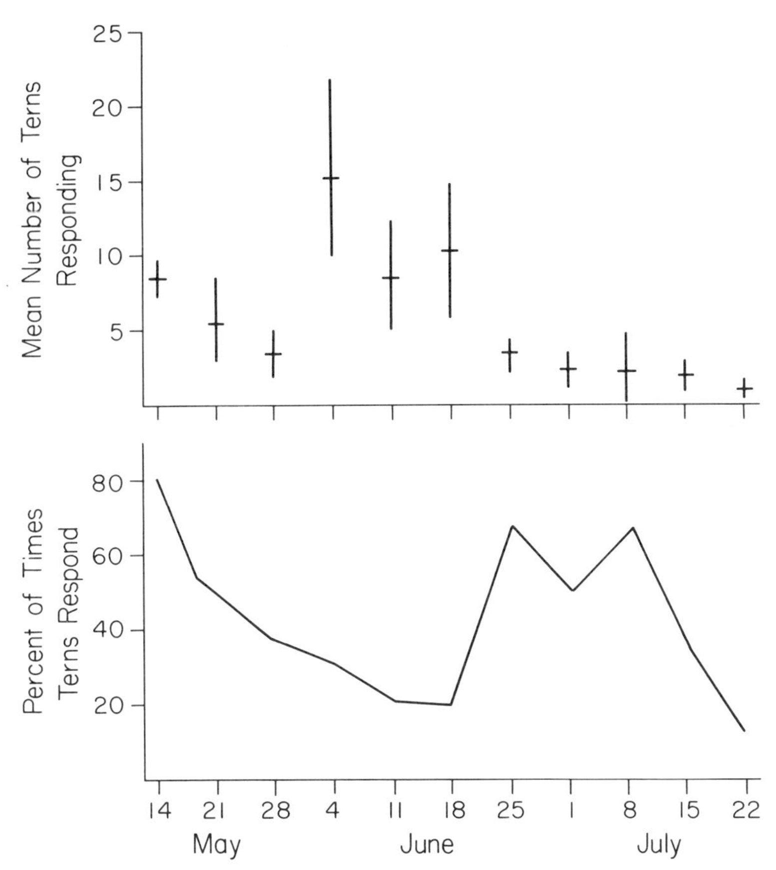

Figure 6.13. Effect of season on tern antipredator responses. Mean (± 1 SD) number of terns responding, by date (top), and percent of intrusions responded to (bottom, N = 284). (Averaged for all species of intruder).

Blackbirds and Gulls

It is instructive to examine tern response to the two most frequent diurnal predators in the sandy beach colony, herring gulls and red-winged blackbirds. The latter are only egg predators whereas the former prey on both eggs and chicks. Thus both the occurrence of the predator and the response of the terns should vary accordingly. We recorded the daily occurence of blackbirds and gulls in a study plot in the West End beach colony (110 m × 30 m) where approximately 320 pairs of terns nested. We noted whenever a predator came into the colony and recorded the terns' responses. Red-winged

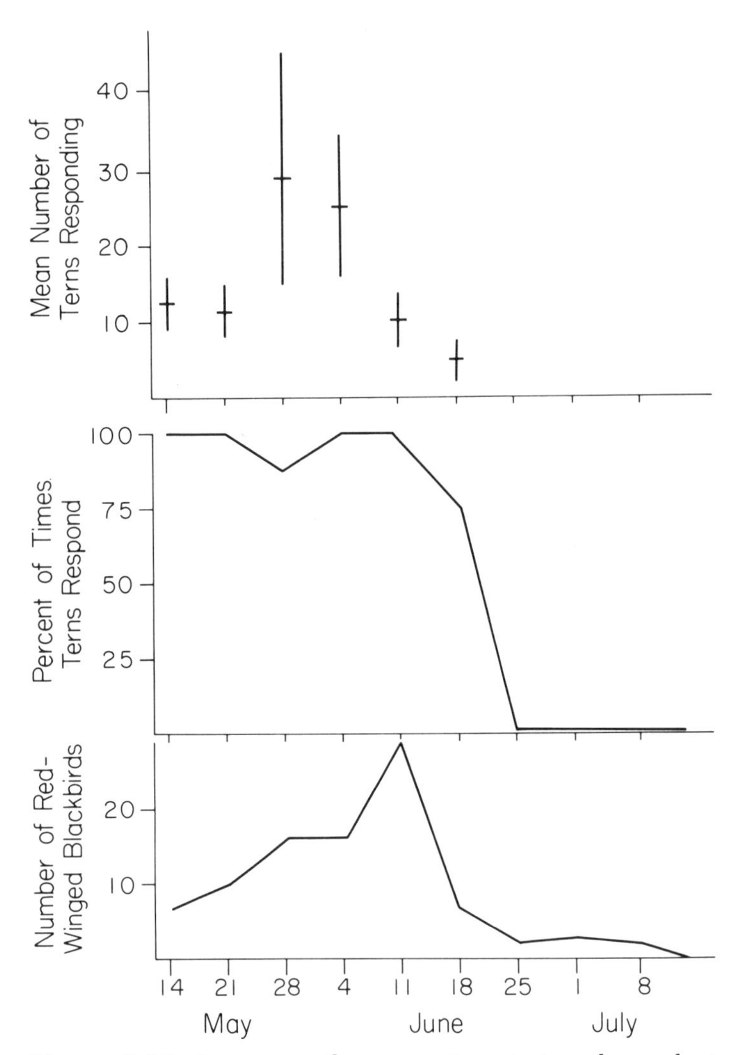

Figure 6.14. Responses of common terns to intruding red-winged blackbirds as a function of season. Frequency of blackbird intrusions per day (bottom). Percent of blackbirds responded to (middle), and mean number of terns responding to the blackbirds (top). (West End 1984).

blackbirds were present throughout the incubation period, but peaked in early June. Figure 6.14 shows the number of red-winged blackbird intrusions per day in the beach colony.

The terns responded to almost all the blackbirds until mid-June when

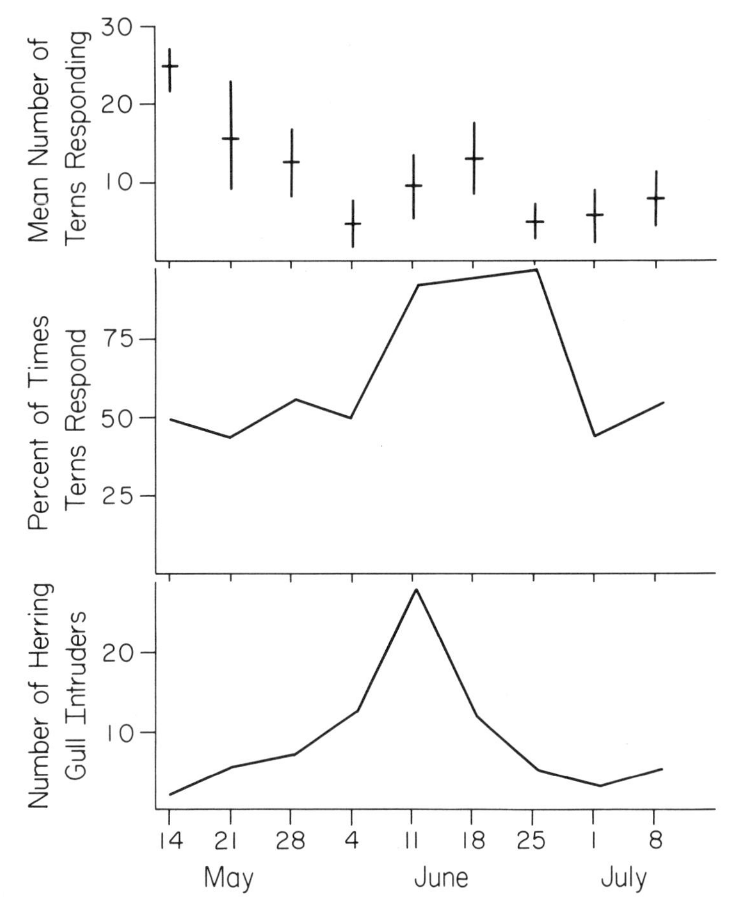

Figure 6.15. Responses of common terns to intruding herring gulls as a function of season. Frequency of herring gull intrusions per day (bottom). Percent of herring gulls responded to (middle), and mean number of terns responding to the gulls (top). (West End 1984).

their response declined markedly. In general, 10 to 30 terns mobbed blackbirds (compare this with the hundreds mobbing harriers), although the greatest number mobbed in late May and early June just before hatching. The terns mobbed, chased, and attacked the red-winged blackbirds singly or in groups of up to 40 terns. In late June and July—when eggs were unavailable—few blackbirds entered the colony, and those that did were ignored.

A few herring gulls were present at the colony throughout the breeding

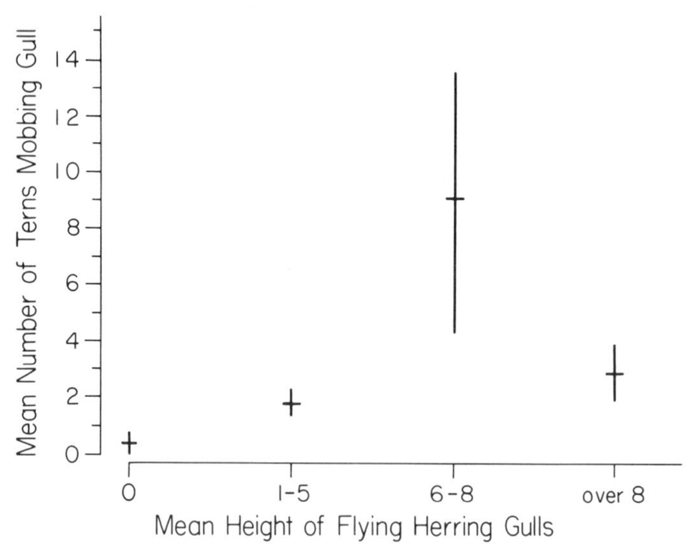

Figure 6.16. Mean (± 1 SD) number of terns mobbing herring gulls as a function of height (in meters) of flying gulls.

cycle of terns, but their numbers increased in mid-June during the peak of hatching. They rested on utility poles, occasionally flying into the colony to take chicks. Terns did not attack the resting gulls, but responded over 50 percent of the time to the actual intrusions. The response rate reached 75 percent in mid- to late June when there were vulnerable chicks. In most cases when terns did not respond, the gulls were flying more than 8 m above the colony, and were flying in a direct line, apparently showing no interest in searching for eggs or chicks (see below). The number of terns in mobbing groups above herring gulls was greater in early May and in early June during the peak of hatching (fig 6.15). Mobbing group size dwindled to 4 to 10 during July, when chicks were large.

We examined the number of terns that mobbed a gull as a function of the height at which the gull flew above the colony. We had predicted that the lower the gull, the more birds would mob it. However, the relationship was complex (fig 6.16). When gulls were on the ground or close to it, they were usually attempting predation on a particular nest and were mobbed only by those terns immediately around them. When the gulls were 5 to 8 m above the colony, they were searching for prey and they were mobbed by dozens of terns. When they flew more than 8 m above the colony, they were usually not looking for prey and so they were largely ignored. Significantly fewer terns mobbed herring gulls that were flying above than below 8 m ($X^2 = 17.5$, $df = 3$, $P < 0.001$).

Responses, Responders, and Season

The graphs (figs. 6.13, 6.14, and 6.15) show no relation between percent of intrusions eliciting a response and number of terns responding. Thus, a high frequency of response may be associated with a low average number of responders. Both factors are influenced by stage of incubation (investment in or vulnerability of offspring) and by behavior of intruder. Gulls plunging to a single nest below their perch elicited mobbing behavior from only a few pairs, but the response occurred many times. Gulls crossing over the colony might be mobbed by many terns or by none at all, depending on their height. At certain stages, an individual tern may have little reason to invest in mobbing, but there may always be a few birds that are in a responsive state.

Turnstones

Both herring gulls and red-winged blackbirds were present in the tern colonies throughout the breeding season. Turnstones represent a different pattern of predator occurrence. They were present only in late May and early June while making a migratory stopover on their way to northern breeding grounds. When they are in the colony, turnstones represent a brief but critical threat as egg predators. They may remain in the colony for hours, damaging many clutches. In 1984, 1985, and 1987, several turnstones were present in the Cedar Beach sand colony for four to six days. They moved around the colony in search of eggs, leaving sections only when they were heavily attacked.

When terns participated in upflights or flew to mob the turnstones, their eggs were exposed to other turnstones which would run toward the nest in a hunched posture and rapidly peck at the eggs. Once one egg of a clutch was pecked open, the terns seemed to become confused and either ate their remaining eggs themselves, or chased one of the turnstones, leaving their eggs exposed to other turnstones.

Terrapins

Diamond-backed terrapin are infrequent visitors to tern colonies, entering from mid-June to early July in their quest for nest sites (Burger and Montevecchi 1975). They do not eat eggs or chicks, but might walk over nests breaking eggs, thereby representing a slight hazard. Nonetheless, the terns mob and attack them vigorously (fig. 6.3). It is apparent in this case that the actual risk and the intensity of mobbing need not be correlated.

Terrapins often are not discovered immediately when they enter the colony, but as soon as one tern responds, a cloud of mobbing birds forms low overhead, and the turtle is vigorously attacked. We examined the terns' response to terrapins in experiments in 1982, by releasing a terrapin in different parts of the colony. We captured a female terrapin that had completed laying her eggs. We covered it with a box, carried it to a release point, and then retreated to our blind, from which we remotely removed the cover so the terrapin could move through the colony. Usually it took 1 to 2 min before the terns responded. When they responded, as many as 100 terns mobbed the terrapin (fig. 6.17). Their response waned after 2 to 3 min and only the terns whose nest the terrapin approached continued to mob it.

Discussion

In tern colonies, "panics" or upflights decreased seasonally as indicated by Marples and Marples (1934) and others. As colony density increased, terns could devote more time to resting and less time to vigilance, indicating an advantage to breeding in large groups. Over the course of the breeding season, many species of potential predators entered the colonies, and the terns responded most strongly to turnstones, gulls, and red-winged blackbirds. These are clearly threats to their eggs and chicks, since these were the species actually observed to prey on them.

Thus the terns' response to potential threats seemed to reflect the actual danger. Overall response varied seasonally, being highest around hatching, and remaining high during the period when the chicks were maximally vulnerable. Terns protected their eggs mainly by continuing to incubate them.

Gulls and red-winged blackbirds posed the biggest threat since they were common residents, particularly on the salt marsh islands, and were frequently seen in or next to the colonies. Both gulls and blackbirds ate eggs,

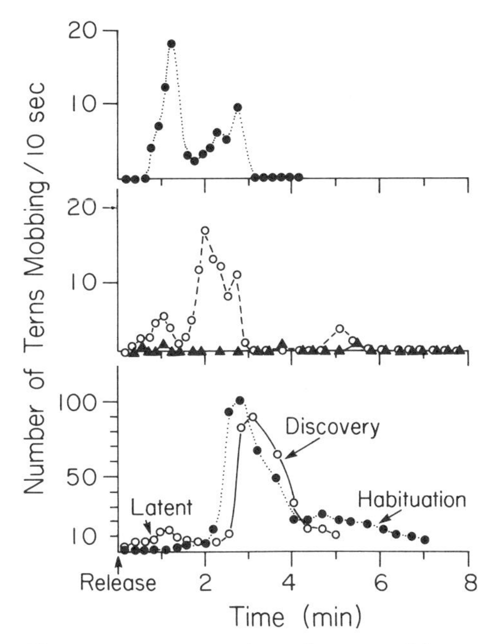

Figure 6.17. Number of terns mobbing a terrapin experimentally released in the colony. Numbers in each 10-sec. interval from time of release through habituation. Shows five different releases. Closed circles represent trials in morning; open circles, trials in evening. Two top graphs for edge of colony; bottom graph for center of colony.

though only gulls ate chicks. The tern response was appropriate in that more terns responded more often during late May and early June, the egg period. After hatching began, the terns were less likely to respond to blackbirds. By contrast, the response to gulls increased after hatching (early June), again an appropriate response considering the vulnerability of small chicks.

6.6 EXPERIMENT 1: ANTIPREDATOR RESPONSES TO A HAWK AND A GULL

In the New York and New Jersey colonies, hawks and gulls are among the most significant avian predators, and they are responded to vigorously by common terns. Gulls are one of the most common intruders into tern colonies and eat eggs and chicks, while hawks take chicks and adults. Gulls and hawks of several species cause substantial disturbance and elicit vigorous mobbing when they enter a tern colony. Such observations delineate the occurrence of potential predation and the range of antipredator responses, and set the stage for controlled experimentation. To examine the effects of type of predator, habitat, density, nest location (center versus edge), and stage of breeding cycle, we undertook a series of experiments.

In 1985 near the Cedar Beach colony, we found an adult herring gull and an immature northern harrier (marsh hawk) both of which were alert, able to walk, but unable to fly. We introduced these birds as live "models" to various parts of several colonies to test the antipredator response of the terns. The experiments, conducted at a time when hatching was just beginning, allowed us to contrast the effects of habitat (marsh versus beach), location (center versus edge), and reproductive stage (late incubation versus chick).

When set in place, the herring gull sat upright with its neck and head erect. It ducked its head slightly at the approach of each passing tern, occasionally making upward jabs with its bill to fend off the approaching birds. We kept the gull for two series of experiments while nursing it back to health, and it was released at the conclusion of the study. The harrier was a recently fledged female with a broken wing. It sat vertically, was alert and responsive, and moved no more than 3 m from the position at which it was placed. We conducted only a limited series of tests with this bird before transporting it to a rehabilitation center from which it was eventually released to the wild.

Many antipredator studies have used real, stuffed, and mounted birds as the model eliciting a response (Lemmetyinen 1971; Curio 1975; Fuchs 1977; Shalter 1978c; Frankenberg 1981; Lowther and Olson 1984). Some studies have employed artificially constructed models (Nice and Ter Pelkwyk 1941; Amlaner and Stout 1978). Nice and Ter Pelkwyk (1941) and Regelmann and Curio (1983) observed the enhanced response to a moving predator, and indeed, use of live "models" has often proven desirable. Our own experience with inanimate or stuffed animals shows that habituation occurs very rapidly, but this did not occur quickly with live models. Some experimenters have found it useful to use the human experimenter as a predator model (Blancher and Robertson 1982; Conover 1987; Erwin 1988).

These birds offered the advantage of being live and responsive "models," yet made little attempt to escape from the position where they were placed. The normal posture and the slight responses they offered greatly enhanced the responses of the nesting birds, compared with their responses to realistic (but nonliving) models we have tried previously (see also Shalter 1978b).

Methods

We compared antipredator behavior at Cedar Beach, at two New York marsh colonies (Seganus E and F), and at two New Jersey marsh colonies (Ham and Pettit). The marsh islands were generally oval in shape, and we made one transect through the long axis of each of the marsh colonies, beginning at a point 10 m outside the colony, and moving inward 10 m at a time (table 6.3; fig 6.18).

At Cedar Beach, we used part of the preexisting 20 m. grid pattern, and ran antipredator trials along several transects from the north edge toward the center, and from the center to the extreme eastern end of the colony (fig. 6.18). We defined "central" as near the intersection of the central transect (east-west) and the central north-south gridline. Birds along the northern border of the colony, but not near the extreme eastern end were called "peripheral," while those at the east end were called "end" nests. Both of the latter two groups correspond to "edge" nests.

The recorder sat 20 m outside the colony, and before the experiment began, recorded the number of birds flying in one minute within 5 m of the future test location (baseline). The experimenter carried the predator (hidden from view) to the test location. We began recording the tern responses for one minute while the experimenter remained standing at the test location

Table 6.3. Experimental design for study of antipredator behavior in response to a gull and hawk model.

	Gull	*Hawk*
Beach colony (Cedar Beach)		
number of transects	10	2
plots in transect	4	4
distance between plots and transects	20 m	20 m
Salt marsh (Seganus A & B, Pettit, East Ham)		
number of transects	1	
points of transect	10	
distance between points	10 m	

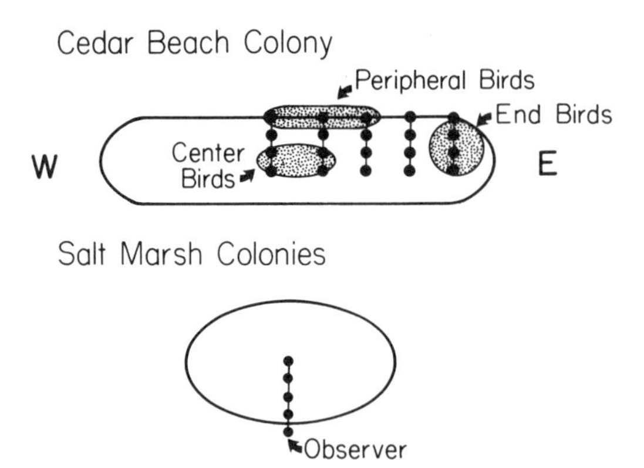

Figure 6.18. Distribution of sampling points for testing tern responses to a hawk and gull. In the marsh, only the gull was used.

with the "predator" hidden. We then placed the predator on the ground and recorded data for one minute while the experimenter was still present. The experimenter then quickly retreated to a point 20 m outside the colony, and we recorded behavior for the next 6 minutes in 1-minute blocks, counting the number of birds and their dives (fig. 6.19). Both the gull and harrier sometimes changed positions, sometimes moving up to three meters toward nearby cover. When the predator was actually moving we treated the time separately. We later noted whether the predator's new position involved more cover than the original test location.

Variables recorded included time of day and distance from the recording position, percent vegetation cover, status and contents of nearby nests, and numbers of nests within 5 m of the test position. For Cedar Beach, we also recorded the number of nests and their contents for all nests in the four 20×20 m quadrats immediately surrounding each test location. In each one-minute sample, we recorded the number of birds overhead, average distance of birds above the predator (the approach or avoidance distance), number of diver per minute, and the number of strikes if any. We initially predicted that density, stage in breeding cycle (egg versus chick), location (center versus edge), predator type (gull versus hawk) and length of time the predator was in the colony (opportunity for habituation) would influence the antipredator defensive behavior of terns.

Results

Factors Affecting Response to Predators

In general, the results showed that reproductive stage, predator type, and time in the colony influenced the number of terns above the predator, the total number of dives and strikes, and the mean distance of the terns above the predator (table 6.4). In addition, nesting density explained variations in the number of terns above the predator and the number of tern strikes. The number of tern dives, but not the mean distance above or the number of terns above the predator varied significantly among colonies (Fig. 6.20).

In most cases, the factor that contributed the most to explaining variations in the number of terns present or their behavior was predator type (gull versus hawk). The only exception was the regression model for the number of strikes, where the number of tern chicks most strongly influenced the variance (table 6.4).

Overall, the behavior of terns to hawks and gulls differed in a number of ways, even though the physical conditions of the test did not differ (table

Table 6.4. Models explaining variations in dependent measures for all tests and habitats. Given are F values (levels of significance).

	Number of Terns Above	*Number of Tern Dives*	*Number of Tern Strikes*	*Mean Distance Above Predator*[a]
Model				
F	19.9	23.4	6.63	9.32
R^2	0.49	0.53	0.25	0.36
df	11,233	11,233	11,225	11,233
P	0.0001	0.0001	0.0001	0.0001
Factors entering F (P)				
number of nests	26.5(0.0001)	47.8(0.0001)	—	—
number of eggs	—	15.1(0.0001)	6.4(0.01)	9.2(0.002)
number of chicks	33.4(0.0001)	—	29.2(0.0001)	7.8(0.005)
density[b]	26.2(0.0001)	—	9.3(0.002)	—
model[c]	97.1(0.0001)	100.7(0.0001)	10.1(0.001)	48.3(0.0001)
Type of test[d]	6.1(0.0001)	16.4(0.0001)	2.4(0.02)	6.1(0.0001)

[a] Mean distance terns flew above predator.
[b] Total number of active nests in plot 100 m x 100 m around experimental plot.
[c] Hawk or gull.
[d] Length of time since model was introduced.

Figure 6.19. Antipredator experiments: single common tern mobbing a harrier (left); group of common terns mobbing herring gull (right).

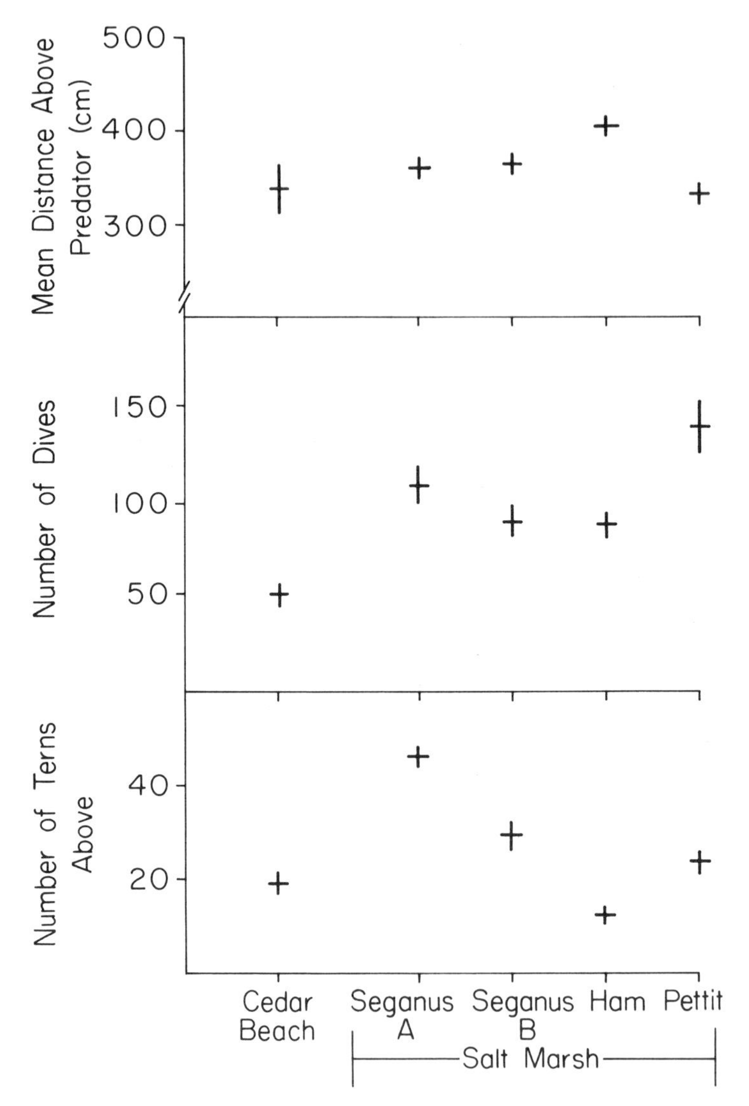

Figure 6.20. Behavior of common terns directed toward a gull model: comparison of beach (Cedar Beach) and salt marsh colonies. Mean (± 1 SD) distance birds circled above predator (avoidance distance, top panel). Mean (± 1 SD) dives per minute (middle) and mean (± 1 SD) number of terns mobbing (bottom).

Table 6.5. Comparison of hawk and gull tests for antipredator behavior at Cedar Beach. Given are means ± one standard deviation.

Characteristic	*Hawk*	*Gull*	*Kruskal-Wallis* X^2	*P*
Independent measures				
cover directly over model (%)	20.8 ± 36.6	32.0 ± 36.3		NS
cover within 1 m (%)	16.1 ± 28.5	19.7 ± 24.2		NS
cover within 5 m (%)	24.2 ± 28.1	34.4 ± 25.0		NS
distance from edge of colony	31.9 ± 16.6	15.2 ± 14.5		NS
Density[a]	6.3 ± 3.3	7.0 ± 6.6		NS
empty nests	2.4 ± 1.5	0.6 ± 1.3		NS
nests with eggs	1.8 ± 1.5	2.4 ± 3.6		NS
nests with eggs and chicks	0	0.4 ± 0.7		NS
nests with chicks	2.1 ± 1.2	2.5 ± 3.4		NS
number of nests[b]	111.3 ± 71.6	101.8 ± 108.2		NS
Behavioral measures				
number of terns above	64.5 ± 58.0	23.6 ± 22.1	85.8	0.0001
number of tern dives	7.8 ± 8.1	77.7 ± 63.4	136.7	0.0001
number of tern strikes	0.1 ± 0.3	0.6 ± 1.4	19.3	0.0001
Mean distance overhead (m)[c]	492.7 ± 140	349.8 ± 127.3	107.9	0.0001

[a] Density = number of nests within 5m.
[b] Number of nests = number of nests within the 100 m^2 quadrat.
[c] Of terns overhead.

6.5). In general, vegetation cover directly over the predator and within 1 and 5 m was similar, and nest density was similar, because the tests were conducted along the same transects.

In general, more terns mobbed the hawk than the gull, but far fewer dove at or struck the hawk, and terns stayed farther above the hawk than the gull (table 6.5). These differences were all highly significant. Although we didn't record it quantitatively, terns also stayed in a tighter, more compact group over the hawk, and remained less organized and more scattered over the gull. In this respect, the group over the hawk was similar to the groups that hovered over dogs.

Behavior Directed Toward Hawk

Because of the major differences in behavior directed toward gulls and hawks, we examined the terns' responses to each predator separately. The variability in tern response toward the hawk was explained by nest density, distance from the edge of the colony, and percent of vegetation cover (table 6.6). In general, more terns responded when the hawk was placed in a dense nesting area within a larger section of the colony that also had a high nest density (table 6.6). Thus, terns came from a great distance to respond to the hawk. Secondly, distance from the periphery and end of the colony was important, partially because these sections had lower overall density.

The percent of cover influenced the number of terns overhead and the number of dives. The hawk frequently hopped to dense vegetation and became motionless, and its crypsis made it more difficult to see. When the hawk hopped to beachgrass or goldenrod, it became less visible and the number of terns mobbing it decreased. Terns responded more to the hawk when it was walking then when it was still, and showed some habituation (fewer responses in the last minute) although they continued to respond (see below).

Behavior Directed Toward Gull

The behavior of terns differed dramatically between salt marsh and beach (table 6.7). In the marsh, nearly twice as many terns flew overhead, dove and struck at the gull, although saltmarsh terns remained higher above the

Table 6.6. Models explaining variation in dependent measures for response to the harrier at Cedar Beach.

	Number of Terns Above	*Number of Tern Dives*	*Number of Tern Strikes*	*Mean Distance Above Hawk*
Overall Model				
F	5.46	5.82	3.92	6.92
r^2	0.63	0.65	0.21	0.30
P	<0.0001	<0.0001	<0.05	<0.0001
Factors entering F (P)				
density	26.8 (<0.0001)	4.5 (<0.04)	3.9 (<0.05)	—
number of pairs	4.5 (<0.04)	9.3 (<0.004)	—	10.9 (<0.002)
distance to edge	—	36.4 (<0.0001)	—	6.0 (<0.01)
distance from east edge	5.0 (<0.03)	4.8 (<0.03)	—	52.2 (<0.0001)
Percent cover	18.9 (<0.0001)	4.2 (<0.04)	—	—

Table 6.7. Behavioral response of common terns to live gull model in salt marsh and beach colonies. Given are means ± one standard deviation.

	Salt Marsh	*Beach*	*P*
Independent measures			
cover directly over model (%)	47.3 ± 39.7	16.1 ± 23.6	NS
distance from edge of colony	9.3 ± 11.2	21.0 ± 15.2	NS
Density[a]			
empty nests	0	1.2 ± 1.7	NS
nests with eggs	4.9 ± 4.1	1.8 ± 2.1	NS
nests with eggs and chicks	0.6 ± 1.00	0.2 ± 0.4	NS
nests with chicks	5.5 ± 4.1	4.4 ± 1.8	NS
Behavioral measures			
number of terns above	27.9 ± 23.8	19.3 ± 19.3	27.4(0.0001)
number of tern dives	105.3 ± 69.1	50.6 ± 42.4	97.3(0.0001)
number of tern strikes	0.7 ± 1.6	0.4 ± 1.0	15.5(0.0001)
Mean distance above (m)	363.2 ± 109.6	335.6 ± 142.6	9.5(0.002)

[a] Total number of terns nests within 5 m of test bird.

gull. The rate of dives per tern overhead was similar in salt marshes (3.75/tern) compared to the beach colony (2.68/tern). The higher level of antigull defensive activity in the salt-marsh colony is extremely interesting because the response to people is opposite; terns at Cedar Beach dive at and strike humans far more than do those in salt marshes.

Effect of Habitat

We examined the factors that contributed significantly to explaining variations in the defensive behavior of terns in the two habitats (table 6.8). Only two factors (nest density and predator activity) were significant contributors to explaining the variability in all the response measures in both salt marsh and beach habitats (fig. 6.21). In general as the number of nests increased, the number of terns overhead increased (fig. 6.22).

Distance from the edge of a colony entered models for three of the four response measures on salt marshes, but for the beach colony influenced only the number of terns overhead. In general, there were more responses in the center and fewer on the edge. The number of nests with chicks present entered two of the models, and was a more important factor in salt marshes than in sand colonies; defense increasing when chicks were present.

In both salt marsh and beach colonies density was clearly the most

Table 6.8. Model explaining variation in dependent measure for the gull model at salt marsh and beach colonies.

	Salt Marsh				*Sandy Beaches*			
	Number of Terns Above	*Number of Tern Dives*	*Number of Tern Strikes*	*Mean Distance Above*	*Number of Terns Above*	*Number of Tern Dives*	*Number of Tern Strikes*	*Mean Distance Above*
Model								
F	20.2	39.0	7.6	13.1	11.1	12.9	4.8	10.3
R^2	0.48	0.64	0.26	0.39	0.53	0.59	0.33	0.56
P	0.0001	0.0001	0.0001	0.0001	0.0001	0.0001	0.0001	0.0001
Factors entering, F (P)								
density	29.4(0.000)	138.9(0.0001)	11.7(0.0007)	81.0(0.0001)	62.4(0.0006)	47.9(0.0001)	15.9(0.0001)	15.5(0.0001)
number of nests	—	—	—	—	7.0(0.0001)	9.4(0.002)	—	—
nests with chicks	30.6(0.0001)	—	30.4(0.0001)	4.2(0.04)	—	—	35.9(0.0001)	—
nests with eggs	19.2(0.0001)	—						
Type of Test	22.8(0.0001)	51.5(0.0001)	4.1(0.0006)	6.7(0.0001)	15.3(0.0001)	20.4(0.0001)	2.6(0.02)	21.7(0.0001)
cover over nest	18.1(0.0001)	—	—	—	—	7.2(0.008)	—	13.9(0.0001)
cover within 1 m	—	—	—	—	—	5.1(0.02)	5.3(0.02)	9.8(0.002)
cover within 5 m	—	—	—	—	—	5.5(0.01)	—	9.4(0.0002)
Distance from								
east end	—	15.1(0.0001)	—	16.7(0.0001)	12.2(0.0006)	15.1(0.0001)	—	—
edge of colony	6.7(0.01)	—	8.2(0.004)	5.9(0.01)	6.1(0.01)	—	—	—

Figure 6.21. Experimental study of common terns mobbing a herring gull in a salt marsh colony, Seganus Thatch (top), and at Cedar Beach (bottom). In general, more birds mobbed the gull in the beach colony and birds tended to approach the gull more closely.

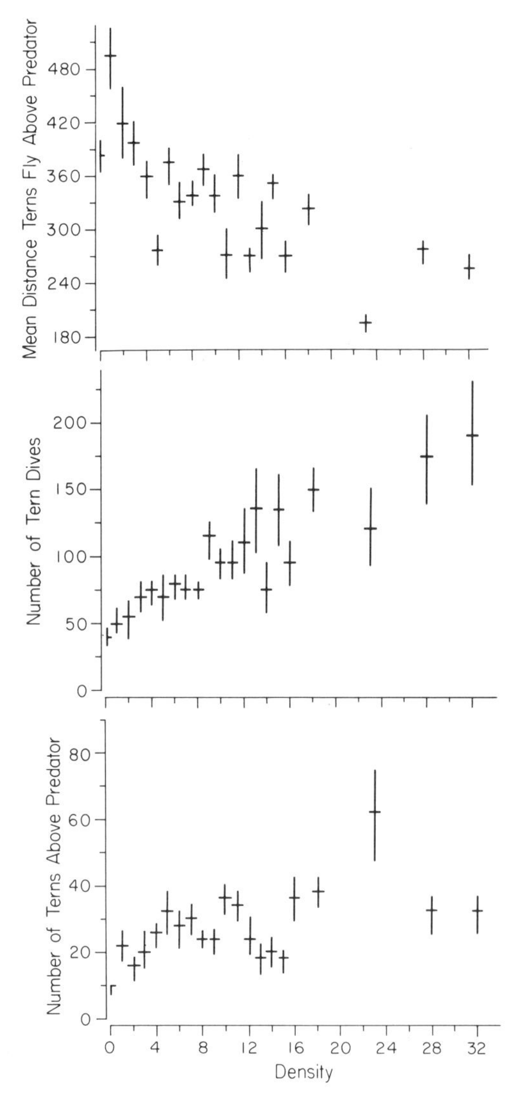

Figure 6.22. Behavior of terns above a herring gull as a function of number of tern nests within 5 m of the gull (density) at Cedar Beach. Shown are means (± 1 SD).

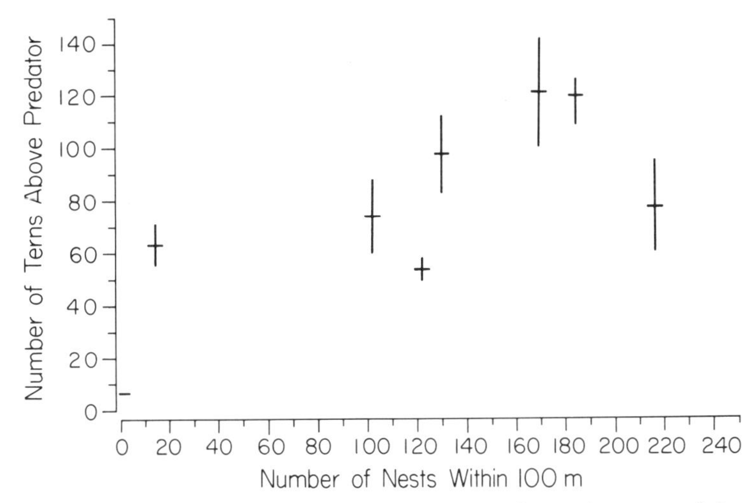

Figure 6.23. Number of terns above the gull as a function of the number of nests within 100 m of the test bird at Cedar Beach sand colony. Shown are means (± 1 SD).

important factor influencing the number of terns overhead, as well as most defense behavior (table 6.8). As density of nests increased, the number of terns above the predator increased, the number of dives increased, and the mean distance of terns above the predator decreased (fig. 6.22). This was also true for the number of nests within 100m of the test situation (fig. 6.23).

Several general conclusions can be drawn: 1) in all habitats, terns responded more to moving—compared to the stationary—gull or hawk; 2) in all cases, terns responded more to the predator alone than to the person; 3) tern response to the person was similar, regardless of whether the person was holding a hawk or a gull; 4) terns responded less to a person standing over the predator than they did in the first minute the predator was alone in the colony; and 5) over 75 terns/min flew over the hawk compared to less than 35 for the gull.

In the absence of disturbance at Cedar Beach, fewer than 10 terns were flying over a 20 × 20 m quadrat in the colony during any one minute. With the person alone, responses increased significantly. Yet when a person with a predator stood in the same place the number of terns overhead and the number of dives increased dramatically (fig cs. 6.24, 6.25). The number of terns overhead and the number of dives was even higher when the predator

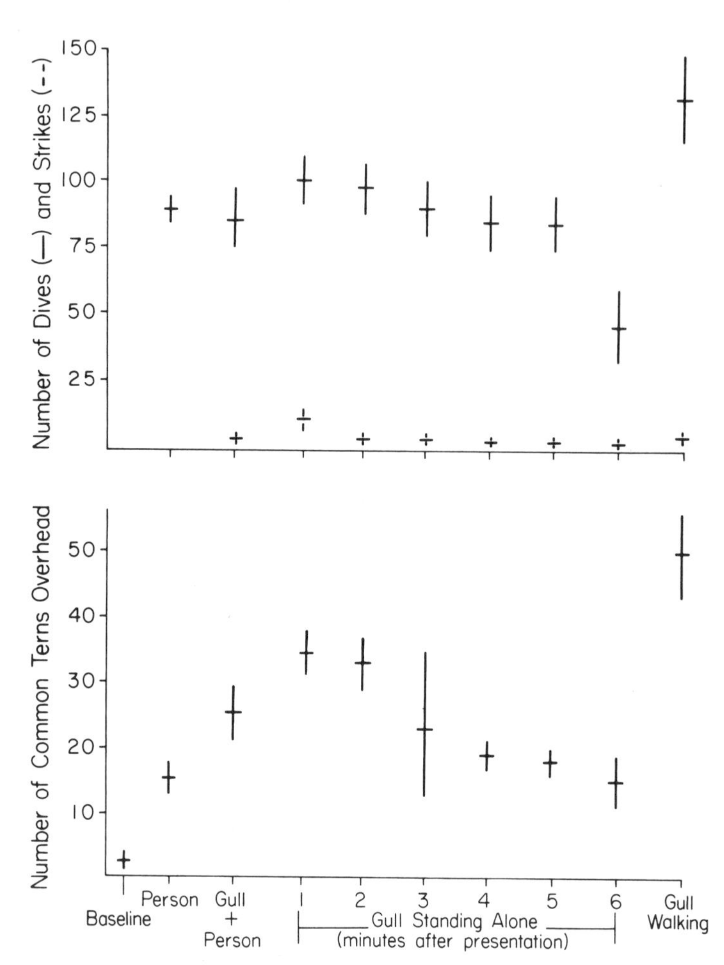

Figure 6.24. Responses of common tern during the Gull Experiment at Cedar Beach comparing the baseline activity, response with person present, and response to the gull alone. Shown are means (± 1 SD) for number of terns overhead (bottom), and number of dives (upper lines in upper panel) and strikes (lower lines in upper panel).

was alone. Thereafter, the birds responding and the number of dives slowly decreased over the next six minutes (habituation).

It is of interest to examine habitat variations in the behavior of the terns toward the person standing alone, and toward the gull when presented for

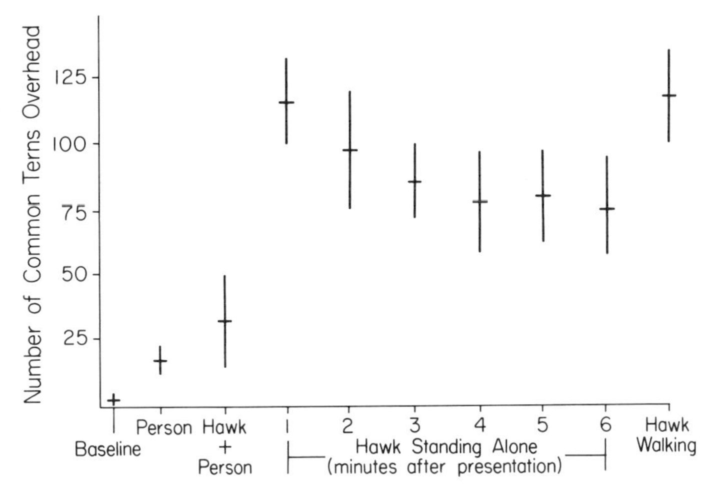

Figure 6.25. Responses of common terns during the Hawk Experiment comparing the baseline activity, response with person present, and response to the hawk alone. Shown are means (± 1 SD) for number of terns overhead.

the first minute (table 6.9). In salt marshes, tern defensive responses were greater toward the gull than toward the person. However, in the beach colony more terns responded to the gull or hawk by flying overhead, but terns made more dives and strikes at the person.

The mean distance terns remained overhead also varied by species of predator, length of stay in the colony for the predator, and whether the predator was moving or stationary (fig. 6.26). Overall, terns remained higher

Table 6.9. Behavior of common terns directed toward a person standing alone and a gull alone (first test minute). Shown are mean ± one standard deviation.

	Salt Marsh		*Beach*	
	Person	*Gull*	*Person*	*Gull*
Terns				
number of terns above	14 ± 8	40 ± 25	15 ± 11	26 ± 23
number of dives	65 ± 27	122 ± 53	94 ± 29	56 ± 41
number of strikes	0.05 ± 0.2	1.4 ± 2.1	7 ± 4	0.4 ± 0.9
mean distance above (cm)	452 ± 98	346 ± 90	369 ± 136	349 ± 114

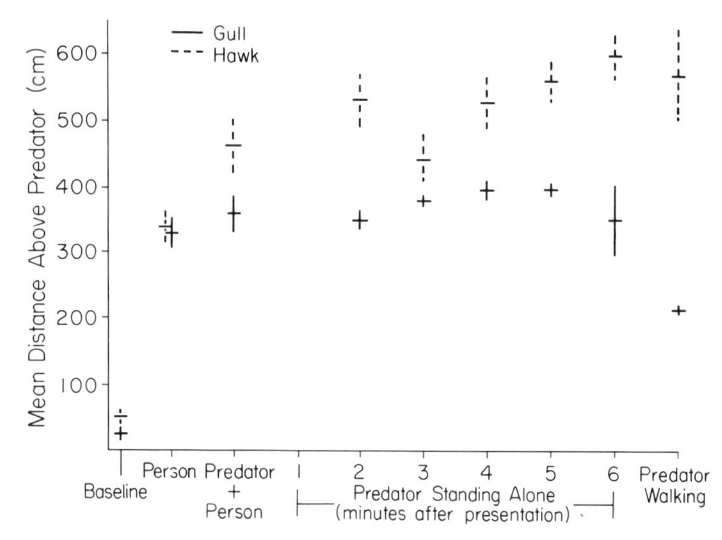

Figure 6.26. Avoidance distance, the mean distance (± 1 SD) terns stayed above a hawk or gull model. Data were collected in the minute before the test, a minute with a person and the predator, and for five minutes with the predator. Data taken when hawk or gull were moving are shown separately. Moving gull elicited increased diving response.

above the hawk than above the gull. Terns were closer to a moving gull than to a stationary gull, yet remained at similar distances or higher when the hawk was moving. The avoidance distance terns stay above predators seems less susceptible to habituation than their other behavioral responses (Fig. 6.26).

Habituation

The process of habituation occurred to both the gull and hawk (fig. 6.26), but habituation was much slower to the hawk. At each point in time, the overall level of response (number of terns overhead) was far greater for the hawk than the gull. The number of terns overhead decreased rapidly over the six minutes, whereas the number of dives decreased slowly until minute 5 for the gull. The number of dives at the hawk, however, did not decrease but remained similar throughout the test period, even though fewer terns were present.

Discussion

Predator Presentation Variations

In these experiments, terns were tested with a person, hawk, and gull. The number of terns that flew overhead and their defensive behavior (number of dives, number of strikes) differed dramatically depending on the predator type. Thus we suggest that studies using only a human as the "model" must be viewed with caution (e.g., Conover 1987; Erwin 1988) since the birds' responses to natural predators might be quite different. Overall, the number of terns flying over the predator was greatest for the hawk, intermediate for the gull, and lowest for the person. However, on the sandy beach, the number of dives and strikes was greatest for the person and lowest for the hawk. Thus, more terns came to mob the hawk, and to a lesser extent the gull, but almost no terns dive-bombed or struck the hawk. Diving and striking were most commonly directed at the person.

We believe the differences between the terns' response to the hawk and gull reflect differences in the threat these predators pose, and the differences between these predators and people reflect experience and habituation (discussed below). Hawks can kill and eat both adult and young common terns. Harriers frequently enter the Cedar Beach colony where the tests were performed, and sometimes kill adults. Terns respond to them by mobbing (fig. 6.19) to encourage the hawk to leave the colony, but they remain at a safe distance above the hawk to avoid any chance of being captured. While mobbing the hawk, terns remained in a tight aerial group.

Gulls, on the contrary, are primarily a threat to eggs and chicks, but not to the adult terns themselves (Veen 1977). Thus, the terns can approach the gull and even strike it in efforts to deter it. Gulls can be an important cause of egg and chick mortality for many species of larids (Tinbergen 1953; Kruuk 1964; Veen 1977; Burger and Lesser 1978).

Habitat Differences

There were habitat differences in the defensive responses of terns to the natural predators. Terns were more responsive to the gull in salt marsh compared to the sandy beach colony, perhaps reflecting differences in density, colony size, vegetation cover or height, behavioral characteristics of different populations, or experience. We do not believe the differences are due to vegetation because the cover is higher, denser, and continuous in salt marshes whereas most of the beach colony was fairly open, and we had expected the converse effect.

The experience of terns in the colonies examined varies with respect to their exposure to gulls, the threats gulls pose, and their exposure to humans. Moreover, the Cedar Beach colony had a much larger population size but was not more dense. Differences could relate to numbers and/or density.

Habituation

Habituation is the process whereby in the face of continued stimulation a response diminishes over time. In the absence of reinforcement (egg or chick loss, adult injury or death), the behavior in question extinguishes and finally disappears. There are two types of experience that can affect habituation in terns: long-term and short-term. Over the days or weeks of the breeding season, exposure to a predator without adverse effects could diminish the response. Secondly, over the six minutes of the test, the lack of a threatening stimulus could decrease the terns' response. The latter clearly occurred and will be discussed below.

However, another distinction must be made for the effect of long-term predator exposure on tern behavior. Habituation can either act to reduce their overall aggression as McNicholl (1973) showed for terns reducing their mobbing of gulls nesting in their colony, or it can act to reduce their fear of the predator, allowing closer approach and attack. Thus, when their fear is reduced, terns may appear more aggressive toward a predator. Repeated exposure to a person or predator reduces fear of that predator through habituation of the fear response. Then tern attacks may increase, revealing a high attack tendency which previously had been masked by a stronger fear response. Increasing exposure to the predator or person might then result in increasing aggression, as the terns experience no ill effects from that aggression. The results of this experiment are consistent with this hypothesis.

Attack as an Expression of Conflict

The evidence that the antipredator response of terns is a compromise between two conflicting tendencies is not novel. The behavior is a composite of approach and avoidance behavior. One primal response to a threat is to increase the distance between self and threat, in other words, to flee. Another response is to approach and thwart it. When the intruder is distant, an incubating bird remains sitting. At a certain point, the response distance, the bird leaves its nest. Marler (1956) pointed out that the response to a threat varies inversely with the distance to the threat.

The response distance depends on the perceived risk and the incubation tendency. The perceived risk depends on the species of predator, its apparent behavior (Schaller 1972), any prior experience with the species or individual, and the response of neighboring birds. Incubation tendency is influenced by: 1) stage of incubation, 2) clutch size, 3) time in this incubation bout, 4) level of fear or stress (through a negative feedback loop), 5) environmental factors (Austin 1946), and 6) by behavior of neighboring conspecifics. The latter factor is important since when one bird responds by flying up, many others follow immediately, as if to say "what does it know that I don't?" Hence, in any colony or section, the response distance is likely to depend on the most timid, apprehensive, or fearful individual, or the lowest common denominator.

One benefit of synchrony would be to place neighbors on a more or less equal footing with regard to incubation tendency, to minimize the risk of having one individual crying "wolf" too often, when its neighbors would otherwise continue to incubate. Austin (1933) and Austin (1946) emphasize that the tendency to "sit tight" and incubate increases during the course of incubation.

Approach or Withdrawal

Once a bird has stopped incubating and flies up, it may move toward or away from the intruder. The resultant distance between bird and intruder is a balance between approach and withdrawal. Approach is a function of the tendency to attack and also the tendency to gather information (Kruuk 1976). The tendency to attack should ultimately depend on some balance between the probability of success (driving the predator away), the need to protect vulnerable eggs or chicks, and the probability of the ultimate failure (getting caught). The perception as to whether the predator can be thwarted may depend on innate recognition and learned response (from previous encounters, Shalter 1978b) and/or the response of neighbors, the distance from intruder to the nest, and the behavior of the intruder (Hinde 1954). The ability of ungulates, for example, to perceive the difference between a hunting and nonhunting lion (Schaller 1972) greatly reduces the energy they would otherwise expend on unnecessary flight. Moreover, we have shown elsewhere (Burger and Gochfeld 1981b) that nesting herring gulls can perceive the difference between a person approaching the nest directly or tangentially.

Predator Recognition

The immediate need to protect the nest depends on the distance of the intruder to the nest, the stage of the cycle, and on the number of other vulnerable nests. The perception of personal risk depends on both innate and learned responses (Kramer and von St. Paul 1951), and on the behavior of the predator (Nice and Ter Pelkwyk 1941; Burger and Gochfeld 1981b). If the predator is obviously hunting, approaches despite mobbing, and acts aggressively towards adults, the avoidance distance will increase. If, on the other hand, it is already leaving the colony under a swirling escort, the birds appear emboldened, and avoidance distances decreases.

The role of previous experience is critical. If birds are familiar with the intruder and have never experienced harm, they may have low avoidance distance (Burger and Gochfeld 1983), whereas a novel intruder engenders more respect. The suddenness and closeness (or perceived size) of the stimulus may initiate initial withdrawal, before approaching (Nice and Ter Pelkwyk 1941). Thus, the tendency of any bird to make a close approach and attack a predator is contingent on many factors, some innate, some learned, and some immediate.

If the perceived risk to self is high and incubation tendency is low, the avoidance distance will be great; conversely, at low risk but high protective tendency, the avoidance distance will be low, and the bird may attack. Over time, the attack tendency may increase as fear is reduced or may decrease through habituation. The effect of time on the response will be influenced by any threats or capture successes on the part of the intruder and by any positive reinforcement which comes from successfully driving it away. If the intruder captures a bird, the other birds are likely to decrease their attack tendency, but may actually increase their mobbing response.

Risk to Adults

Kruuk (1976) distinguished the "nervous" approach behavior for the purpose of gathering information from attack behavior. Smythe (1970) suggested that mobbing elicits responses from the intruder which signal its position, capability, and intention. Thus, the intruder may prematurely signal its intention or "tip its hand." Schaller (1972) described the tendency of grazing ungulates to monitor potential predators, Kruuk (1972) observed predators killing ungulates that approached too closely, and Smith (1969) suggested that forest falcons might elicit mobbing responses from small birds in order to attract them as potential prey. We have seen a coyote attempt to capture a mobbing magpie, and have ourself learned to snatch a mobbing tern by hand (Fig. 6.6). We have seen a peregrine and a harrier catch mobbing terns and shorebirds, and Myers (1978) reported a caracara capturing a mobbing lapwing. Close approach can indeed be dangerous.

Fear

We tend to infer fear or fearfulness indirectly by measuring the animal's behavior. We recognize that in birds with a high attack tendency but a high degree of fear, their avoidance distance may be so great as to preclude effective attack. Thus mobbing birds will hover low over and attack an inoffensive rabbit, terrapin, or biologist, or injured fledgling, while milling ineffectually high over a dog, or fox, or hawk. Russell (1979) reviewed the general classes of fear-evoking stimuli. Fear functions by enabling an organism to avoid harmful stimuli or situations. There is no assurance that a fearful response is appropriate, for the animal must judge on incomplete data.

6.7 EXPERIMENT 2: SEASONAL AND COLONY SIZE EFFECTS ON MOBBING

In the previous experiment, we showed that mobbing behavior varies by habitat, species of predator, density, and the length of time the predator is in the colony. In this section, we examine how mobbing behavior changes seasonally and as a function of colony size. We examined salt marsh colonies, because we could study 30 colonies of varying sizes (pairs of terns) in one year.

The benefits that a colony member derives from nesting in a colony should relate to the number of birds that engage in mobbing a potential predator (see also Wittenberger and Hunt 1986). Presumably, if more birds mob a predator, there is a greater chance that the predator will depart before

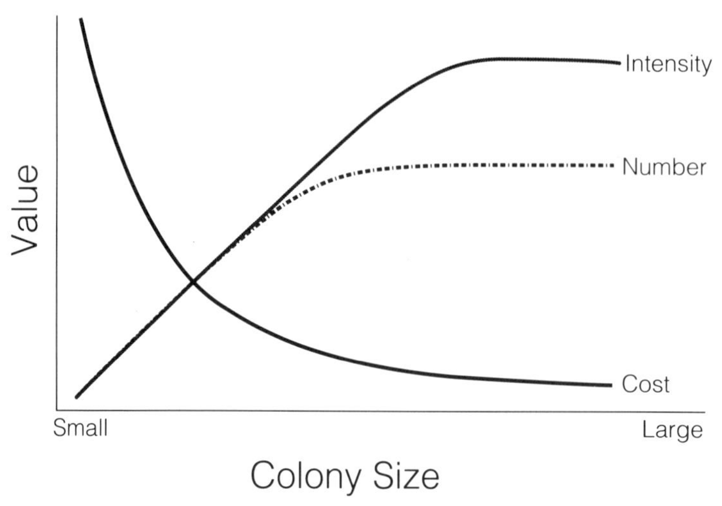

Figure 6.27. Proposed model relating the intensity of mobbing/attack, number participating, and cost per individual as a function of number of birds nesting in a colony.

eating any eggs or young. Andersson (1976) noted that during unsuccessful predation attempts on kittiwake nests there were more birds mobbing the predator than during successful attempts.

Another possible benefit is the quality of the mobbing. Mobbing birds can merely circle overhead, or birds can attack and even strike the predator. The intensity of the mobbing behavior should increase as a function of colony size (fig. 6.27). Both frequency and intensity of mobbing may be influenced by social facilitation where birds increase a particular behavior in the presence of other birds performing that behavior.

One of the costs of nesting in colonies relates to the energy and time devoted to mobbing predators, whether they are over the mobber's nest or elsewhere in the colony. Secondly, mobbing can result in direct damage to adults if they get too close to a large predator capable of killing them (i.e., foxes, Kruuk 1964; or hawks, Myers 1978). In order for large colony size to be selected for, with respect to mobbing behavior, the cost of the mobbing behavior (or the cost/benefit ratio) should be less for individuals nesting within larger colonies (see fig. 6.27). At a certain point, however, one reaches a plateau of maximum benefit. If colonies differ in density, it will be important to examine the effects of density independently from the effects of size.

The relative costs and benefits illustrated in figure 6.27 are expected to vary for different species. The colony size at which the benefits cease to increase, presumably also varies with the type of predator as well as the species of defender.

The model we propose (fig. 6.27) suggests that the cost of coloniality should decrease as the advantages of antipredator behavior increase, assuming that larger colonies don't attract more predators. The advantages of antipredator behavior can be separated into at least two components: an advantage due to the numbers of birds which mob a predator and swirl overhead, and the intensity of the mobbing by those individuals (the number of dive-attacks, and actual hits). We predicted that the intensity of mobbing should increase beyond the point where the number of birds mobbing no longer increases. This increase, which we attribute to social facilitation, should be greater in larger colonies.

These ideas generate a number of hypotheses which we tested in the field:

1. For a predator in the colony, mobbing behavior should be a monotonic function of colony size (number of breeding pairs), increasing in frequency and intensity in larger colonies.

2. For the individual nesting within 5 meters of the predator, the cost of directly protecting their nest or eggs should relate inversely to colony size. That is, each individual should mob less.

3. For all individuals in the colony regardless of the location of the predator, the cost (amount of time or energy used) of mobbing should relate inversely to colony size.

We tested these predications by examining the mobbing behavior in 30 colonies of common terns nesting in New Jersey. A human was used as the stimulus in order to control for speed of movement, and to insure that the predator behaved the same way under all conditions (see Yasukawa 1978). Several authors (Kruuk 1964; Veen 1977) have noted that gulls and terns respond to man in a similar manner as to mammalian predators such as fox, but this was clearly not the case in the New York-New Jersey tern colonies.

Methods

From May 1 until July 30, 1979, we visited 30 colonies (6–500 pairs) of common terns at 4- to 5-day intervals. Predominant vegetation on the islands was *Spartina alterniflora* and *S. patens*, although some of the islands have a few *Phragmites* and bushes. Except for one colony, all colony sites had been occupied for at least 4 years. The terns nested mainly on mats of dead *Zostera*.

We approached each colony by boat, and one person walked slowly from the edge of the island (usually only 3 to 7 m from the nesting terns) into the center of the colony. The person walking through the colony did not wave his arms, nor did he carry anything larger than a clipboard and a white helmet. The clothing worn was similar from day to day. The second observer remained in the boat and recorded data for 1-min. sample periods. From three to five samples were taken in each colony on every visit, depending on the size of the colony. During each 1-min. sample period, we recorded the following information: colony, date, location (edge, intermediate, and center of the colony), time of day, weather variables, total number of birds visible in the colony, number of birds flying overhead, and number of birds diving, number of dives, and the number of times the intruder was hit by a diving tern. Nests were individually marked, and checked on each visit so that the number of active nests of various clutch sizes was known.

The number overhead, the number of dives, and the number of hits are an indication of the quality of mobbing a particular intruder or predator will experience, for presumably the discouraging effect on any predator increases as the number of mobbers, dives, attacks, and hits increases. We used these data to estimate the costs of mobbing to an individual nesting in a colony, and the advantage to those individuals nesting within 5 m of the predator (or human intruder in this case). To determine the cost of the former, we divided the mobbing measures (number of terns overhead, number of terns diving, number of dives, and number of hits) by the number of birds present at the colony. To determine the advantage to the tern nesting within 5 m of the predator, we divided the mobbing measures by the number of nests within 5 m of the intruder. To derive a benefit from nesting in a larger colony, these cost measures should decrease in larger colonies.

Social facilitation implies the enhancement of activity by the mere presence of numbers of birds performing the activity. To test for the existence of social facilitation we divided the mobbing behaviors (divers, dives, hits) by the number overhead; and the dives and hits by the number of divers. For social facilitation to occur, these values should be higher in larger colonies and when more birds are mobbing.

We computed the mean values (± SD) for all the above measures as a function of colony size. We arbitrarily designated four colony sizes: tiny (0–25 pairs), small (25–100 pairs), medium (100–250 pairs), and large (over 250 pairs). Each colony was assigned to the appropriate category at the end of each sampling period.

Results

Seasonal Effects

There was a significant seasonal effect on mobbing behavior. When all the data are pooled across colonies there was no significant seasonal effect for the number of terns overhead mobbing, but there was a significant difference in the number of dives and the number of strikes (fig. 6.28). The increase in dives and hits occured in mid-June when eggs were hatching. It is remarkable that across colonies a seasonal effect is apparent, given the lack of synchrony across colonies.

Likewise, there was a difference in the number of strikes per diver and the number of strikes per tern overhead (fig. 6.29). This indicates that the birds overhead increased the intensity of their attack behavior, suggesting the presence of social facilitation. The increase in intensity corresponded to hatching.

Quality of Mobbing

The number of birds overhead and diving, and the number of dives and hits generally increased with increasing colony size (summarized in table 6.10, fig. 6.30). Significant differences occurred in all measures, and thereby the potential effectiveness increased with colony size. We also examined three breaking points (at 25, 100, and 250 pairs) for the effectiveness of mobbing by colony size. For all comparisons, the mobbing measures (number overhead and diving; number of dives and hits) were greater in the larger colonies (table 6.11). Thus, as predicted, predators entering larger colonies are mobbed, dive-bombed, and struck by more birds at any one time than are predators in small colonies, even though the likelihood of any single tern's losing its chick is smaller in a large colony.

Advantage to the individual nesting within 5 meters

We computed the advantage to an individual nesting within 5 m of the potential predator as a function of colony size (table 6.11). For all measures,

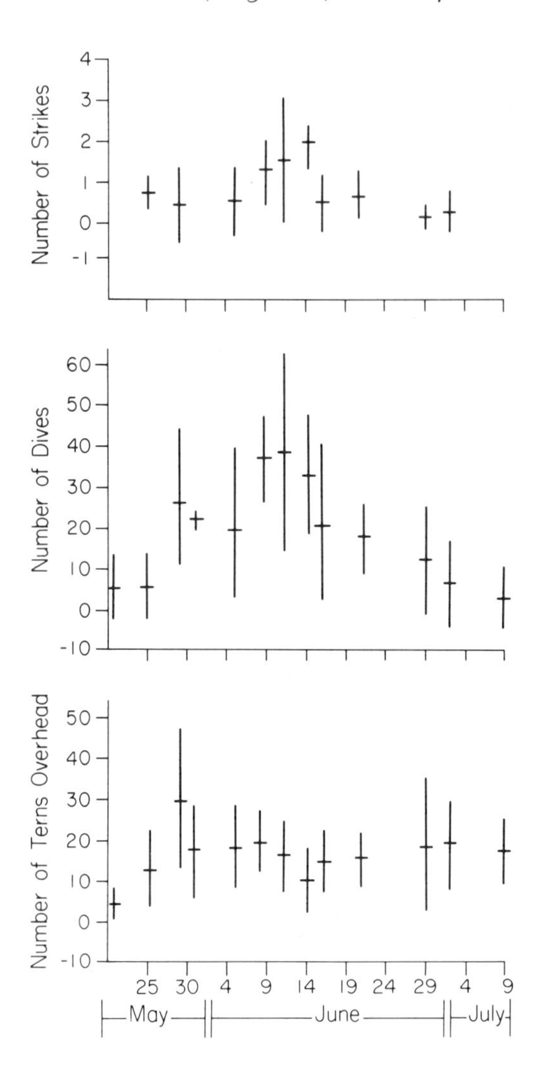

Figure 6.28. Common terns mobbing human intruder: effect of season. Shown are mean (± 1 SD) number of birds overhead mobbing (bottom), number of dives per minute (middle) and number of strikes per min. (top).

except hits per pair, significant differences occurred with colony size. There were more divers, dives, and number of birds overhead per pair in large than in small colonies. The values of these measures, however, did not continue to increase with colony size endlessly (fig. 6.31). The number overhead was larger for colonies over 25 compared to those below 25, but no differences

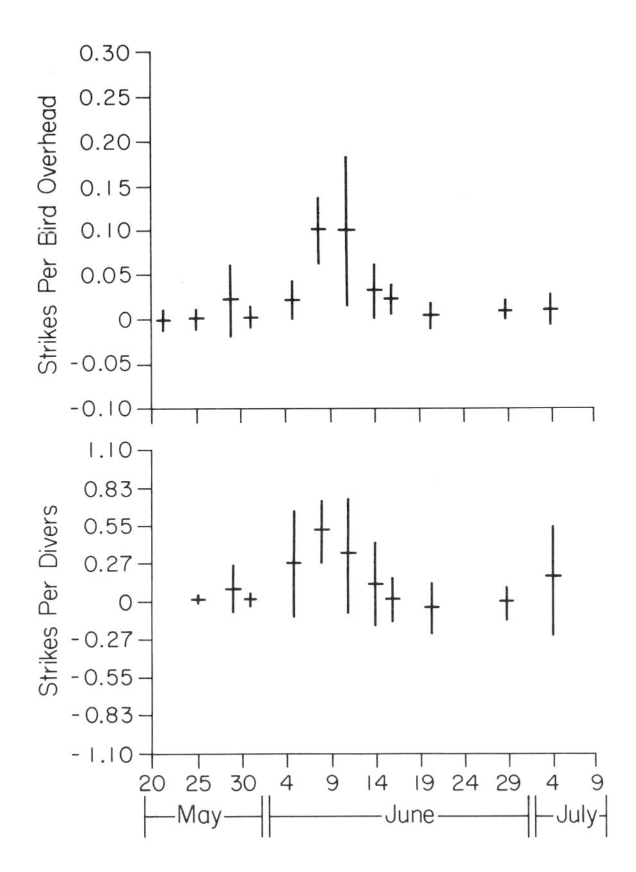

Figure 6.29. Seasonal variation in the intensity of terns mobbing a human intruder. Number of strikes per bird overhead (top) and per number actually diving (bottom) are shown with means (± 1 SD). If there was no seasonal increase in mobbing intensity, values shown remained similar throughout the cycle.

were noted thereafter (table 6.11). Similarly, no significant differences were noted for colony sizes of 100 or more pairs for the other measures.

Cost to the individual nesting in the colony

Time and energy invested in mobbing is a cost, quite apart from the risk to the adult or the exposure of eggs and young. To evaluate the cost to an individual nesting in that colony, we divided the mobbing measures by the

Table 6.10. Differences in behavior of mobbing terns as a function of colony size. Given are the *t* values and levels of significance (* = $P<0.05$, ** = $P<0.01$, *** = $P<0.0001$).

	Colony Size[a]		
	<25 versus >25	*<100 versus >100*	*<250 versus >250*
Colony characteristics			
clutch number	2.50*	1.97*	1.82 NS
area	6.3	13.2***	21.0***
density peak	1.9 NS	2.20*	0.24 NS
Effect on predator			
number overhead	10.1***	5.4***	5.2***
number diving	9.5***	5.64***	6.4***
number dives	9.5***	7.1***	7.4***
number hits	4.6***	3.0**	2.8**
Advantage to individual nesting within 5m			
divers/pair	4.81***	3.03**	0.84 NS
divers/pair	4.47***	3.16***	1.43 NS
hits/pair	2.90**	1.97*	0.8 NS
overhead/pair	4.43***	1.69 NS	1.3 NS
Cost to nesting in colony			
dives/number present	0.4 NS	2.71**	3.6***
hits/number present	——[b]	1.34 NS	1.4 NS
overhead/number present	4.7***	9.1***	13.1***
divers/number present	1.01 NS	4.49***	5.6***
Social facilitation			
dives/number diving	2.64*	2.1*	3.03**
divers/number overhead	2.32**	0.93 NS	1.8 NS
dives/number overhead	2.84**	3.5***	4.2***
hits/number overhead	4.8***	1.8 NS	1.94 NS
hits/diver	3.96***	0.58 NS	0.5 NS

[a]Colony size based on number of pairs.
[b]There were no hits in small colonies.

number present in the colony. For all measures, the cost decreased with colony size (table 6.11, fig. 6.32), however, significant differences between groups occurred only for the number overhead and the number diving. This is partially due to the maximum number of birds that can dive or hit within any minute without interfering with each other.

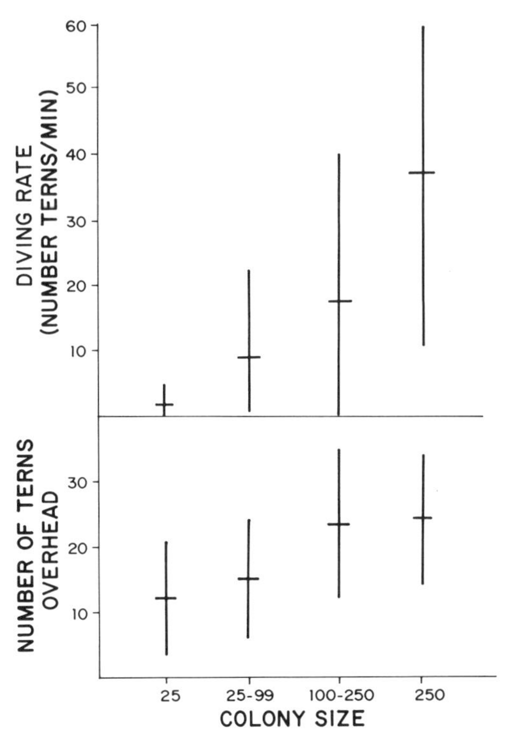

Figure 6.30. Response to human intruders as a function of colony size. Shown are the number of terns mobbing overhead (bottom) and the rate of dives/minute (mean ± 1 SD). This shows an increase in the quality of mobbing with colony size.

Social Facilitation

We accept that social facilitation is occurring when the frequency of a behavior increases disproportionally to group size. In this study, the rate of mobbing behaviors increased with colony size for all measures (table 6.10, fig. 6.33). However, significant differences among groups occurred only for dives/number overhead, hits/number overhead, and dives/number diving. For these measures, being in a larger group resulted in a higher relative rate. These results are consistent with the occurrence of social facilitation. The *t* tests for differences among colony sizes (table 6.11) corroborate these findings.

Table 6.11. Effect of colony size on mobbing behavior of common terns. Given are means ± one standard deviation.

	Tiny	*Small*	*Medium*	*Large*	*Trends*[a]	F[b]	$P<$[c]
Colony characteristics							
number of birds	0–25	25–100	100–250	over 250			
number present	11.7 ± 6.3	56.1 ± 20.4	145.0 ± 32.2	446.4 ± 1.48	+		
number nesting	10.8 ± 11.7	39.5 ± 34.8	75.0 ± 66.0	257.4 ± 1.48	+		
clutch size	1.95 ± 0.69	2.30 ± 0.53	2.29 ± 0.45	2.43 ± 0.42	+		
effect on predator							
number overhead	12.4 ± 9.6	15.6 ± 9.6	23.4 ± 22.0	24.7 ± 10.7	+	21.9	0.0001
number diving	0.5 ± 0.8	3.2 ± 5.5	4.1 ± 4.2	7.2 ± 4.6	+	21.4	0.0001
number of dives	1.3 ± 2.3	9.0 ± 13.7	17.3 ± 24.0	37.8 ± 26.1	+	46.4	0.0001
number hits	0.04 ± 0.3	0.02 ± 0.08	0.29 ± 0.66	0.74 ± 1.3	+	7.8	0.0001
Advantage to individual nesting within 5m							
overhead/pair	2.8 ± 3.2	5.2 ± 4.8	6.8 ± 6.2	3.8 ± 4.8	+, −	6.7	0.0003
divers/pair	0.25 ± 0.45	0.86 ± 1.22	1.6 ± 2.8	1.2 ± 1.3	+	6.26	0.0005
dives/pair	0.7 ± 1.9	3.3 ± 5.5	7.3 ± 14.8	6.3 ± 14.8	+, −	6.11	0.0006
hits/pair	0.02 ± 0.15	0.07 ± 0.24	0.12 ± 0.28	0.11 ± 0.23	+, 0	2.02	NS
Cost to individual nesting in colony							
Overhead/ number present	0.49 ± 0.48	0.31 ± 0.22	0.17 ± 0.18	0.06 ± 0.03	−	26.9	0.0001

divers/number present	0.05 ± 0.10	0.05 ± 0.09	0.03 ± 0.02	0.02 ± 0.01	−	3.9	0.01
dives/number present	0.15 ± 0.3	0.14 ± 0.26	0.11 ± 0.15	0.08 ± 0.06	−	1.7	NS
hits/number present	0.008 ± 0.06	0.004 ± 0.02	0.003 ± 0.004	0.002 ± 0.003	−	3.8	NS
Social facilitation							
divers/number overhead	0.15 ± 0.2	0.28 ± 0.61	0.28 ± 0.27	0.33 ± 0.28	+	1.4	NS
dives/number overhead	0.41 ± 1.1	0.79 ± 1.42	0.97 ± 1.2	1.6 ± 1.1	+	7.2	0.0001
hits/number overhead	0.000	0.02 ± 0.07	0.018 ± 0.05	0.029 ± 0.053	+	2.6	0.05
dives/diver	2.9 ± 0.25	4.08 ± 3.05	4.2 ± 3.1	5.1 ± 1.8	+	2.9	0.03
hits/diver	0.000	0.012 ± 0.36	0.151 ± 0.543	0.086 ± 0.15	+, −	1.0	NS

[a] + = increase with colony size, − = decrease with colony size.
[b] df = 3,260.
[c] NS = not significant.

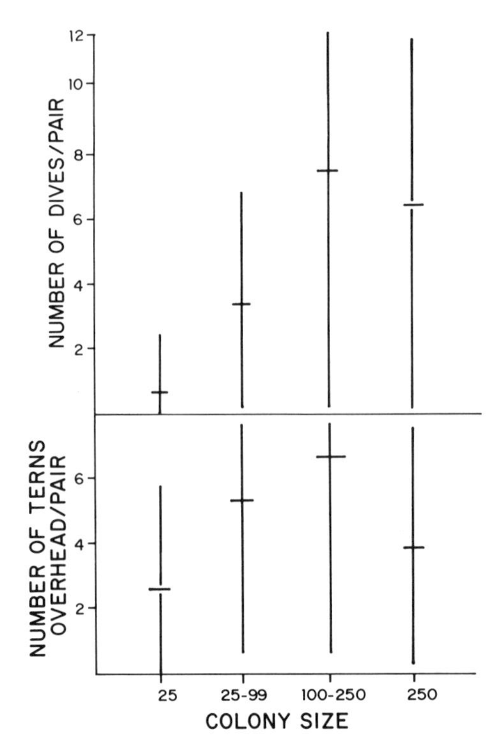

Figure 6.31. Defense advantage to terns nesting within a colony. Shown are number of terns overhead per pair within 5 m and number of dives per minute per pair within 5 m for different sizes of colonies. When controlled for density, the trend shows more birds mobbing and diving for larger than for smaller colonies. The difference is only partly obscured by large variance. Shown are mean (± 1 SD).

Discussion

Seasonal Effect

In this experiment, terns exhibited a clear seasonal effect. Their mobbing increased in frequency and intensity at hatching and early in the chick phase, then declined somewhat late in the chick stage. We suggest that the terns are responding to the increased vulnerability of the chicks. Eggs are less vulnerable, because the terns can prevent predation merely by sitting on them. Older chicks are less vulnerable because they are too large to be taken by some predators, and can run or hide from others. Yet very young chicks

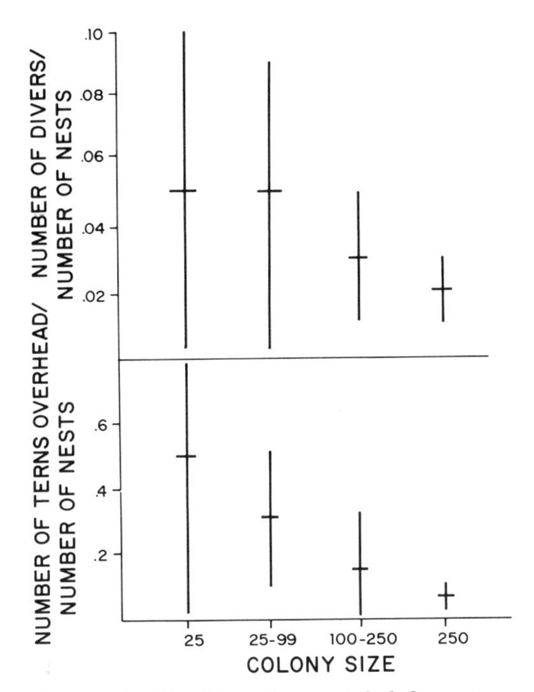

Figure 6.32. Cost of colonial defense to a tern nesting within a colony. Shown are means (± 1 SD) for responses divided by number of nests in the colony.

wander about, are exposed, and are sufficiently small to be easily consumed. Seasonal differences in mobbing have been reported in a variety of species including Eastern kingbirds (Blancher and Roberston 1982), barn swallows (Shields 1984), herring gulls (Burger 1984c), and Arctic terns (Lemmetyinen 1971).

Investment Versus Vulnerability

Andersson et al. (1980) suggested that parental defense should increase directly with increased investment from which it is apparent that investment will increase approximately linearly through the season. Our data show a decline in mobbing during the prefledging period. Thus, older chicks do not appear to elicit more protection than young ones. In fact, at individual nests, we can already detect a decline in mobbing when the chicks are about one week old. We argue, therefore, that the parental response is influenced also

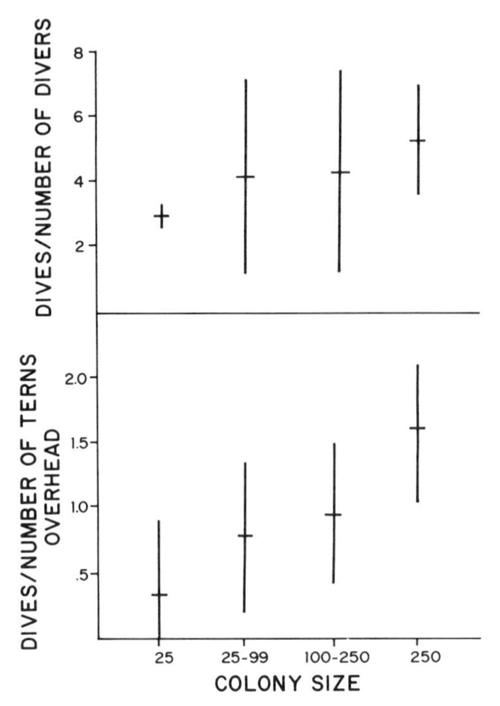

Figure 6.33. Social facilitation of group defense as a function of group size. Shown are means (± 1 SD) for dives per minute per number of birds mobbing (bottom) and number diving (top). Increased intensity, manifested by increased amount of diving, occurs in larger colonies.

by chick vulnerability, rather than simply by the amount of previous investment.

This is not a simple relationship in any case. The number of birds and intensity of mobbing is not a linear function of the season. Our results show a dramatic increase around hatching, as does Shields' (1984) work with barn swallows. In the case of the swallows, there would not seem to be a tremendous increase in risk at the time of hatching. One wonders, therefore, whether the process of hatching, the stimulus of begging chicks, produces a categorical change in parental behavior of which increased mobbing tendency is only one aspect.

Cost-Benefit Analysis

We predicted that one adaptive advantage of nesting in colonies relates to increased benefits from antipredator behavior. To the predator, the frequency and intensity of antipredator behavior should be greater as a function of colony size. For the individual nesting in a colony, the cost should decrease while the benefits increase as a function of colony size. An individual nesting in a large colony should spend less time mobbing while a predator is mobbed by more birds. The field data supported these predictions. Further, these results were not a function of density among colony sizes, as significant differences in density did not occur.

Austin (1946) observed that mobbing was stronger in large tern colonies. Erwin (1988) recently examined mobbing in seven common tern colonies in southern New Jersey and did not find a relationship between attack frequency and colony size or density. However, seven colonies may not be a sufficient sample size for this analysis, and the number of active nests only ranged from 3 to 87. By contrast, in our study of 30 colonies ranging in size from 6 to 500 active nests, we did find significant differences in nest defense as a function of colony size. The optimal colony size at which antipredator behavior, such as attacks, were most frequent (i.e., costs were lowest, and benefits were highest), varied.

In terms of potential impact on a predator, the number of birds overhead did not change from medium to large colonies although the number diving and the number of dives and hits did continue to increase. In terms of advantage to the bird nesting in the colony, the mobbing advantages were optimium in medium colonies, and decreased slightly in large colonies. Thus, over 250 pairs, the advantages did not continue to increase. This may relate to the functional colony size. That is, if a predator enters a colony of fewer than 250 birds, all terns leave their nests and respond to the predator. In much larger colonies, birds no longer act as one unit, for only part of the colony (where the predator is) responds. A similar effect has been noted by Burger (1979b) and Gochfeld (1979a) with respect to synchrony of egg-laying. Beyond a particular colony size (species dependent), the birds began to act as discrete subcolonies rather than as one unit.

The cost of nesting in the colony in terms of contribution to the mobbing effort likewise decreased for all measures as a function of colony size. Thus, although in large colonies more birds engaged in antipredator behavior over the predator at any one time, the cost to each individual was less. The optimal colony size in terms of cost to the individual varied among measures, but generally the cost was significantly higher in smaller colonies.

Overall, terns nesting in large colonies were protected by more frequent and intense antipredator behavior, but each individual contributed less to that defense. Thus, in terms of antipredator behavior, the results of this study are consistent with our initial predictions: One advantage of coloniality relates to increased predator protection at a decreased cost to the individual. It is important, however, to remember that there are other possible advantages of coloniality such as breeding synchrony, identification of safe nesting sites, and information transfer relative to food finding. The existence of an antipredator advantage does not preclude an interaction of several beneficial factors selecting for coloniality.

Social Facilitation

Darling (1938) first proposed that large colonies of gulls were more successful because they laid eggs more synchronously than smaller colonies; synchrony increased reproductive success by decreasing predation, because predators were "swamped" by the overabundance of available food during a brief period of time, thereby reducing the relative risk for any individual chick. Since then, many authors have argued about the occurrence of synchrony and social facilitation (see Gochfeld 1980b for review). Authors frequently have argued about the cumulative effects of numbers of birds on breeding success. Southern (1974) showed social facilitation in ring-billed gulls with respect to copulatory wing-flagging. Similarly, Gochfeld (1980b) showed social facilitation in wing-flagging and other copulatory behavior in common terns. However, this phenomenon is limited to a brief time period during the reproductive cycle. We considered social facilitation of antipredator behavior for the entire breeding cycle, and found that behavior rates were relatively higher when more birds mobbed than when fewer mobbed.

Birds may respond to the sight of an intruder or to the alarm calls of other birds by flying up from the nest, circling over, and even diving at it. If the intruder moves away from the nest, the bird continues to attack the intruder, and the tendency to attack may be enhanced, thereby increasing the frequency, intensity, and duration of the attack any individual makes. Our data are consistent with this model for social facilitation of antipredator behavior.

As individuals persist in their attacks on the intruder, even as it moves away from the nest, one sees more birds circling and diving. This effect is out of proportion to the number of birds in a colony, which leads us to conclude that social factors apart from mere numbers are conducive to effective attacks on intruders or predators.

SEVEN

Piracy in Common Tern Colonies

Piracy, also called kleptoparasitism, is the theft of already procured food from one animal by another, either a conspecific or another species. Piracy may be a severe cost of colonial nesting for some species, because it reduces reproductive success by decreasing food available for their young (Nettleship 1972). Generally, the pirate pursues or ambushes the victim and steals the food from the victim's grasp before it can be swallowed or fed to a mate or young. With so many adult terns returning with fish visible in their beaks (fig. 7.1), the opportunity for piracy is substantial, and some individual birds become specialists at stealing fish, ambushing adults as they fly into the colony or making sudden attacks at the moment a fish is being fed to a young chick.

Piracy is widespread among birds (Brockman and Barnard 1979), particularly among frigatebirds (Nelson 1975; Ashmole 1971), jaegers and skuas (Grant 1971; Andersson 1976; Furness 1977, 1978; Taylor 1979; Furness and Furness 1980; Maxson and Bernstein 1982), gulls (Spaans 1971; Hatch 1975; Fuchs 1977a,b; Pierotti 1980, 1981; Burger and Gochfeld 1981a; Schnell et al. 1983; Hulsman 1984; Panov and Zikova 1987), and terns (Hopkins and Wiley 1972; Hulsman 1976). Some individuals of generally nonpiratic species may specialize in piracy.

In general, the victims of avian pirates are other birds, but piracy has also

Figure 7.1. Common tern adult returning to colony with sand eel in its beak.

been reported on river otters (Kilham 1982) and sharks (French 1982). Seabirds generally pirate from other seabirds, but other species are not immune. For example, gulls will rob herons (Quinney et al. 1981), mergansers (Lamore 1953), whooper swans (Kallander 1975), lesser scaup (Siegfried 1972), American coot (Bartlett 1957), eiders (Ingolfsson 1969), grebes and loons (Meinertzhagen 1959), starlings (Burger and Gochfeld 1979) and shorebirds such as dunlin (Payne and Howe 1976), Eurasian lapwing (Barnard and Stephens 1981; Kallander 1977, 1979), African black oystercatcher (Hockey 1980), Eurasian oystercatcher (Dummigan 1977), and plovers (Vader 1979a,b; Barnard et al. 1982, Thompson 1983). Occasionally, the tables are turned. Shorebirds such as the lesser sheathbill regularly kleptoparasitize seabirds such as rockhopper penguins (Burger, A. E. 1981), and white pelicans have been reported to pirate from gulls (O'Malley and Evans 1983). Interspecific piracy can have important potential negative impacts on bird populations, as Nettleship (1972) pointed out for herring gulls at a Newfoundland puffin colony and Veen (1977) reported for black-headed gulls on Netherlands sandwich tern population.

To be cost-effective, piracy should occur when and where potential victims are abundant. Otherwise, the cost of attempted piracy, in terms of time and energy, would be high, and might exceed what the pirate would expend foraging directly on its own.

Piracy behavior has been extensively studied at seabird colonies where pirates steal from adults bringing food to young (Hulsman 1976; Andersson 1976; Arnason and Grant 1978; Gauzer and Ter Mikhaelyan 1987; Delude et al. 1987; Panov and Zikova 1987). Pirates sometimes breed in the same colonies with their victims. However, piracy can also occur in feeding flocks far from the colony (Safina and Burger 1985). In general, these studies have dealt with interspecific rather than intraspecific piracy (Burger and Gochfeld 1981a, Goss-Custard et al. 1982, Rockwell 1982). Skuas, gulls, and terns may pirate from one or several species breeding in a colony. These pirates often concentrate near the edge of the colony and wait for adults to bring back large, visible food items to feed their young. Rockwell (1982) performed a cost-benefit analysis of intraspecific piracy of glaucous-winged gulls on tidal flats away from the breeding colonies.

In common tern colonies, both inter- and intra-specific piracy are prevalent. The two have been compared for black-naped, roseate, and crested terns in Australia (Hulsman 1976), and common terns in Maine (Hopkins and Wiley 1972) and New York (Hays 1970). In these studies, the relative frequency and success of intraspecific piracy was low. Indeed, it may not even occur every year (Hays 1970).

In contrast, intraspecific piracy in ring-billed gulls and Western gulls is an important component of food acquisition by some parents and of food loss by other parents (Elston et al. 1978; Pierotti 1980, 1981; Elston and Southern 1983). In ring-billed gulls, the feeding of chicks is initiated by begging from chicks, and the parents respond by regurgitating fish. Pirates steal fish when it is being transferred from parents to young. The presence of pirates affects parental behavior, and parents devote more time to the initial phases of regurgitation, often reswallowing the food if pirates are near. The percent of successful piracy attempts ranged from 18 to 51 percent, but it was difficult to determine the percent of total food destined for chicks that was lost to pirates. Partially, this is due to the regurgitation method of feeding in Gulls. Gulls can regurgitate different quantities of food at any feeding and can control the amount and timing of the food provisioning. In contrast, most terns bring back a single intact fish in their bills where it is clearly visible to pirates and scientists alike. Thus, terns provide ideal examples to study intraspecific piracy as a cost of colonial nesting.

Ecological Correlates of Piracy

Brockman and Barnard (1979) reviewed the ecological correlates of avian piracy, arguing that for piracy to become common several conditions need apply: 1) large concentrations of hosts, 2) either large quantities of food, or a food shortage, 3) large, high quality food items, 4) predictable food supply, and 5) visible food. Piracy is indeed more common where there are concentrations of hosts with large food items, as at breeding colonies or in foraging flocks. Large food items are more subject to piracy than small food items (Hopkins and Wiley 1972; Dunn 1973; Veen 1977; Gochfeld and Burger 1981b). Brockman and Barnard (1979) suggest that piracy should be more common where food supply is predictable, although we suggest that in breeding colonies where conspecifics are always present, parents can easily switch to piracy when potential victims are abundant. Parents that are guarding chicks can watch for food items to steal from the neighbors without abandoning their own chicks. Furthermore, chicks can steal from neighboring chicks.

Several authors have noted that kleptoparasitism is more common during periods of food shortage, during inclement weather, or for younger birds (Munro 1949; Zusi 1958; Bergman 1960; Veen 1977; Burger and Gochfeld 1981a). We suggest food would not necessarily have to be in short supply, only that there be a higher benefit than cost to piracy versus obtaining food directly.

7.1 MODELS FOR PIRACY BEHAVIOR IN A TERN COLONY

The ecological correlates of piracy behavior can be used to construct models suggesting how the frequency of piracy should vary over time in common tern colonies. Food, generally fish, is brought back to the colony in abundance during the courtship feeding and chick feeding stages (fig. 7.2). There continues to be a low level of courtship feeding during most of the breeding season, not only through incubation (Nisbet 1973a), but because some pairs lose mates and pair again and there are new recruits to the colony who might be seeking mates. Once chicks hatch, the quantity of fish brought back to the colony increases rapidly, and this continues at an increasing rate due to increasing needs and numbers of growing chicks.

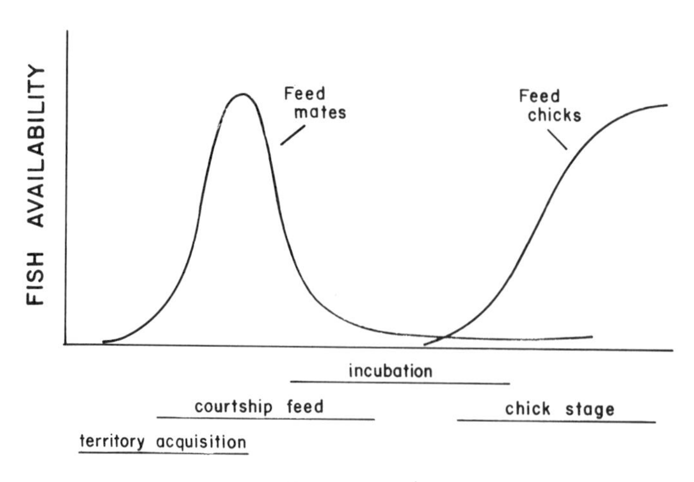

Figure 7.2. Model of fish availability to pirates in common tern colonies as a function of breeding season and activities.

A number of different factors might interact to produce opportunities for piracy (defined as piracy potential). The curve for a colony is determined by increasing numbers of chicks entering the log phase of growth. Thus food available for piracy increases during the courtship feeding phase, decreases during incubation, and then rises dramatically during the chick rearing phase. For individual parents, the piracy potential curve is similar to the colony curve during the courtship stage, but during the chick stage we suggest a number of factors determine piracy potential including: age of chicks, number of chicks, size of fish being brought back, and age of neighboring chicks (who might themselves be pirates). These factors all interact in a largely additive fashion resulting in a piracy potential increase for individual parents as the age of their chicks increases (fig. 7.3).

Presumably parents should increase the size of the fish they bring back as their chicks get older, an assumption we confirmed during this study. However, fish brought back for courtship feeding were also larger later in the season. Small chicks may have difficulty swallowing even small fish (fig. 7.4), and large fish may protrude from their beaks for hours as digestion proceeds (fig. 7.4). Annett and Pierotti (1989) found that hatching of chicks caused Western gulls to shift their feeding from garbage to fish. Veen (1977) reported that Sandwich terns bring fewer small fish to older chicks and Cairns

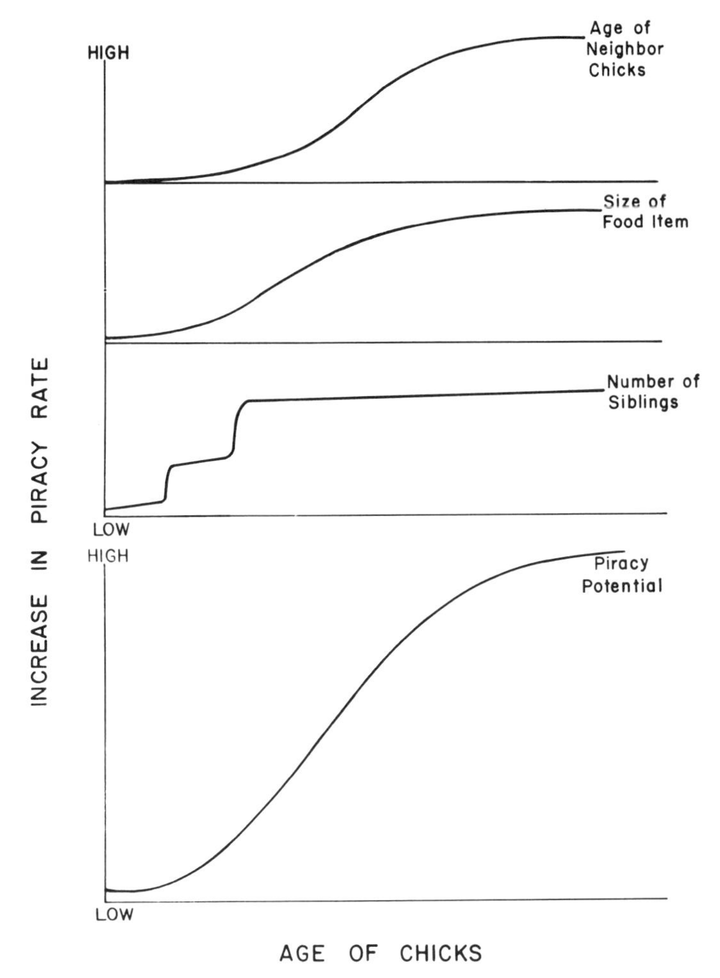

Figure 7.3. Factors affecting variation in the potential piracy rate on a hypothetical parent-chick family as a function of chick age. The top three factors are additive, producing a summation curve showing increased piracy potential with age of the chicks in the bottom graph. Age of neighbor chicks contributes most when neighboring chicks are older than chicks in the family. Size of food item is inversely related to chick handling times. Number of siblings contributes most when family is older and larger.

Figure 7.4. (top) Day-old common tern chick swallowing a medium-size sand eel. This may take many seconds and may require several attempts during which piracy may occur. (bottom) Five-day old chick with large sand eel protruding from its mouth may invite piracy if it fails to hide while digesting its meal. (top-facing) Two week-old siblings struggling over a large sand eel which was eventually swallowed by the chick on the left.

(1987) found black guillemots increased food size brought to chicks as they grew. Wiggins and Morris (1987) reported that with increasing age of the chicks, male common terns increased the size of the prey they brought back, but females did not. We believe the age of a parent's chicks influences the risk of piracy, because as chicks get older not only do they require more (and larger) food items, but presumably there is increased sibling competition for food. When siblings fight over a particular food item, there is an increased opportunity for piracy by adults or neighboring chicks (fig. 7.4). At this time, both parents and chicks may be less vigilant for potential pirates since their attention is diverted to the dispute over the food.

We also suggest that piracy is a function of parental vigilance as well as piracy potential (fig. 7.5). Presumably as piracy potential—and the resultant piracy attempts—increase, parents will become more vigilant and develop behaviors to reduce piracy attempts and piracy success. Common terns almost always bring back a single fish transported in their bill. Thus, the fish is conspicuous, and losing a single fish necessitates a return to the feeding grounds.

Food supply affects piracy rates. Piracy attempts should increase when food supply is low (Brockman and Barnard 1979; Pierotti 1980). With food difficult to find, some individuals may switch to piracy as the method of food acquisition requiring the least time and energy. However, if food being

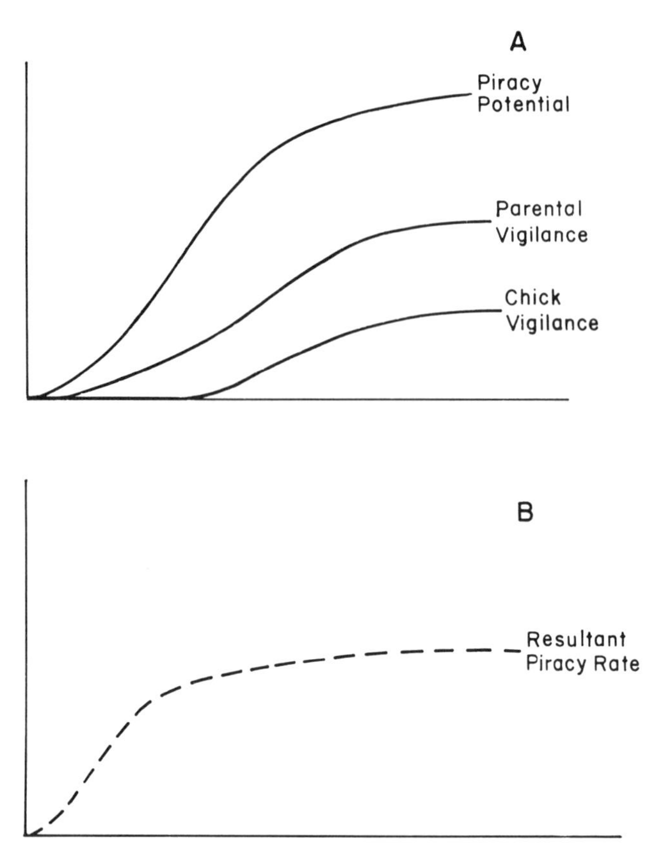

Figure 7.5. Relationship of piracy potential, and parental and chick vigilance as a function of stage in breeding season. (A) When piracy potential is high, vigilance increases—first for parents and then for chicks—leading to a lower curve for resultant piracy rate (B).

brought back to the colony is very scarce, opportunities for successful piracy would be few, and parents may choose to follow successful terns to the foraging grounds (passive information transfer). Conversely, when food supply increases, the opportunities for piracy increase, and parents may be less vigilant when the time required to replace a lost fish is low.

We predicted that food size would influence piracy, much as Veen (1977) indicated for Sandwich terns, where black-headed gull pirates stole larger fish preferentially. If fish size increases during the season, this will confound any other seasonal effect.

7.2 FOOD AVAILABILITY TO PIRATES

In this chapter, we are not primarily concerned with feeding behavior of common terns, but with piracy as a cost of coloniality for terns nesting in colonies. Whereas most studies of piracy have focused on the strategies of the pirates, we were more interested in the avoidance strategies of the potential victims. In the following sections, we will discuss fish availability in the colony (fish brought back by terns), piracy rates, and the factors affecting piracy. We use the following definitions:

Percent Pirated: The proportion of fish brought to the colony, study area, or nest that is pirated.

Success Rate: The proportion of piracy attempts that end with successful thefts.

Piracy Rate: The number of successful piracy events that occur per unit time.

Methods

We examined food availability and piracy behavior of common terns in 1982 and 1984 at West End Beach and at the Pettit and Carvel marsh colonies. The frequency of piracy varies dramatically from year to year, but in both 1982 and 1984, piracy occurred commonly in these colonies, and we were able to examine piracy behavior from early May through July. In 1982, we made observations on groups of terns and, in 1984, we followed individual pairs in each study site (throughout the season at West End and intermittently at Pettit and Carvel). For all fish brought into the study sites during our observations we recorded: date, time of day, fish length (relative to bill length), and age of chicks. Terns returning with fish fly closer to the dunes, and are thus easy to observe. With binoculars, we could readily estimate fish size relative to the terns' bill length (ca 3.0 cm). In 1984, we also recorded tidal stage.

To evaluate factors contributing to the frequency of piracy attempts or the piracy percent, we used multivariate statistics (SAS 1982). An independent variable is said to contribute to explaining the variance if it enters the general linear model at the $P < 0.05$ level.

Models for Fish Delivery and Piracy

Table 7.1 shows the multivariate models for a number of important variables related to fish transport and piracy. We used the variable season (corresponding to date) as the surrogate for stage in reproductive cycle. Because phenomena vary with date in a curvilinear fashion we also constructed the variable date^2 (date $\times$ date). We also constructed other synthetic variables such as date^3 and tide time^2 and tide time^3 in order to estimate the polynomial regression of piracy on these variables.

Over the entire reproductive season the variance in number of fish brought back by common tern parents was explained by time of day, season, time since the last low tide (lag), and date^2 $\times$ lag intereaction (table 7.1). Generally the number of fish per unit time fed to mates or young is influenced by the same variables.

The variability in piracy attempts (all piracy tries, regardless of their fate) was explained by date^2, time of day, number of fish brought back, fish size, and date^2 $\times$ lag (table 7.1). Fish size is thus a significant variable influencing piracy attempts. The percent of total fish brought back that are pirated is affected by season, mean fish length and date $\times$ tide interactions. Variability in piracy success is explained only by fish size. This indicates that piracy success itself doesn't vary linearly by date, but is directly related to the proportion of large fish. Since fish length was an important component of piracy behavior, we also examined the factors influencing variability in mean fish length (table 7.1) and found fish size varied by date, time, season, and tidal state.

We also examined the effect of time, date, age of chicks, and date $\times$ age on feeding and piracy behavior of common terns with the use of linear regression models on raw date (table 7.2). Again, the number of fish brought back by parents (and the number of fish eaten by their chicks) was influenced by these three variables. Interestingly, date and age also influenced the number of fish eaten by chicks when sibling fights were involved. Thus, the effects of season and age are additive (see below).

The number of fish lost to pirates was influenced by time and chick age, but date $\times$ age also entered the model for fish lost to neighbor chicks (table 7.2). This reflects the ability of older neighbor chicks to pirate food from younger chicks, and this can occur only later in the season (see below). The models discussed above serve as a basis for our discussion of food availability and piracy behavior in the rest of this chapter.

Table 7.1. Factors affecting fish delivery, piracy, and fish sizes for common terns during the reproductive season (West End 1984).

	Total Fish Brought Back	*Fish Eaten by Mate or Young*	*Piracy Attempts*	*Percent Pirated*	*Success Rate of Pirates*	*Mean Fish Length*
Model[a]						
R^2	0.22	0.16	0.47	0.39	0.32	0.36
F	6.76	4.81	8.72	5.23	14.99	33.6
P	0.0001	0.001	0.0001	0.002	0.001	0.001
df	5125	4126	5125	4126	3127	4126
Factors entering, F (P)						
Date[2,e]	NS	NS	6.37(0.01)	NS	NS	55.9(0.001)
Time	5.21(0.02)	12.87(.0005)	8.89(0.003)	NS	NS	13.30(0.0004)
Season	6.93(0.009)	3.04(.08)	NS	7.18(0.001)	NS	16.82(0.0001)
Total fish	—[d]	NS	59.27(0.0001)	NS	NS	NS
Mean fish length	NS	NS	NS	4.62(0.03)	3.37(0.06)	—[d]
Total small fish	—[e]	NS	17.31(0.0001)	10.49(0.001)	33.83(0.0001)	NS
Total large fish	—[e]	NS	NS	NS	20.02(0.0001)	NS
Time from low tide (Lag)	20.55(0.0001)	NS	NS	NS	NS	5.97(0.01)
Time X lag	3.01(0.08)	16.02(.0001)	NS	NS	NS	NS
Date X lag	20.93(0.0001)	12.95(.0005)	12.16(0.0004)	12.22(0.0007)	NS	72.05(0.0001)

[a] Models constructed by using stepwise multiple regression with a Max R procedure. The best variables were then entered in order in a GLM Procedure using the variables that contributed significantly to the model.

[b] Fish under 2.5 cm.

[c] Fish over 7.5 cm.

[d] Not included in model.

[e] Date squared is a polynomial term showing a curvilinear change.

Table 7.2. Models explaining the fate of fish brought back to feed young common terns at the nest. Data are from 1982 for chicks 0–28 days old.

	Model[a]			*Factors, F (P)*			
	F	R^2	P	*Time*	*Date*	*Age of Chick*	*Date × Age*
Fish brought back by parents/hr	13.75	0.36	0.0001	15.65(0.0001)	23.1(0.0001)	11.9(0.0001)	NS
Fish eaten by parent's chicks/hr							
without sibling fights	3.14	0.08	0.0001	3.24(0.001)	10.2(0.001)	2.31(0.03)	NS
after sibling fights	13.9	0.28	0.0001	12.9(0.0001)	21.9(0.0001)	13.8(0.0001)	14.7(0.0001)
Lost to piracy/hr							
taken by neighbor's chicks	5.49	0.13	0.0001	2.65(0.001)	NS	10.3(0.0001)	4.88(0.02)
taken by adults	3.76	0.09	0.0001	5.69(0.0001)	NS	2.42(0.02)	NS
unsuccessful attempts	16.90	0.32	0.0001	27.9(0.0001)	NS	7.85(0.0001)	NS
Fate of fish							
eaten by parent's chicks	3.03	0.11	0.0001	NS	3.24(0.07)	4.53(0.0002)	7.57(0.0006)
lost to pirates	3.03	0.11	0.0001	NS	3.24(0.07)	4.53(0.0002)	7.57(0.0006)

[a] *df* = 16,571. Based on 587 feedings (80 hrs observation, 1984).

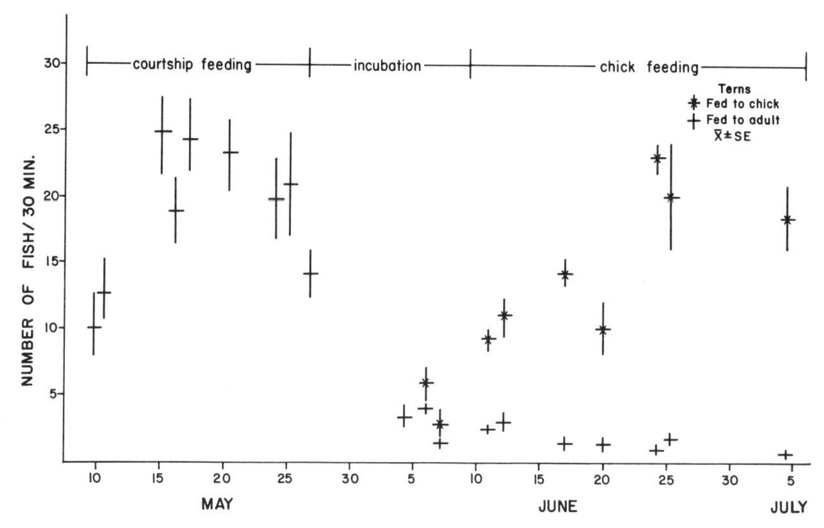

Figure 7.6. Seasonal availability to pirates of fish brought back by common terns (May 10–July 5, 1982, West End) (mean ± 1 S.E. shown).

Seasonal Fish Delivery Rates

We counted and estimated the size of fish brought back to the colony for twenty pairs nesting in the center of the West End colony (N = 4523 fish), twenty pairs at Pettit (N = 284 fish), and for 12 pairs at East Carvel (N = 126 fish). Observations were made in three- to five-hour watches at different times of the day from early May (before laying) through early July (fledging) at West End and intermittently at the other colonies. As expected, large numbers of fish were brought back to the colony only during the courtship feeding and chick stages (fig. 7.6). During incubation fish brought back to the colony decreased.

The period when few fish are being brought back to the study site is short, because not all pairs initiate incubation at the same time. Thus most pairs initiated incubation in late May, but some began in mid-May or early June. A low level of courtship feeding continues during incubation, and there is also courtship feeding involving either newly established pairs or renesting birds who lost their eggs and chicks and are beginning a new cycle (fig. 7.6). Newly established pairs may be young birds whose breeding activities have been delayed due to late arrival at the colony, difficulty in finding a mate or establishing a territory, or inexperience in obtaining enough food to adequately provision the female.

Colony Variations in Fish Availability

The rate at which fish are carried into the colony depends not only on food availability, but on the demand created by females (Smith 1980) and growing young. We compared the mean number of fish per time brought back to the three colonies (table 7.3), and found no differences in the number of fish brought back per breeding pair between West End and Pettit. There was a strikingly lower rate of fish brought back to Carvel, though it did not reach significance ($P > 0.05$). Differences might be expected if fish size varies and if terns in one colony consistently brought back significantly larger than average fish. They might therefore bring back fewer fish per hour. This was not apparent in our data, where if anything, the mean size of fish at Carvel was slightly smaller.

Fish availability to pirates may have two components: quantity and size. Small fish may be less available because they are too small to steal, or conversely, too easy for potential victims to retain. We determined fish size by comparing fish lengths to the size of the tern's bill (about 3.0 cm). Common terns brought back fish that ranged from 1.3 to 12.0 cm, with most fish being from 4 to 7.5 cm in length (fig. 7.7).

Seasonal Changes in Fish Size

Overall, during the breeding season, terns brought back fish that averaged 5 cm. Presumably terns should bring back larger fish to females or large chicks

Table 7.3. Variations in fish delivered to common terns as a function of colony (summed for season). Data are means ± 1 SD.

	West End	*Pettit*	*East Carvel*
Habitat	Sand	Salt marsh	Salt marsh
Number of pairs in study area	20	20	12
Number of samples[a]	263	18	18
Mean number of fish brought back per half hour (±SD)	17.2±11.7	15.8±6.5	7.0±3.8
Mean size of fish (cm)	5.5±1.6	4.5±0.5	4.4±1.2
Mean percent of fish (±SD) that were pirated	7±8	1±2	2±7
Piracy success (percent) (successes/attempts)	52±39	29±32	25±35

[a] Half-hour samples.

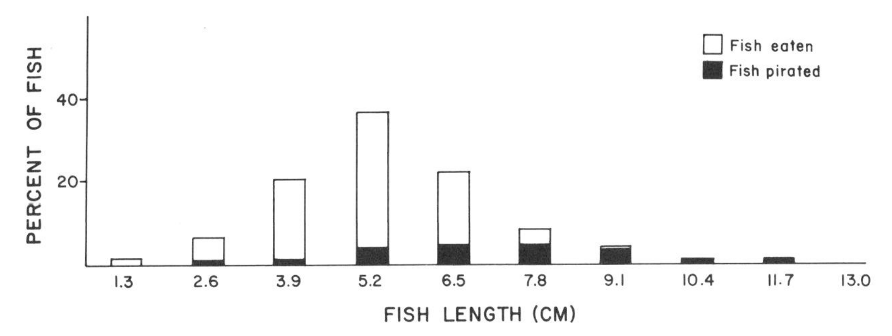

Figure 7.7. Length distribution for fish brought back by terns to the West End colony (N = 2750). Open bars show fish eaten by mates or chicks; shaded bars show fish pirated. The data were summed over the period from May 10–July 5, 1982, and the total height of all bars = 100 percent.

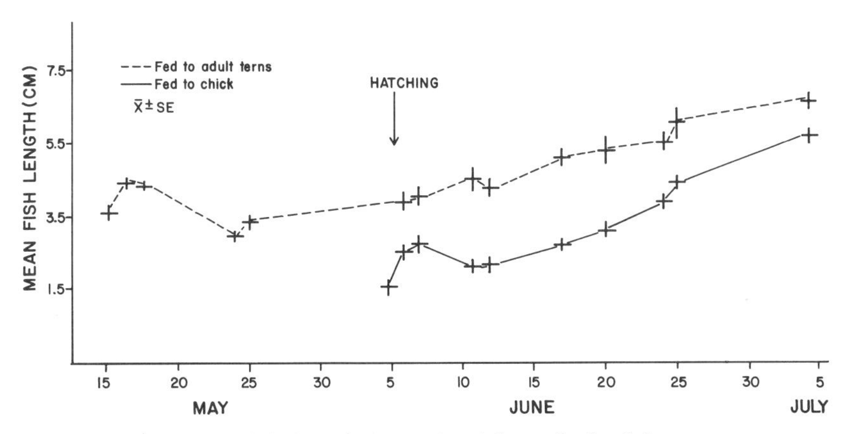

Figure 7.8. Mean fish length (± 1 S.E.) brought back by common terns for mates and chicks as a function of date (West End 1982 for 20 pairs, N = 4530).

than to small chicks. Field data support these hypotheses (fig. 7.8). Fish fed to females are larger than those fed to young, at least until early July. Very young chicks can only swallow very small fish. They have trouble manipulating larger fish, and even if they can swallow them the tail protrudes from the bill for an hour or more while digestion proceeds (fig. 7.4). Larger chicks can swallow larger fish more quickly and completely, and parents with older chicks tend to bring back larger fish, thereby necessitating fewer trips. Although courtship feeding decreases once pairs have eggs, some pairs in the colony are still courtship feeding. Moreover, females guarding chicks sometimes eat the fish that males bring back for the chicks.

7.3 BEHAVIOR OF COMMON TERN PIRATES AND VICTIMS

Courtship activity, including pair formation (if it is a new pair), pair-bond maintainance, and precopulatory displays all begin or take place on the territory, although terns performing fish flights cover a wide area over the colony (Palmer 1941a). Common terns bring back fish in their beak, usually one at a time, to the female who remains on the territory. Often, the males leave again immediately or within a minute to search for another fish. Potential pirates cruise over the colony and chase males returning with fish. Occasionally, pirates attempt to steal fish when males are offering the fish to their mates.

During the chick phase, both parents bring back fish. Fish size is initially small and increases as chicks get larger. During this phase, pirates attempt to steal fish from flying parents, but some also attempt piracy during the transfer of fish from parent to young and again while the chick is trying to swallow the fish (see fig. 7.4).

As chicks get older the piracy interactions become very complex (table 7.4). Fish can be eaten by the young with no mishap, or there can be several piracy attempts before the food is consumed. Piracy attempts can be by adults or by neighboring chicks. If neighboring chicks are large, their chances of success are high, because very young chicks often give up the fish easily. We have even seen chicks carried away and dropped when a pirate seizes the tail of a half-swallowed fish.

In the West End study area, courtship feeding accounted for 21 percent of the fish brought back to the colony over the season, and 16 percent of all fish were pirated during courtship feeding (table 7.5). During the chick phase, fish were stolen by both adults and neighboring chicks. Indeed, for some locations, neighbor chicks accounted for considerable losses. In 1984, when we examined piracy intensively during the chick phase, neighbor chicks pirated up to 10 percent of the fish being brought back to chicks 21 to 28 days old (fig. 7.9). This equaled the loss to adult pirates.

Table 7.4. Complexity of food delivery. Fate of fish delivered to chicks in five nests with flying young (July 15, 1984).

	Initial Fate	*Secondary Fate*	*Tertiary Fate*				*Number*
Fish	Eaten by young						38
	Sibling fight	Eaten by young					22
		Piracy attempt	Eaten by young				3
			Eaten by pirate				14
	Piracy attempt	Eaten by young					9
		Eaten by pirate					12
		Piracy attempt	Eaten by young				6
			Eaten by pirate				4
			Piracy attempt	Eaten by young			2
				Eaten by pirate			3
				Piracy attempt	Eaten by young		1
					Eaten by pirate		3
					Piracy attempt	Eaten by young	3
						Eaten by pirate	4

Table 7.5. Comparison of fish eaten by adult, young and pirates for common terns at West End Beach study area.

	Adult[a]	*Young*[b]	*Pirates*[c]
Overall percent of fish eaten	21 ± 21	63 ± 10	16 ± 21
Mean number fish eaten per 30 min.	10.0 ± 12.8	8.6 ± 6.8	2.9 ± 2.3
Mean (± SD) fish length (cm)	4.9 ± 2.2	4.5 ± 1.5	6.3 ± 2.7

[a] During courtship feeding (May).
[b] During chick feeding stage (June–July).
[c] Throughout season (May–July, combined data 1982 and 1984).

7.4 FACTORS AFFECTING PIRACY BEHAVIOR IN COMMON TERNS

Seasonal Effects

We chose an area of the West End colony where piracy occurred. For the 20 pairs in this study area the percent of piracy varied seasonally (fig. 7.10), as did the piracy rate (fish per half hour; fig. 7.11). The piracy rate was high during courtship feeding, low during incubation, and highest during the chick stage. During the chick stage, the rate was high initially when chicks were very young, then it dropped in mid-June, and then slowly increased as chicks became larger (and ate larger fish).

Because of the increase in piracy during the chick phase (fig. 7.11), we examined this phase of the cycle in detail in 1984 to determine the factors affecting piracy at that time. Absolute piracy rates varied depending on the time of day (fig. 7.12), but generally accounted for 0.2 to 0.8 fish per hour per pair. The percent of fish that were eaten by the parent's own chicks decreased seasonally from late June through mid-July in 1984 (fig. 7.13). Thus, for older chicks in 1984, fish loss was quite high, and inflicted a considerable cost on parents. The number of sibling fights over fish is not a direct cost of coloniality, and is discussed here only in passing.

Fish Delivery Rate and Piracy

Initially, we had predicted that piracy rates should be related to the number of fish delivered per hour. We used this as our measure of fish available for pirates to steal. However, when we examined the data from May to July, 1982 (fig. 7.14) it was clear that the percent of fish pirated did not relate to

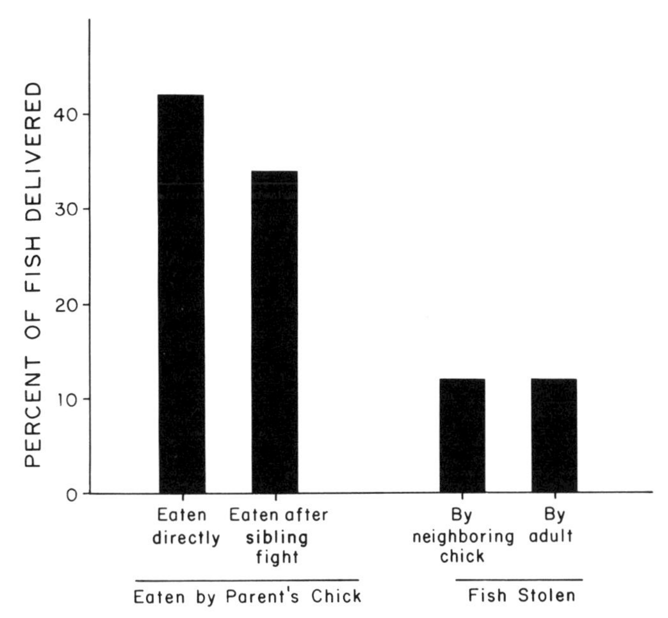

Figure 7.9. Percent of fish (N = 3576) that were eaten by parent's chick with or without a sibling fight, pirated by common terns, and the percent stolen by neighboring chicks or by adults.

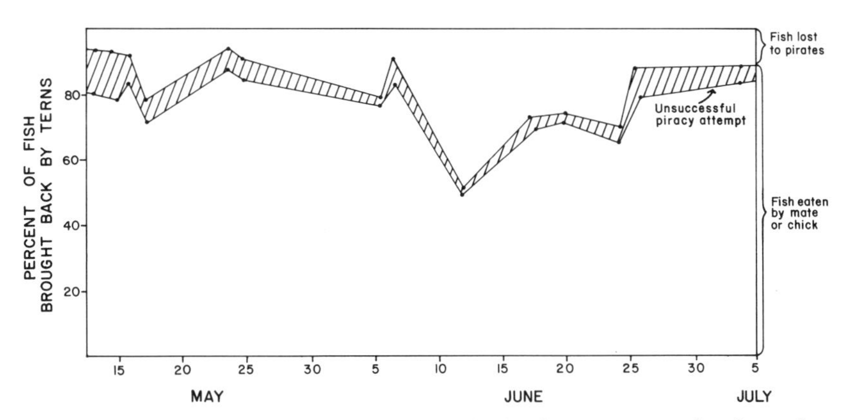

Figure 7.10. Percent of total fish brought back to West End colony that were pirated (N = 715) as a function of season.

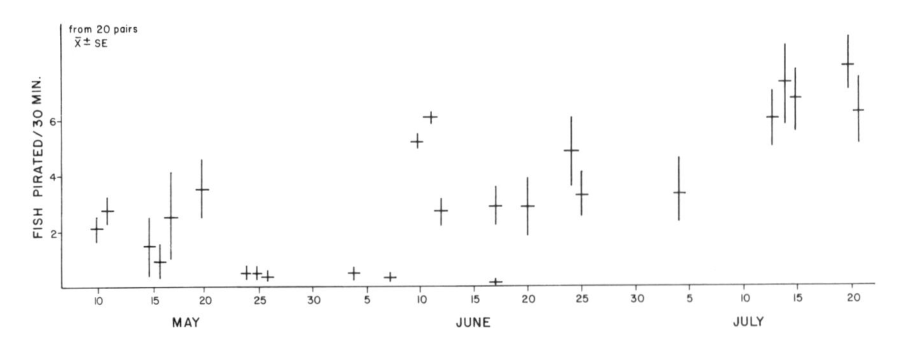

Figure 7.11. Piracy rate (fish pirated/30 min) in common terns as a function of season (N = 715). Given are means ± standard error for 30-min. time blocks.

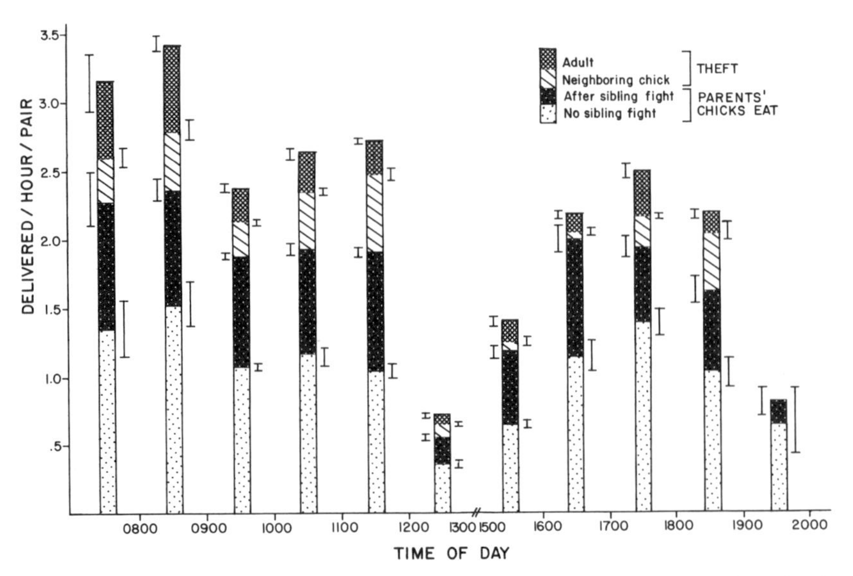

Figure 7.12. Fate of fish delivered by common terns during the chick phase as a function of time of day (N = 2816). Total food delivery rates (fish per hour per pair) is shown by total bar, food pirated by adults (netted bar), food pirated by neighboring chicks (hatched bar), food eaten by the parent's chick after a sibling fight (white dots on black) and food eaten by parent's chick without a fight (dotted bar). Lines alongside bars indicate SD associated with each mean value.

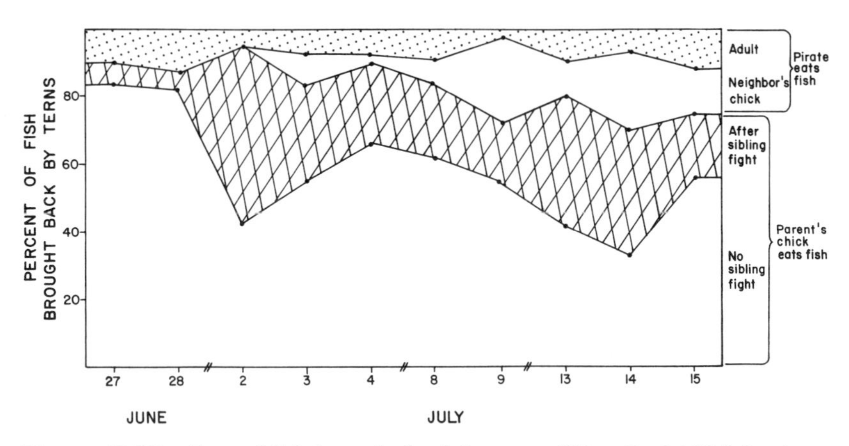

Figure 7.13. Fate of fish brought back by terns (West End 1984) by date (N = 4530). Percent of fish eaten by a parent's chick with and without a sibling fight, and percent lost to pirates (adults and young). Shows seasonal increase in importance of chicks as pirates.

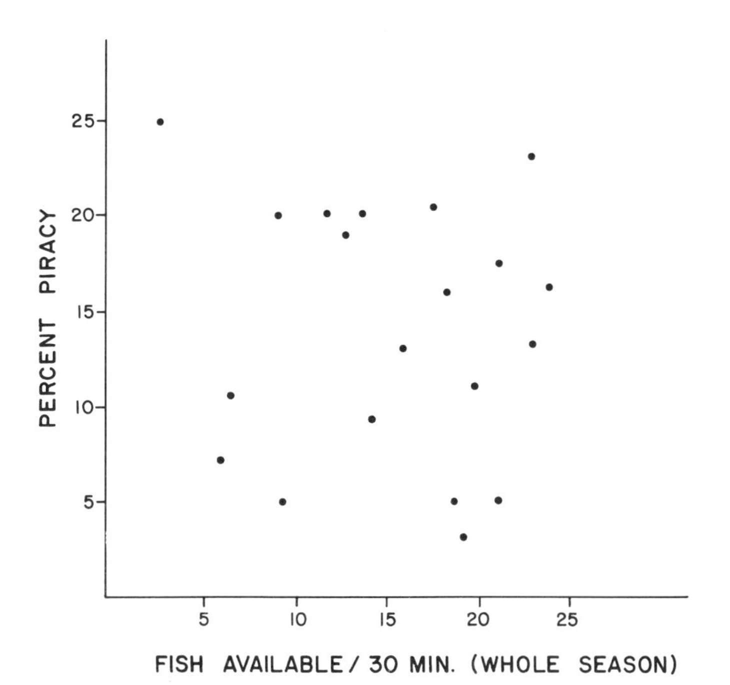

Figure 7.14. Percent piracy as a function of fish delivered per 30 min. for common terns (entire 1982 season), showing lack of correlation.

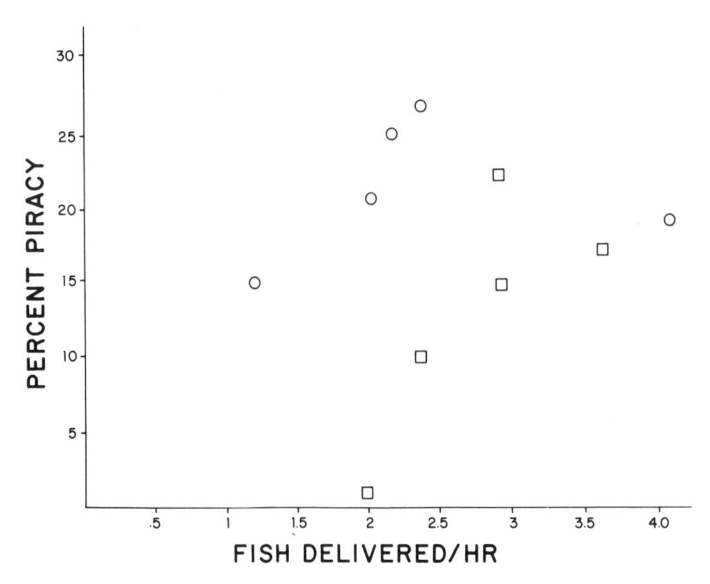

Figure 7.15. Percent piracy as a function of fish delivered to common terns during the chick phase (1984). Squares represent piracy from chicks under 15 days; circles, from chicks over 15 days.

fish availability per se either linearly (*tau* = 0.12, $P > 0.05$) or curvilinearly.

We found evidence that piracy was dependent on other factors such as fish size (see below) that might vary seasonally. Therefore, during 1984, we examined piracy rates only during the chick stage, and found that the percent of fish pirated showed no better correlation with fish availability (fig. 7.15, tau = 0.16, $P > 0.05$). However, chicks under 15 days of age that were fed smaller fish generally had fewer piracy losses than chicks over 15 days of age, even though the latter were more proficient at swallowing quickly.

Daily Variations

In 1982, we examined fish delivery and piracy rates in terns by time of day and found that both varied significantly (fig. 7.16). For the colony as a whole, fish delivery rates were high very early in the morning, in midmorning, and again around 13:00. Piracy rates in terns were low until almost noon; then piracy increased markedly and remained high for the rest of the day. These morning versus afternoon differences were significant (table 7.6).

However, once chicks were about ten days of age, the piracy pattern

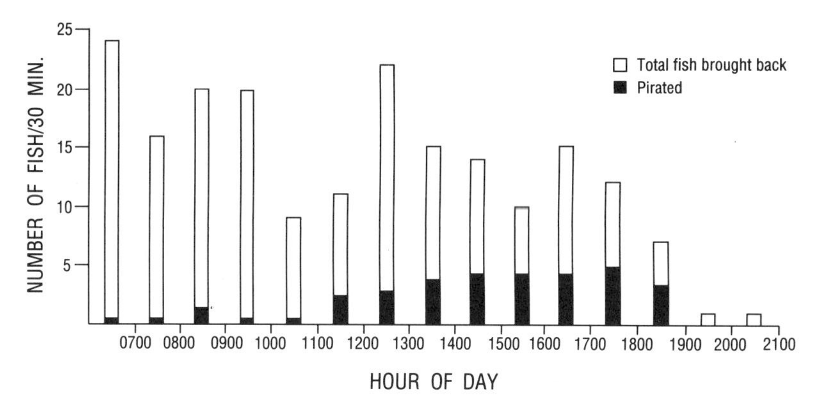

Figure 7.16. Fish delivery and piracy rates in common terns by time of day (1982, West End, N = 4530).

Table 7.6. Comparison of morning and afternoon feeding bouts and piracy in common terns. Kruskal-Wallis test used to contrast A.M. versus P.M. (see fig. 7.16).

	Kruskal-Wallis X^2	P	*Time with Higher Value*
Total fish brought back by parents/hr	4.04	0.05	morning
Fish eaten by parent's chicks/hr	3.56	0.05	morning
Fate of fish			
percent eaten by parent's chicks	7.82	0.005	morning
percent lost to neighbor's chicks	4.41	0.03	afternoon
percent lost to adults	7.56	0.006	afternoon

changed and the pirates became active all day. Figure 7.17 shows the results for the 1984 data. Piracy rates increased by date (and thus with age of the chicks). Again, sibling fights were much more common in older chicks, and contributed to some of the increased piracy because the attention of the chicks was diverted to fighting with their sibling rather than avoiding piracy (see below).

Habitat Variations

Since terns flying with fish in their bills within the colony are conspicuous whether they are flying over open sand or vegetation, aerial piracy rates

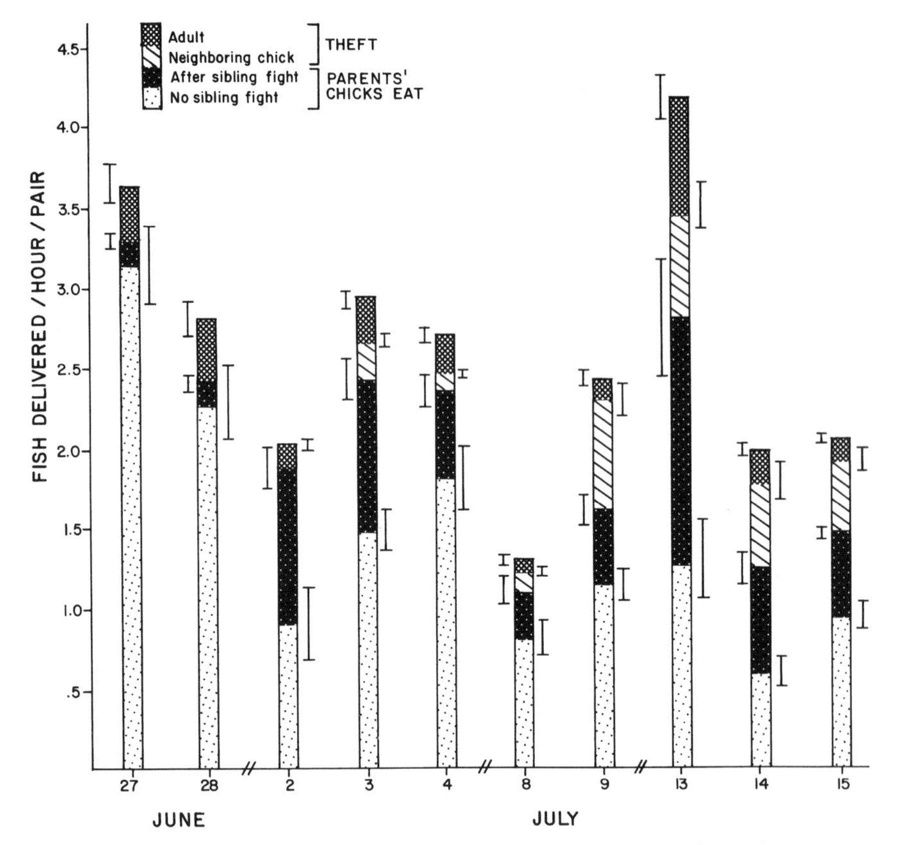

Figure 7.17. Fate of fish brought back by terns (West End 1984, N = 2370) from June 27–July 15. Mean fish delivery and piracy rates in common terns when their chicks are over 10 days of age. Netted bar = piracy by adults; hatched bar = piracy by neighbor chicks; dotted black bar = eaten by parents chick after sibling fight; dotted white bar = eaten by parent's chick directly (West End, 1984). Lines alongside bars show SD associated with each mean value.

should be similar. Our data confirm this hypothesis. However, piracy rates may be expected to vary by habitat since vegetation cover may make it more difficult for aerial pirates to see chicks engaged in fights with siblings over food or to see small chicks that might be having difficulty swallowing large fish.

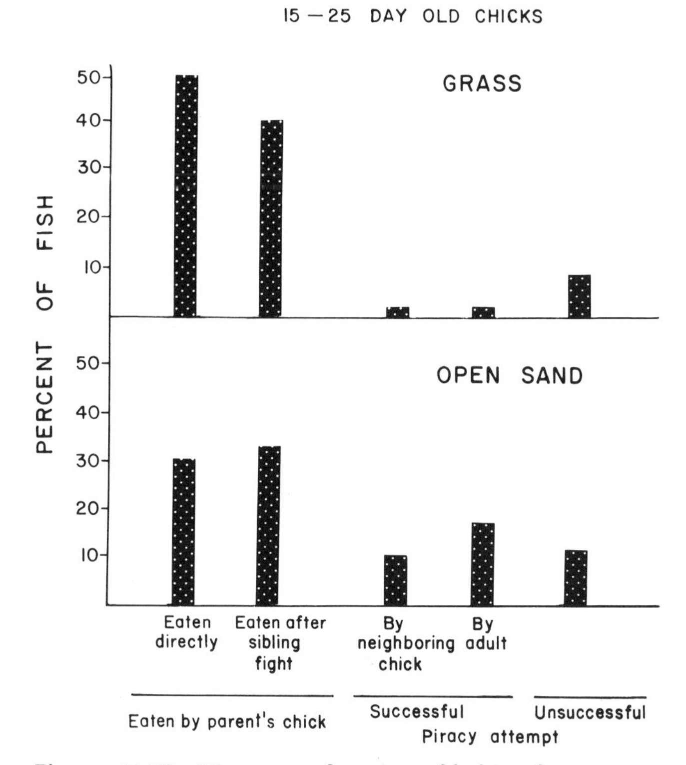

Figure 7.18. Piracy as a function of habitat for common terns nesting in grass (> 20 percent cover) or on open sand (West End, 1984). Each bar represents a percent of total fish delivered.

In 1984, we examined piracy occurring on the ground in open sand and in habitats with vegetation (> 20 percent grass). We chose sites where chicks were over 15 days of age. A significantly higher proportion of fish were eaten by chicks in grass habitats compared to open sand habitats (fig. 7.18; Yates $X^2 = 12.5$, $df = 4$, $P < 0.02$). Both neighbor chicks and adults succeeded in stealing more fish from chicks standing in open habitats, although aerial piracy rates were similar over both habitats. Thus habitat differences impose costs in terms of increased piracy rates in open habitats. The benefit of a lower piracy rate in vegetation is balanced, however, by the potential costs of increased mammalian predation.

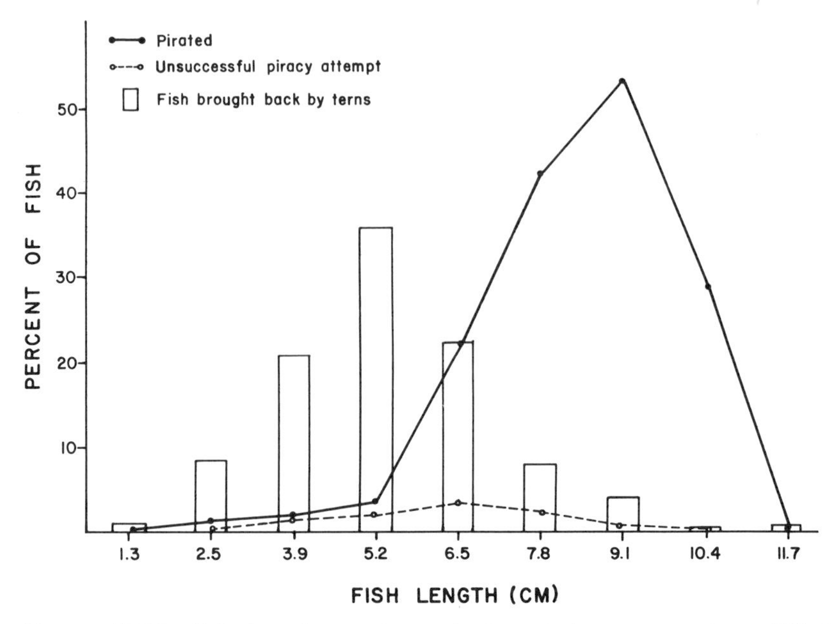

Figure 7.19. Relative piracy rates and attempts in common terns (West End, 1984) as a function of fish size (N = 3576). Histograms show size of fish brought back to colony. Dashed line represents piracy attempts; solid line represents successful piracy.

Food Size

Common terns bring back fish that vary in size from about 1.3 cm in length to rarely over 12 cm. We initially predicted that piracy rates would increase with increased fish length because large fish are easier for pirates to see, are more difficult for terns to carry, and are easier to seize from a chick's bill. Our observations confirmed this, as piracy rates exceeded 50 percent for fish that were 8–10 cm long (fig. 7.19). Further, figure 7.19 also shows that the fish size distribution for unsuccessful piracy attempts was shifted to the left. Thus, pirates attempt to steal fish of all sizes, but they concentrate their efforts and are more successful with the larger fish.

Tide

Tidal variables affect the number of fish brought back by foraging terns. Terns bring back more fish at low tide and again four to six hours after low

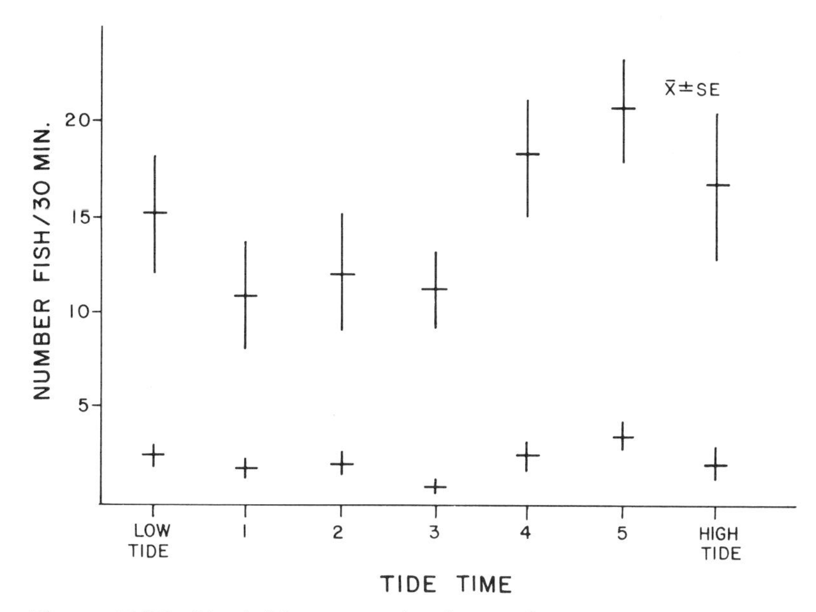

Figure 7.20. Food delivery rate (top bar) and piracy rate (bottom bars) for common terns as a function of tide. Given are the hours between low and high tide. Shown are mean ± 1 standard error.

tide compared to other tide states (fig. 7.20). Piracy rates also vary, and the highest piracy rates are close to high tide.

To examine specifically what tidal factors may affect total number of fish brought back and piracy behavior, we entered several potential tidal variables into a general linear models procedure to determine which tidal variables (defined in table 7.7) influenced variability in the dependent measures. Clearly, the time since the last low tide had the greatest effect on total fish brought back, and we used this in our models (see table 7.7). Surprisingly, tidal factors alone did not enter any of the piracy models, although it did in combination with other factors (refer to table 7.1).

Age of Chicks

In 1982, we found that piracy rates varied seasonally and hypothesized that some of these differences were related to the age of chicks. Chicks increase

Table 7.7. Tidal factors affecting fish delivery and piracy rates for common terns. Regression procedures were used to determine which tidal factor(s) influenced the outcome. *F* values and significance levels are given. Variables with numerical superscripts are squared or cubed to explore curvilinear relationships. (If a squared term contributes significantly, it shows a relationship that increases and then decreases or vice versa).

	Total Fish Brought Back	*Piracy Attempts*[a]	*Percent Pirated*[b]	*Success Rate of Pirates*[c]
Factors				
Lag[d]	22.76 (0.0001)	0.55 (NS)	2.48 (NS)	2.21 (NS)
Pre or Post Low[e]	1.27 (NS)	0.21 (NS)	0.56 (NS)	0.48 (NS)
New Time[f]	12.07 (0.0008)	0.19 (NS)	0.31 (NS)	2.21 (NS)
Tide Time[g]	4.64 (0.03)	0.26 (NS)	2.48 (NS)	1.41 (NS)
New Time^2	6.71 (0.01)	21 (NS)	0.17 (NS)	0.35 (NS)
New Time^3	9.46 (0.002)	33 (NS)	0.05 (NS)	0.74 (NS)
Tide Time^2	5.43 (0.02)	2.13 (NS)	1.87 (NS)	0.11 (NS)
Tide Time^3	1.76 (NS)	0.03 (NS)	0.00 (NS)	0.15 (NS)

[a] Piracy rate + unsuccessful piracy attempts.
[b] Percent of total fish that were eaten by pirates.
[c] For piracy attempts only.
[d] Time since previous low tide.
[e] Tide is rising (+) or falling (−).
[f] Time until nearest low tide using + or −.
[g] 0–5 values where 0 = low, 5 = high tide.

in size and weight with age and require more food. Parents can meet these increased food demands either by increasing their delivery rate (number of fish per hour) or by bringing back larger fish.

Parents increase the food size as the chicks get older (see fig. 7.8). Food delivery rates, however, varied markedly, no doubt depending on the fish available to foraging terns. Generally, parents feeding older chicks brought back more fish, particularly when the chicks were learning to fly. As might be expected, sibling fights increased in frequency with the age of the chicks (fig. 7.21), and for some older broods most fish were contested.

The piracy percent also varies with the age of chicks (fig. 7.22). The percent of fish lost to pirates is high for the youngest and for older chicks, and lower for chicks between 11 and 25 days old. The increase in fish lost by

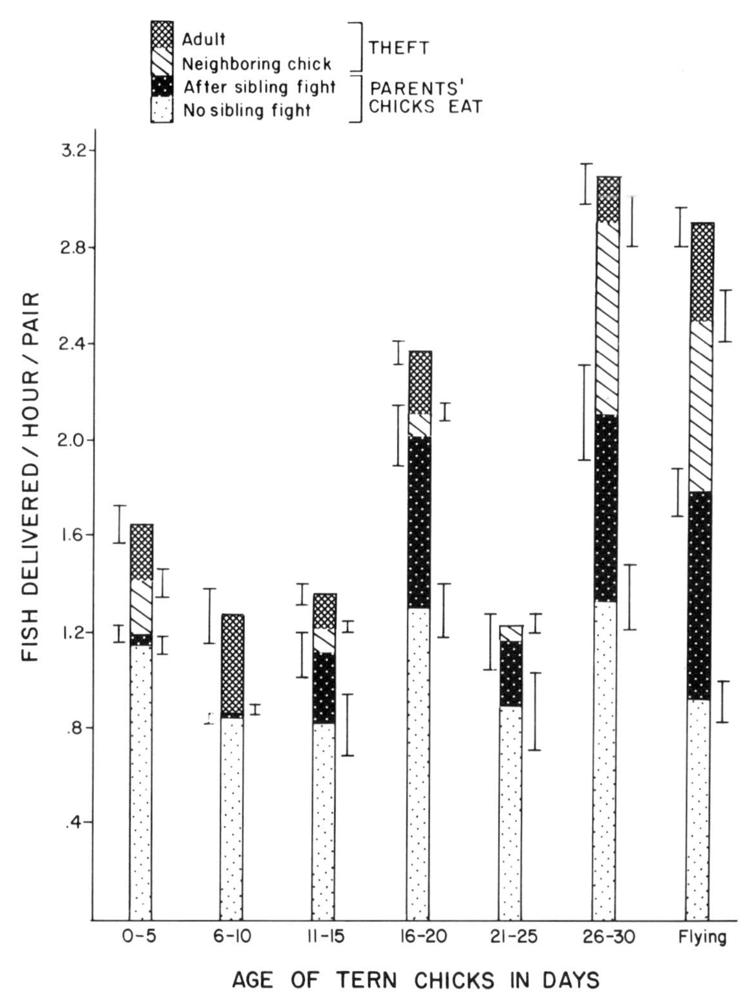

Figure 7.21. Fate of fish (N = 3576) brought back by common terns as a function of age of the chick (legend same as Fig. 7.17).

older chicks is due to increased sibling fights, making the fish particularly vulnerable to neighboring chicks that can rush in and pirate the fish. Very young chicks lose fish to larger neighbor chicks and to adults because they cannot protect their fish.

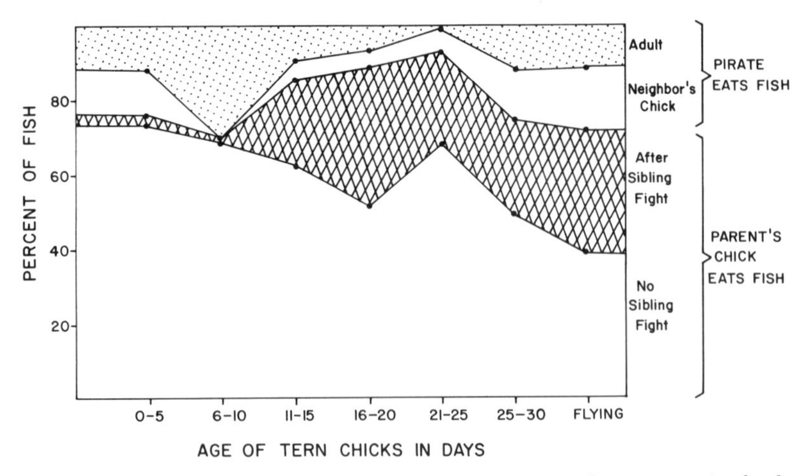

Figure 7.22. Percent of fish lost to pirates or eaten by a parent's chick as a function of age of the chick in common terns (same data as fig. 7.21 shown as percentages).

7.5 BEHAVIORAL RESPONSE OF PARENTS AND CHICKS TO THREAT OF PIRACY

Piracy behavior is an interactive process with pirates assessing their chances of success and their own requirement for food, and parents and chicks attempting to decrease piracy attempts and successes.

Chick Behavior

The behavior of chicks directly affects piracy rates in that the opportunity for piracy depends on the length of time a fish is visible before being swallowed. Common tern chicks swallow fish head first (Gochfeld 1975), and they have more difficulty swallowing large, fat, or long fish. Newly hatched chicks require several days to master this task (Gochfeld 1980d), after which most fish are swallowed in one or two seconds, but about 15 percent of the fish require over 5 seconds to swallow (fig. 7.23). Most piracies by neighbor chicks occurred when a chick experienced a delay in swallowing.

Some species of fish take longer to swallow than others. Pipefish are particularly difficult for common tern chicks to swallow. We watched young try to swallow 32 pipefish, and the mean time to swallow the fish was 6.4 ± 6.6 minutes. Such a long swallowing period clearly exposes the chick to

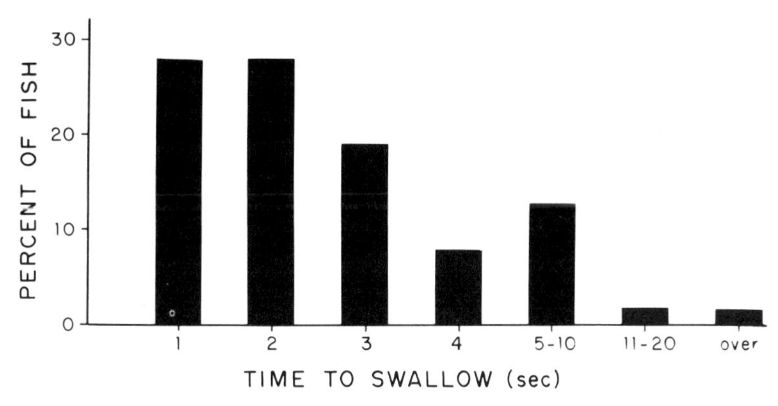

Figure 7.23. Time required for common tern chicks to swallow fish (West End, 1984, N = 270 fish).

piracy. Fortunately pipefish are not preferred food items and adults almost never try to steal them, chicks don't usually try to steal them, and sibling fights are less common. Indeed, some chicks refuse to eat pipefish and a number of nests had 30 or more uneaten pipefish scattered around.

Parental Avoidance Behavior

Parents can reduce piracy by bringing back small fish, by feeding their chicks near or in vegetation (obscuring the fish and making piracy difficult), by not landing if a piracy attempt is imminent, or by flying low over the colony when carrying fish—thereby making themselves less conspicuous to pirates. Often, when parents start to land by their chick, a neighboring chick or adult rushes up beside the parent's chick to grab the fish. Parents actively avoid this by aborting their landing attempt and circling around to land elsewhere. During an avoidance circle, the parent calls to the chick and the chick moves to another part of the territory. The behavior is particularly frequent when all the chicks in an area are over 20 days of age, and are running all over their territories. During this time, parents may make as many as ten circles before succeeding in feeding their chick, although over 40 percent land the first time and successfully feed their chicks (fig. 7.24).

Physically, pirates can steal fish only when they have clear access to their intended victim. Victims lose their fish to pirates when their flight patterns (or behavior on the ground) provide pirates with clear access to the fish. Our observations indicated that piracy attempts were less frequent when terns flew

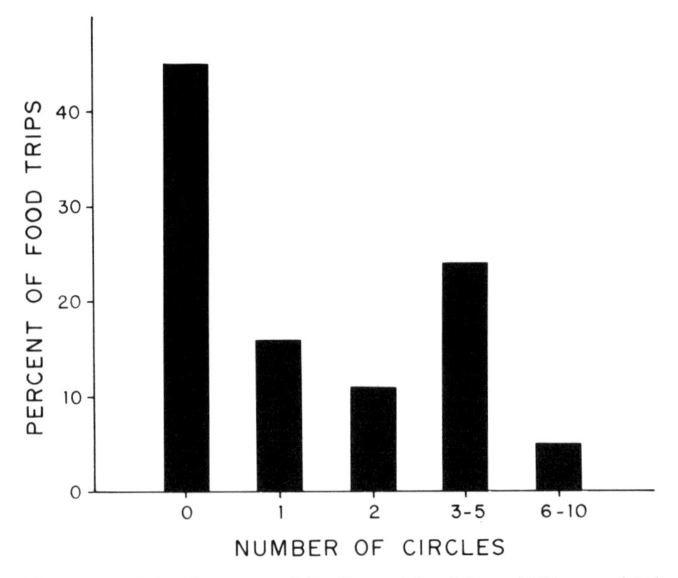

Figure 7.24. Percent of feeding visits (N = 270) on which adult common terns carrying fish for chicks made 0, 1, 2, or more circles over the territory before landing to feed their chicks. Parents are avoiding piracy by neighboring chicks when the chicks are over 20 days of age.

close to the ground. Above, we demonstrated that terns attempt piracy on large fish, but not on small fish.

In addition to piracy at nests, many pirates ambush terns as they enter the colony. It is therefore advantageous for an incoming tern to "hide" as long as possible. Taking a lesson from aircraft that must fly low to evade enemy radar, we predicted that

1. Terns flying into the colony with fish should fly lower over the colony than terns departing without fish.
2. Terns flying with fish at some distance from the colony will be higher than terns over the colony.
3. Terns carrying fish should fly high near the colony only when they have small fish; and, conversely, terns carrying large fish should fly low over the colony.
4. Piracy rates should be highest when terns carrying large fish fly high above the colony.

We examined these hypotheses at Cedar Beach in 1985 by recording the height above the ground, estimated to the nearest meter, for 223 terns flying out to fish, for 249 terns returning with fish when they reached 100 m from the colony, and for 725 terns with fish flying across the colony. It was possible to estimate the height of flying terns by using as reference points features of known height such as dunes, fences, lifeguard towers, and utility poles. For incoming terns carrying fish, we also recorded fish size (relative to bill length) and piracy attempts.

Common terns entering the colony with fish flew lower (mean height = 3.1 ± 1.3 m.) than did terns that were a hundred meters away from the colony (mean height = 7.6 ± 2.7 m, $t = 37.6$, $df = 972$, $P < 0.0001$) or than terns leaving the colony (mean height = 7.9 ± 2.3 m, $t = 39.1$, $df = 956$, $P < 0.0001$; fig. 7.25). There was no height difference for terns flying 100 m from the colony versus those leaving the colony to forage for fish ($t = 1.1$, $df = 480$, NS).

We recorded almost no piracy attempts on terns at 100 meters from the colony, but considerable piracy on terns entering the colony with fish. Pirates concentrated efforts at the edge of the colony where they ambushed potential victims. Terns flying with fish at 100 m from the colony flew 4–10 m above the ground or water. As they reached the dunes on the colony edge they descended abruptly and flew low over the colony to their nest sites and chicks. Terns carrying large fish are more susceptible to piracy attempts than those with smaller fish, suggesting that mean flying height could decrease with fish size. Our data are consistent with this hypothesis (figs. 7.26, 7.27), showing a clear trend, although the sample was too small to achieve significance.

It is also clear that not only do potential victims modify their flight heights to avoid piracy, but pirates do pursue terns that fly high carrying large fish (figs. 7.27, 7.28). This results in there being few very large fish that are carried at heights over 2 m (fig. 7.28). Terns returning with small fish regularly flew 3 to 8 m above the colony, whereas those with large fish almost always were below 4 m (fig. 7.27). For the 7.5 cm size class, terns flying below 1 m suffered significantly less piracy than those flying at 2.5 m ($p < 0.05$). Further, birds leaving the colony fly farther above the colony (90 percent over 3m) than those returning with fish (70 percent below 3m).

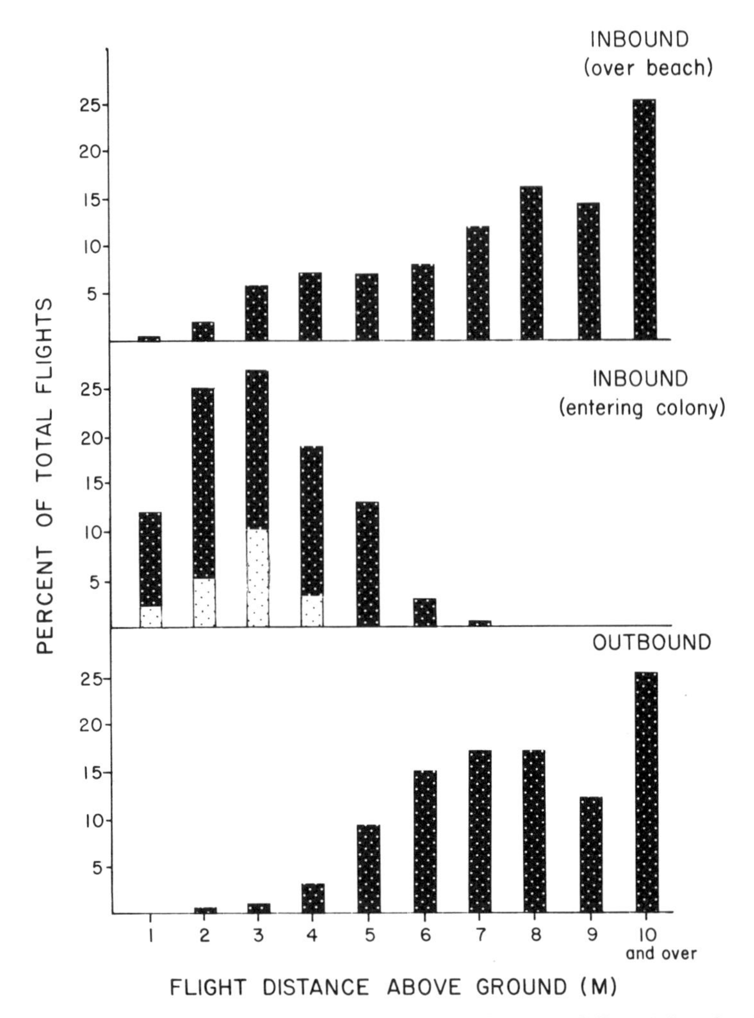

Figure 7.25. Percent of common terns flying at different heights for those that are about 100 meters from the colony carrying fish (top), terns that are entering the colony carrying fish (middle), and those that are leaving the colony without fish. The dotted white bar shows percent of fish pirated at each flight height.

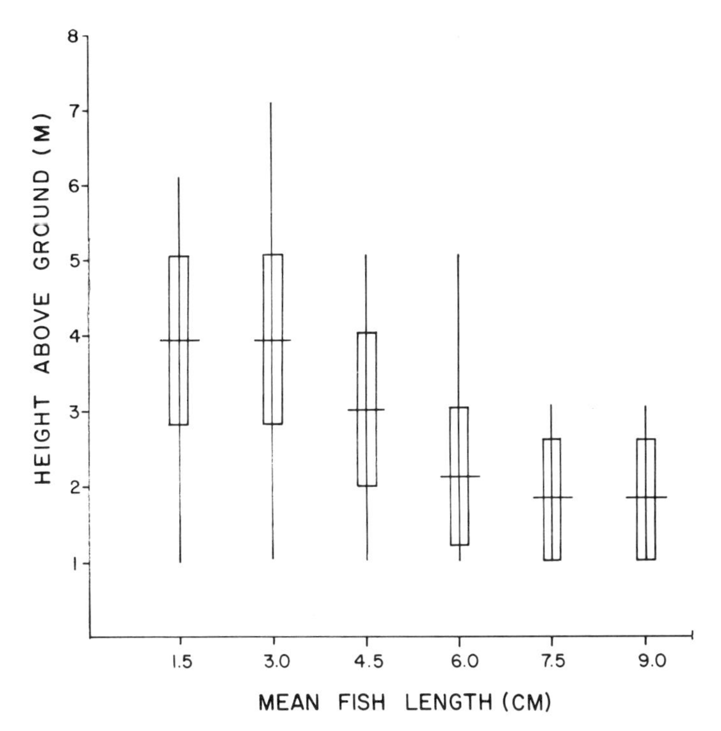

Figure 7.26. Mean flying height of common terns carrying fish back to the colony, as a function of fish size. Vertical line shows range of heights (estimated to nearest 1 m), box indicates ± 1 SD.

Parental Flight Costs

It is difficult to examine quantitatively the costs of piracy in terms of increased time to reach feeding grounds and capture additional fish to replace those lost to pirates. However, within the colony, it is clear that victims increase their flight speed when pursued by pirates. Increased flight speed would increase the cost of that flight, but not nearly as much as losing a fish. As a rough estimate of flight costs, we counted wing beats per second for terns engaged in different flight activities involved with foraging (fig. 7.29).

Parents leaving the colony to forage had fewer wing beats than those returning with fish. The difference persists when corrected for wind speed and direction, indicating that carrying a fish imposes a cost on flight. Terns circling high above the colony with fish intended for mates flew with slower

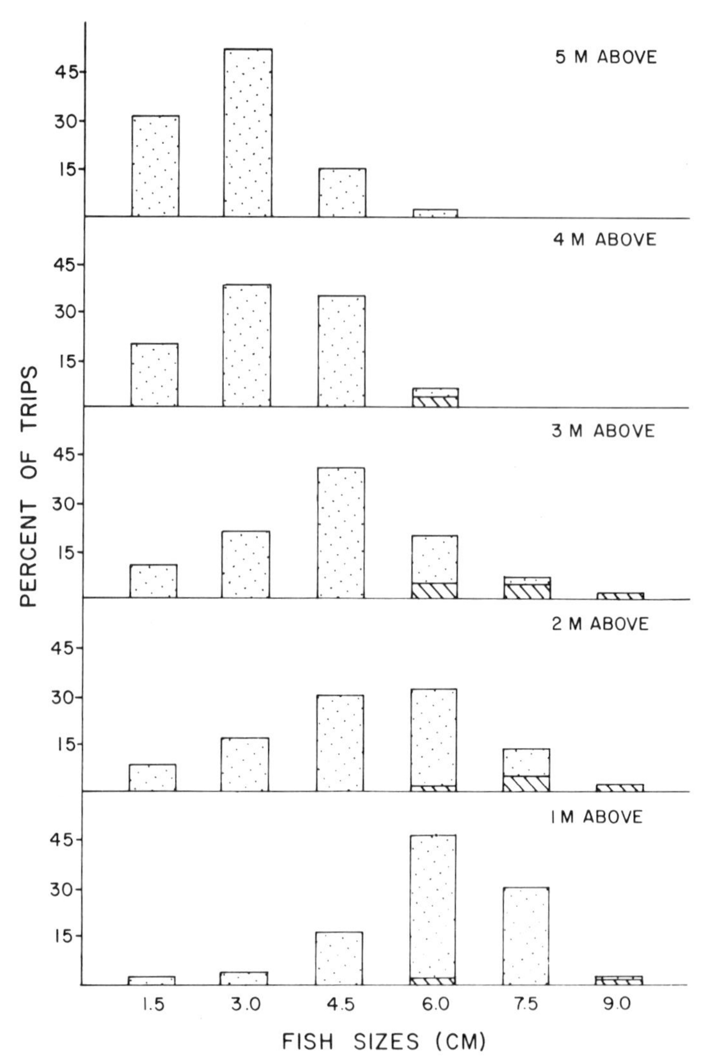

Figure 7.27. Flight heights of common terns returning to the colony with fish as a function of fish size. Hatched bar = piracy attempt.

wing beats than birds circling their territories with fish for their chicks (fig. 7.29). During a piracy attempt, flight speed (and wing beats) increased significantly, and during aerial fights over fish, wing-beat rate increased to over 6 per sec. This is even faster than that employed in dive-bombing a human intruder in the colony. More importantly, the terns had to engage in

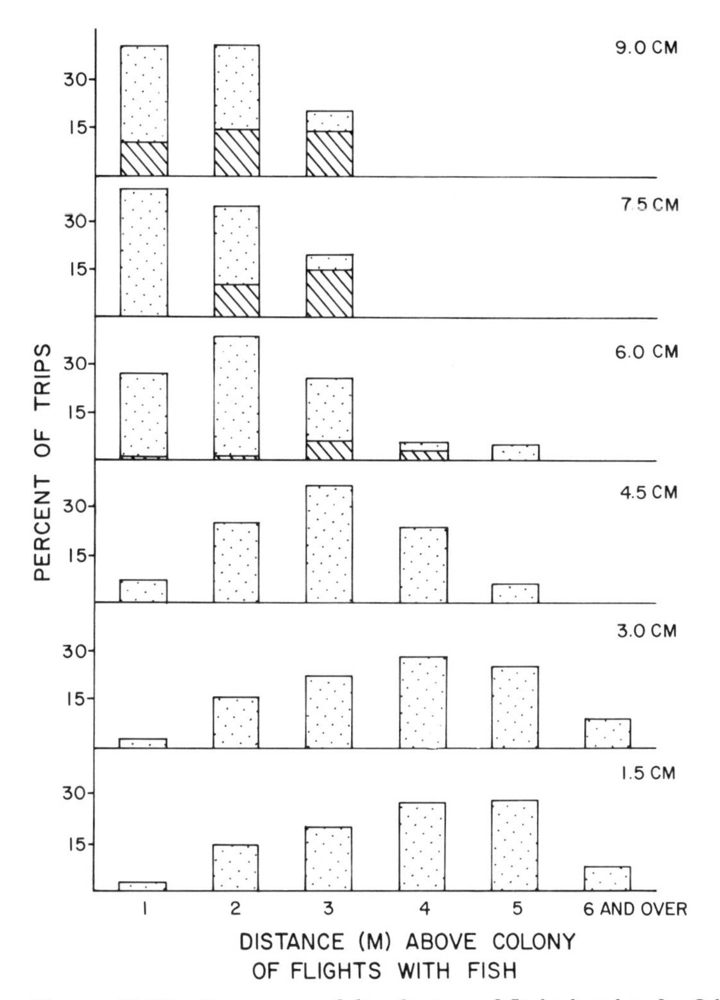

Figure 7.28. Frequency of distribution of flight heights for fish of varying sizes (from 1.5 to 9 cm in length). Hatched bar = piracy attempt (shows same data as fig. 7.25, rearranged to compare flight heights).

evasive action or circle repeatedly, which was not only energenettically costly, but increased the time between feeding for chicks.

In conclusion, adults can avoid piracy by altering their flight path and flying low over the colony. In general, pirates pursue from above and behind, although some pirates chase from below. Most piracy occurs at the colony. The hypothesis that terns that are most vulnerable (those carrying the largest

Figure 7.29. Relative flight costs (wing beats/sec) for common terns engaged in activities related to food acquisition, piracy avoidance, and aggression.

fish) should fly low over the colony is borne out by the present data set. The dramatic and characteristic drop in height as birds flying in from the ocean cross the beach, can reflect aerodynamic considerations as well, but the potential for thwarting piracy is considerable.

Discussion

In common terns, fish are generally brought back to the colony throughout the breeding season, because some birds are always courting or feeding chicks while others are incubating. When the late-courting pairs begin to incubate, the early nesting pairs are already feeding chicks. Fish available to potential pirates also varies by time of day and tidal conditions. For terns, more fish are generally brought back to the colony in the morning and at high tide.

Tidal effects on foraging behavior have been noted for seabirds including gulls (Drent 1967; Delius 1970; Spaans 1971; Galusha and Amlaner 1978; Burger 1976, 1980a), terns and skimmers (Hulsman 1976; Erwin 1977a,b; Black and Harris 1981; Burger 1982b). Erwin (1977b) reported that common terns were unaffected by tide, but that black skimmers fed at low tide. Similarly Black and Harris (1981) found that black skimmers fed at low and incoming tides. In Jamaica Bay, Long Island, Burger (1982b) found that tide significantly affected the activities of both species during the migration period. Common terns fed at high tide and skimmers fed at low tide. We found that the number of fish brought back to the colony clearly related to tides (refer to fig. 7.20), with terns bringing more fish at low tide and at about four to six hours after low tide. Fish might be concentrated in shallower water. Thus common terns may feed in different areas of the bay or over mudflats as tidal conditions shift.

Piracy Rates

Common terns were exposed to conspecific piracy which varied from year to year, but occasionally approached 50 percent of all fish brought back to the colony, depending on the season, time of day, age of chicks, and size of fish. Piracy rates increased with increasing fish availability, size of prey being brought back, and age of the chicks. Although these factors are partially related, each exerts an independent and statistically significant effect. These rates did not occur in all years or all colonies, but are generally higher than those reported in the literature.

Holding fish size constant, increasing prey availability increased the piracy rate. We believe this result reflects the ability of terns within the colony to rapidly assess fish availability as well as prey size. A common tern in the colony can observe whether few or many fish are being brought back, and whether they are of sufficient size to encourage success or a high benefit/cost ratio if the piracy is successful. Thus, a given tern can quickly switch opportunistically to piracy tactics when a large number of terns are bringing back large fish to the colony, and return to independent foraging when fish are less available. Further, a tern that is not otherwise engaged in courtship, foraging, incubation, or chick care activities may engage opportunistically in piracy if the number of opportunities is large. Cost would be low if such an individual were otherwise only loafing and were not interrupting other, high priority activities.

We also suggest that when large numbers of fish are being brought back to the colony foraging conditions may be optimal, and it may take less time and energy for the potential victim to replace the fish lost to piracy than when fish are scarce. Thus, potential victims might theoretically spend less energy avoiding pirates when food is readily available, though we found no evidence for this.

Piracy rates increased sharply for large prey items (refer to fig. 7.19), and prey size available to a potential pirate varied depending on whether males were feeding females, whether adults were feeding chicks, and on the size of the chicks. We believe that in common terns the length of fish being brought back to feed mates or growing young is constrained in part not only by fish availability on the feeding grounds, but by the ability of chicks to swallow fish of certain sizes and by piracy pressures. Presumably, it is optimal for males to bring back the largest possible fish for females, both as courtship inducement and to provide the highest energy levels for egg formation. Similarly, as chicks grow, parents should bring back the largest fish they can carry and that the chick can eat quickly. Both assumptions suggest that adults should bring back large fish except when they are feeding very young chicks incapable of swallowing them. However, large fish are subject to considerably higher piracy pressures. Thus, the optimum fish size to bring back is just below the size that induces increased piracy attempts (about 6 to 7 cm; see fig. 7.7).

Chick Age, Chick Behavior, and Piracy Rates

In common terns, piracy (both as a rate and a percent of total fish, figs. 7.10, 7.11) increased with chick age. As chicks increase in size, they: 1) are able to swallow larger and larger fish, 2) increase in mobility, 3) increasingly compete with siblings for fish, and 4) increasingly attempt to pirate fish from neighboring chicks. All of these factors are additive in increasing the opportunities for piracy.

As they grow, young chicks require more total energy, either as increased number of fish, increased size of fish, or both. These factors are associated with increased piracy rates. By increasing their mobility chicks may also make it more difficult for parents to locate them, or may increase the time necessary to run toward their parent. These two actions provide visual cues to potential aerial pirates or to neighboring chicks that a feeding is about to occur.

Chick mobility is also an advantage since chicks can avoid the piracy attempts of neighboring chicks by running to the opposite edge of the territory. Similarly, an individual chick can sometimes avoid sibling competition by running toward the approaching parent faster than its sibling. Delude et al. (1987), however, found that increases in piracy attempts and parental territorial defense increased with increasing chick mobility.

As chicks get older, they engage in more direct competition with siblings over food. This takes the form of scrambling for the food item and fighting over a particular fish. During these sibling clashes, both chicks and their parents may direct their attention toward the sibling fight, and thus fail to note an approaching pirate (often a larger neighboring chick).

Synchrony and Piracy Rates

In a large colony of common terns, such as West End, there are always pairs that are initiating courtship. Late-nesting pairs may be young birds or failed breeders from this or another colony. However, within subsections of a colony, there is a higher degree of synchrony. One interesting side effect of synchrony is to decrease piracy rates by decreasing the disparity in ages of neighboring chicks. When neighboring chicks are the same age the probability of piracy by a neighboring chick decreases since chicks can defend themselves and their fish from a chick of their own size.

Conversely, the greater the degree of asynchrony the higher the piracy attempts by the larger neighboring chicks. indeed, the high rate of piracy on very young chicks was largely due to older chicks from neighboring nests rushing over to steal fish from the young chicks' bills. In any case, only a small proportion of chicks become pirates. Thus, the first chicks to hatch in any area have lower piracy rates than later-hatched chicks, because there are no older neighbor chicks to act as pirates.

Parental Behavior and Piracy Rates

Parents can lower the risk of piracy by: 1) bringing back intermediate-sized fish and avoiding large fish, 2) decreasing their flight time within the colony, 3) flying low over the colony, 4) aborting landings when a potential pirate is hovering nearby, 5) avoiding landing when a neighboring chick is a potential pirate, and 6) reducing sibling competition.

Most piracy attempts take place within the colony or at its edge, suggesting that adults bringing back fish should minimize their flight path over the

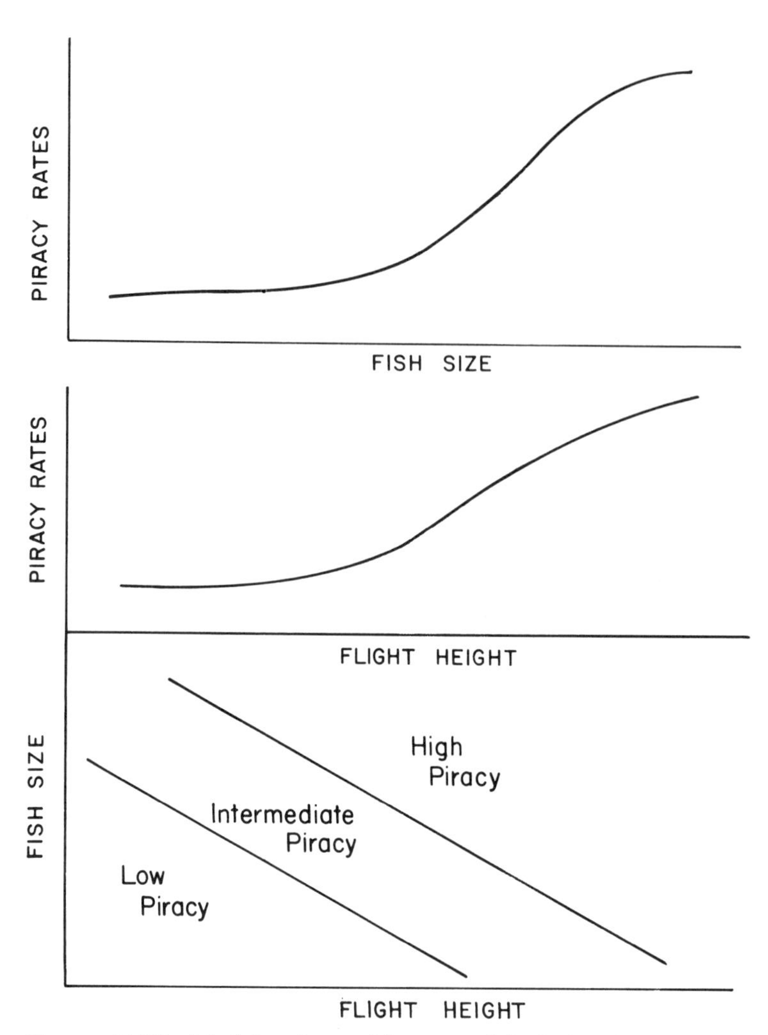

Figure 7.30. Models relating fish size and flight height to piracy, and the interactions of fish size and flight height to indicate zones of low and high piracy potential.

colony. Indeed, this is often the case, and some common terns fly along the water's edge and veer into the colony only when they are near their nests.

Secondly, terns returning with fish, particularly large fish, fly low over the colony to thwart piracy attempts. By flying low, they decrease their conspic-

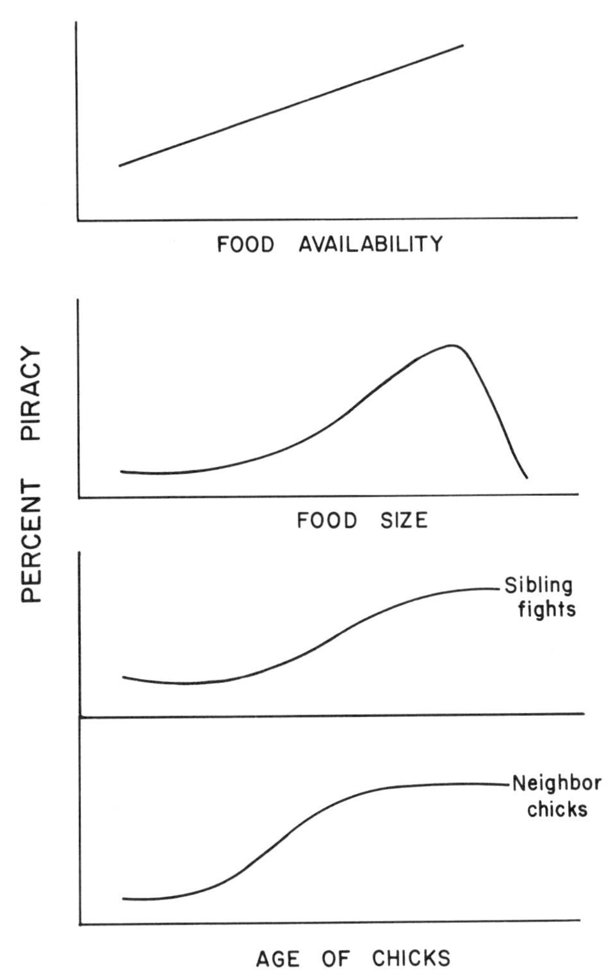

Figure 7.31. Model for relationship of piracy to food availability, food size, and age of siblings and neighboring chicks.

uousness and reduce the maneuverability space for pirates, making them less likely to attempt piracy, and less likely to be successful. Thus, terns carrying small fish did not especially lower their flight heights, and small fish are rarely pirated.

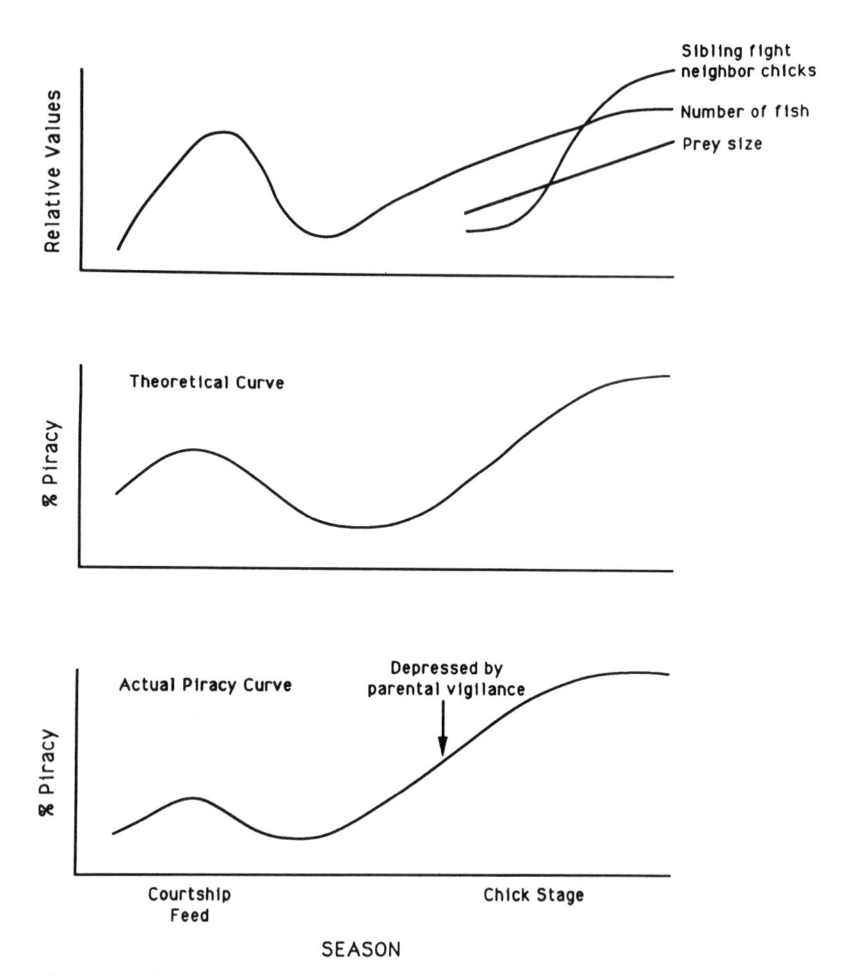

Figure 7.32. Seasonal relationship of number of fish brought back, size of fish and age of chick (sibling fights and neighbor chick piracy) on piracy rate (top). Theoretical piracy curve derived from adding piracy potentials (middle). Actual piracy rate curve observed in common terns, adjusted for parental vigilance and avoidance behavior (bottom).

Overall Model for Piracy as a Cost of Coloniality

The results presented above suggest that the cost of piracy in common tern colonies relates to food availability, food size, sibling competition, age of neighboring chicks, and behavior of the chicks and parents (fig. 7.30). When these constraints are imposed on the seasonal availability of fish, the resultant predicted piracy curve is similar to our results, but lower (figs. 7.31, 7.32). The piracy rate was lowered only by parental vigilance and avoidance behavior. Nesting early and in synchronous subcolonies further reduces the potential of piracy from neighboring chicks.

EIGHT

Floods, Heavy Rains, and Human Disturbance

Common terns, like other colonial birds, are exposed to inclement weather, flooding, and human disturbance. Inclement weather includes heavy and prolonged rain, fog, strong winds, or excessive heat (fig. 8.1) or cold. In addition, species nesting in coastal areas are exposed to tidal flooding as a result of normal tidal cycles and in conjunction with storm winds or hurricanes which may coincide with normal high tides.

Human disturbance is a cost for most nesting birds except those that nest in inaccessible locations such as remote islands (Lack 1968; Kaverkina 1986b). Even steep cliffs can be scaled by motivated indigenous people with ropes and a tradition of egging. Human disturbance can take the direct form of injury to eggs, chicks, or adults or the indirect form of harassing birds, interruption of incubation or parental care, or exposure to the elements or predators. In this chapter, we discuss the effects of floods, heavy rains, and human disturbance and examine the responses of the terns to these threats.

8.1 TIDAL FLOODING

Tidal flooding poses a threat to terns in many habitats, but particularly those nesting on salt marshes, low-lying sandbars, and rocky islets in the Great

Figure 8.1. Incubating adult common tern responds to heat stress by gular fluttering, a slow vibration of the floor of the mouth that increases evaporative heat loss.

Lakes. By their very nature, salt marsh islands are low-lying, subject to frequent flooding, and often the highest parts are less than 30 cm above mean high water.

Common terns can avoid tidal flooding by selecting nest sites that are on high parts of islands or by behavioral means such as use of floating mats, construction of tall nests, or movement of vulnerable eggs and chicks. Nest construction is a critical factor affecting reproductive success (Collias and Collias 1977; Burger 1979b). Figure 8.2 shows a pair of common terns that originally laid three eggs on marsh vegetation and then constructed a higher nest as water levels rose, but were only able to retain two of their eggs in this enlarging nest. Common terns nesting on the barrier beaches seldom experience flooding because the average height of such sites is well above the highest tides (4 *m* above sea level at Cedar Beach). Distance alone is not sufficient to avoid tidal flooding because overwash areas can be several hundred meters back from the surf line. In the New York beach colonies, common terns lost nests to tidal flooding only once, in 1972, at the Wantagh colony during a July 4th hurricane (Gochfeld and Ford 1974). Otherwise,

Figure 8.2. Joanna Burger holds a tern nest with two eggs in one hand and points to an abandoned egg left behind as the incubating tern built up its nest to escape tidal flooding.

common terns at Cedar Beach and West End have not suffered losses due to high tides or flooding.

Sandy Beaches

In New Jersey, the barrier beach colony at Holgate has lost nests to tidal flooding in only two years, and the losses were small. However, on the sandy fringe of Mordecai, nests have been completely washed out in 7 of 11 years. Most of the nests on the island, however, were located on wrack or salt marsh vegetation, higher on the marsh, where they usually escaped flooding. Thus, on the barrier beaches and sand spits on salt marsh islands, the primary defense of the terns against flooding is selection of higher parts of the island well above the high tide line. Consequently, nests are merely depressions on the vegetation or in the sand, with no added material.

Salt Marshes

Common terns nesting in salt marshes can similarly avoid tidal flooding by their choice of colony and nest sites. However, since most marshes are vulnerable to flooding at least once during a nesting season, additional behavioral means can spell the difference between success and failure. Colony site choice involves selecting islands which are high enough to minimize tidal flooding, but not so high as to support abundant dense bushy vegetation or *Phragmites* which might harbor mammalian predators or attract nesting gulls. Height of the island alone is not sufficient to determine the risk of tidal flooding. Normal high tides are augmented by storm winds that can force water toward one side of the bay. Also the location, number, and size of the inlets will affect how much water can flow into and out of the bay at each tide.

To examine the effect of tide on colony location, we divided the Barnegat Bay area into three sections corresponding to the northern, middle, and lower thirds (table 8.1). The upper third has no direct inlet and has only four tern colonies. The middle third has 17 colony sites north and south of Barnegat Inlet. The lower third has the very wide Beach Haven Inlet and 12 colony sites, including the large Holgate colony.

Winds blowing from the south should pile up water on the north end of the bay, while winds from the north force water toward the south. Assuming equal likelihood of predominant winds, colony sites should be equally subject

Table 8.1. Effect of tidal flooding on percent of colonies with more than 33 percent of nests washed out, with respect to location in Barnegat Bay.

		Number of			*Percent*	
Location in Bay	*Width of Inlet*	*Sites Occupied*	*Actual Occupancies*	*Possible Occupancies*[a]	*Occupied*	*Washed Out*
Northern	none	4	60	60	100	7
Central[b]	240 m	17	148	255	58	25
Southern[c]	2 km	12	107	180	59	36

[a] Occupied sites × 15 years of study (1976–1990).
[b] Area around Barnegat Inlet.
[c] Area around Beach Haven Inlet.

to flooding or equally suitable for tern occupation. However, suitable colony sites, defined as a physically suitable site that was used at least once, were not equally distributed (table 8.1, contingency table $X^2 = 10.9$, df $= 2$, $p < 0.01$). There were more islands in the southern section of the bay. All the northern colony sites were used every year, whereas only 58 percent and 59 percent of the suitable sites in the middle and lower thirds of the bay were used.

Washouts, defined here as colonies in which at least one third of all nests were washed out, were less severe in the northern section compared with the others in every year except one (table 8.2). Washouts in the other two sections, however, varied, but were usually more severe in the area of Beach Haven Inlet (lower section, 10 of 15 years, table 8.2). As might be expected from the distribution of washouts, terns had the highest occupancy rate in the north (even though there were only a few suitable islands), and the lowest occupancy in the south. Their choice of colony sites corresponded to the proportion of colony washouts (table 8.1).

Once having selected a colony site, common terns then select a nest site, and these nest sites are also differentially vulnerable to high tides because of elevation differences. Most tern nests on salt marshes are either on mat of *Zostera* or dead grasses or are on flattened living stems of *Spartina patens* or *S. alterniflora* (see chapter 4). For several years, we monitored nest loss due to flooding as a function of habitat (table 8.3). In all years and on all colonies, flood losses were least on wrack, intermediate in *S. patens*, and greatest in *S. alterniflora*. Severe high tides, of course, resulted in total losses of all nests (as at West Vol and West Ham in 1982).

Having selected a nest site, common terns can still reduce the risk of flood

Table 8.2. Percent of colonies in Barnegat Bay that suffered washouts resulting in loss of more than 33 percent of nests due to flood tides.

Year	*North* %	*Central* %	*Southern* %	*Overall* %
1976	0	27	70	38
1977	0	15	0	4
1978	25	80	100	77
1979	50	46	38	44
1980	25	9	29	18
1981	0	38	57	35
1982	0	90	100	77
1983	0	83	75	61
1984	0	55	43	38
1985	0	30	71	40
1986	0	0	75	28
1987	0	29	60	33
1988	0	38	33	11
1989	0	29	43	32
1990	0	0	11	8

damage by constructing a tall nest. In the New Jersey salt marshes, common terns nesting in *S. alterniflora* usually build substantial nests, while those in *S. patens* build smaller nests or none at all. Nest construction varies for terns nesting on wrack, but usually they build no nests. Nests are usually augmented during the rising tides, as the waters get closer to the nests, or when eggs get wet sitting on the ground (see Burger 1980a,b). Both sexes add nest material, and the nests may gain as much as 5 cm in height during one high tide.

With minor tidal flooding, eggs may be displaced only a few cm from the nest, and such eggs are normally retrieved by the incubating adults. Eggs that remain wet for several hours develop a chalky appearance, and the adults are likely to desert such clutches. When tidal waters completely inundate nests and eggs, the parents fly to higher ground, returning as the waters recede, when they may attempt to resume incubation.

Discussion

Tidal flooding is always a threat for ground-nesting species in coastal habitats. Terns can avoid flood damage by appropriate colony and nest site selection,

Table 8.3. Percent of nests washed out as a function of nest location on salt marsh islands in Barnegat Bay. Shown are the percent of nests in each habitat that were washed out.

Colony	*Number of Nests*	*Wrack*	S. patens	S. alterniflora
1976				
Lavallettes	886	0	0	5
West Vol	55	6	—[a]	27
Carvel	91	0	—[a]	16
1977				
Lavallettes	850	3	2	12
West Vol	50	10	—[a]	10
Carvel	123	0	—[a]	16
West Ham	40	0	0	12
Mordecai	30	0	10	20
1982				
Lavallettes	590	0	2	8
West Vol	61	100	100	100
Carvel	319	27	5	0
West Ham	51	100	100	100
1985				
Lavallettes	1610	4	7	18
West Vol	64	73	—[a]	92
Carvel	116	98	—[a]	98
West Ham	55	60	100	75

[a] No *patens* available on these islands.

and by the behavioral response of nest-building. In the New York and New Jersey colonies we studied, terns avoided flooding by nesting high on the barrier beach, beyond the reaches of the highest spring and summer tides. Flooding of such habitat was extremely rare, although winter overwash occurred in two years at Cedar Beach. The West End II site was more than 350 m from the surf and did not flood. However, Austin (1929) reported tidal flooding of beach colonies.

In the salt marshes of Barnegat Bay, flooding is a problem in some colonies almost every year. A combination of a high nest site and a high nest reduces the likelihood of a washout. In our analysis of data for all of Barnegat Bay, we found more consistent island occupancy in the northern part of the bay where flooding was rare. This difference was not due to the height of the islands, for the northern islands, like most of the bay islands, were also covered with S. *alterniflora,* indicative of low elevation. The difference lay in the movement of water, the northern part of the bay being remote from the influx of ocean water through the narrow Barnegat Inlet.

In the northern colonies, the Lavallettes, a higher proportion of the terns nested in the S. *alterniflora* (up to 50 percent of nests). These islands have very little wrack available, partly because of the buffering of the high storm tides which would deposit the dead eelgrass in winter, forming the mats or wrack on the marsh. Both the lack of mat and the relative scarcity of damaging flood tides reflect the distance of these northern islands from the nearest inlet. Although there are only a few salt marsh islands—the Lavallettes—in the northern part of the bay, compared with elsewhere, a disproportionate number of terns (over 1600 in some years) nested there. In some years, more than half of the terns nested in these four colonies which were occupied in every year (table 8.1). Thus, the terns were showing a consistent tendency to return to colonies that had the lowest washout rate. In the south, the tern colonies are close to the 2-km-wide Beach Haven Inlet. On an incoming tide running in front of a wind, water rapidly rushes into the bay, rising over the southern islands.

One must question why more terns don't nest on these northern marshy islands which still have much unused area. One possible explanation is that food is more available near inlets, and that the distance to the inlet which protects the islands from flooding, also entails an added travel cost for foraging terns. If we had begun our study in the 1980s, we would have invoked the thriving herring gull colony on one of the Lavallettes as a deterrent to further growth of these colonies. Yet, although there were no gulls nesting there in the 1970s, the colonies on these islands have remained fairly stable during both periods. On all islands, the terns had differential washout rates depending on their nest site choices (table 8.3). Terns nesting on wrack had lower rates than those nesting in S. *alterniflora*. The fact that the wrack is deposited by the highest winter storm tides explains why this habitat is usually secure from most spring or summer high tides. On most islands, the terns choose to nest on wrack before nesting in the *Spartina*, indicating a preference for that habitat (see chapter 4). Selecting the highest sites thus mitigates against tidal flooding. Birds nesting in S. *alterniflora* can still avoid some tidal flooding by building high and sturdy nests—platforms of eelgrass and cordgrass piled high on the depressed grass. In some cases, nearby grass stems are incorporated in the nest, thus providing some sort of anchor in the event that the mat and nest begin to float. Similar behavioral adaptation for nesting in salt marshes has been reported for laughing gulls (Bongiorno 1970; Montevechhi 1977, 1978; Burger and Shisler 1980), herring gulls (Burger 1979b, 1980b), common and Forster's terns (Burger and Lesser 1978; Burger 1979b; Storey 1987a,b), and black skimmers (Burger 1982b). In general, these authors describe colony and nest site selection as a mechanism for avoiding tidal flooding. Furthermore, nest building behavior in these species in New Jersey has been experimentally studied, and all species respond to rising tidal waters with increased nest building activity (Burger 1979b).

The tidal dynamics of Barnegat Bay are unusual in that islands of equal elevation are not equally likely to get flooded. This study afforded the first opportunity to study a coastal system showing how several colony sites are used by terns in response to tidal condition.

8.2 HEAVY RAIN AND FOG

Heavy rains, wind, and fog present additional difficulties for nesting terns, primarily because they interfere with foraging. Dunn (1973) showed that tern diving success is greatly reduced when the water surface is rippled by wind. Rain and fog also reduce visibility. Although adults can survive for several days without food, decreases in feeding success can delay egg-laying or lead to smaller clutches, may increase the length of incubation (if females must leave to find food for themselves), or the length of incubation bouts (if one mate stays away longer than otherwise), or may decrease chick growth and survival. If adults leave to find food and stay away for long periods, chicks will be exposed to the elements or to predation. Thus food shortage, or increased intervals between feedings, can have profound effects on many aspects of tern breeding behavior.

Although many authors have mentioned the impact of food shortage or inclement weather, relatively few studies have documented the resulting mortality. However, extreme weather conditions are encountered by nesting seabirds in many parts of the world. Howell et al. (1974) discuss the harsh environment faced by nesting gray gulls. Burger (1980b) discussed the impact of weather on herring and laughing gulls. Tomkovich (1986) reported the effects of wind on nesting ivory gulls. Austin (1929, 1933) discussed weather impacts on common terns. Parson (1985) documented weather effects on

Figure 8.3. Joanna Burger and Michael Gochfeld checking common terns for general condition and body mass following excessively heavy rains and hail storms.

growth and survival of heron chicks. Egg and chick losses due to heavy rains have also been reported for jackass penguins (Randall et al. 1986).

In New Jersey, heavy rains in late May 1978 delayed egg-laying by a week in the salt marsh colonies (mean egg-laying was pushed into early June). In 1985, egg-laying was also delayed by a week, because of heavy rains in mid-May, and in 1990 cold wet spring weather delayed egg laying by two weeks (refer to chapter 3, fig. 3.5). Chick survival can be affected if parents cannot bring back sufficient food, particularly during the first week of life, and even adult condition may be compromised by inability to feed for several days. However, unlike black skimmers (Burger and Gochfeld 1990a), common terns brood their chicks effectively during heavy storms, thus reducing thermal or cold stress. As an example, a severe hailstorm in July 1977, with 2 cm hailstones lasting for 90 sec, killed one adult on a nest, injured two others (fig. 8.3), but did not kill any chicks, presumably because all were protected by adults or vegetation. High temperatures can also produce thermal stress (Austin 1933; Hand et al. 1981), particularly for chicks that cannot hide in vegetation or reach the water.

8.3 HUMAN DISTURBANCE

Human disturbance varies dramatically among habitats and colonies. For many people, the marsh is a forbidding environment. In late spring and summer, the salt marsh is often wet and sometimes muddy. On some days, mosquitoes and green-head flies are severe, while cordgrass scratches the unprepared visitor. Moreover, these islands usually have no high ground or sand for picnics or sunbathing. Thus, except for occasional visits by fishermen, the salt marsh islands are usually left alone, and human disturbance of the nesting birds is minimal. In the fifteen years of our New Jersey salt marsh studies, only two colonies suffered severe losses due to people, in both cases because of accompanying dogs. Two islands, Mordecai and East Carvel, have narrow sandy beaches which occasionally attract swimmers or campers.

By contrast, barrier island colonies adjacent to large, popular bathing beaches used by thousands of people on hot summer days, are subject to frequent human visitation. Holgate, Cedar Beach, and West End Beach are heavily used through the season. The frequent intrusion of people at the Holgate colony lead U.S. Fish and Wildlife personnel in 1986 to close the beach to all people and vehicles for the duration of the nesting season.

On summer weekends, human disturbance at West End II was a daily occurrence, although in most cases people walked along the road shoulders through the colony, without actually entering nesting areas. When, in 1987, the terns began to relocate to the West End I site (1 km to the east), the potential for human disturbance was increased because of the closer proximity to the recreational-use areas. In 1988, the colony was fenced, but we still saw people cross the fence and run through the colony. In recent years, the Cedar Beach colony has been almost completely fenced. At first, people would break down the fence, sometimes using off-road vehicles. After several years, the fencing coupled with the presence of researchers effectively reduced human transgressions, and the fact that the beach was closed for much of the 1988 and 1989 seasons due to beach pollution greatly reduced human disturbance.

At several points during our study of the terns, we have kept detailed records on the frequency of human intrusion (table 8.4). Visitation rates in the colony vary seasonally (lower before May 30 and greater after July 1), by day of the week (more on weekends), and with weather. Certain events have favored intrusion for some years and discouraged it in others. Close surveil-

Table 8.4. Human disturbance and intrusion at Cedar Beach and West End Beach (not including visits by investigators). Given are the average number of people/day at the edge of and entering the colony (exclusive of researchers).

	West End Beach			*Cedar Beach*		
		Average No. of Visitors			*Average No. of Visitors*	
	No. of Days	*to Edge*	*to Colony*	*No. of Days*	*to Edge*	*to Colony*
1964	16	1.5	1.2			
1969	23	2.4	1.4			
1975	16	2.1	3.0	24	4.3	3.5
1977				15	5.8	4.3
1978	17	3.4	1.3	29	16.2	24.4
1986				19	11.8	22.7
1987				14	4.2	1.7
1988	2	3.8	3.0	13	1.4	0.7
1989	6	0.8	0.3	10	1.1	0.6

lance by park police during the mid-1970s kept cars off the lawns and intruders out of the colony. Interest in surveillance waned during the late 1970s and the size of the police force was reduced. Construction at Cedar Beach and beach nourishment projects greatly increased the intrusion rate (table 8.4) in the 1977–1980 seasons, as did the favorable wave conditions that attracted numerous surfers. In the mid-1980s, the surfers were attracted elsewhere, there was no construction, and fences were erected. Table 8.4 shows the average number per eight-hour day of persons standing at the edge of the colony and actually entering the colony for several years. An added component may be the increased awareness fostered by the media and by sign posting that nesting birds are vulnerable.

Occasionally, people deliberately damage eggs or chicks. They may break them, play catch with them, or make designs in the sand with eggs from many surrounding nests. Such vandalism has not occurred in the salt marsh colonies and is, in the large beach colonies, a relatively insignificant cause of mortality. In over twenty years, we have observed four instances of vandalism such as deliberate egg or chick destruction at West End and two at Cedar Beach. We twice found evidence of people actually shooting at adult terns, once at Wantagh and once at Cedar Beach. Egging, the collecting of bird eggs for food, is still practiced in Europe and elsewhere, and although illegal in the United States, is a potentially serious matter. We have seen two instances of egging. At Meadow Island, eggers took over 120 common tern eggs on one morning in 1971, and in 1977 we found eggers at work at Cedar Beach. Much more significant human disturbance occurs without a deliberate intention to damage the birds. Most dramatic was the case of a pilot in training who continually landed and took off for about 45 min. from the Cedar Beach site early in the 1970 breeding season. At least 40 nests were destroyed, and many neighboring ones were abandoned. Instances of direct damage, both intentional and nonintentional, have decreased during recent years, probably due to the increased journalistic coverage of endangered and threatened species, improved sign-posting, and increased public awareness of threats to colonial birds. Programs involving signs, leaflets, television programs, and increased protection with fences and even on-site wardens have made a large difference in protecting breeding tern populations (Burger 1989). Even well-intentioned people can cause damage. Biologists who study birds must carefully plan their work so as to cause minimal interference or disturbance of the colony. Nature photographers and bird watchers frequently visit the tern colonies (table 8.5). Although such visits do not directly injure the adults, they may keep the adults disturbed and off their nests, and may unnecessarily expose eggs or chicks to thermal stress from cold or hot temperatures. Chicks exposed to direct sunlight on the hot sand in the open, without vegetation cover, will die in about 25 min.

Although not creating a direct effect, humans can cause a variety of problems through pollution of the environment with a variety of heavy metals and organic chemicals (Gochfeld 1971, 1973, 1975b, 1981; Hays and Risebrough 1972; Connors et al. 1975; Burger and Gochfeld 1985). Oil pollution is another problem for terns (Duffy 1977; Butler and Lukasiewicz 1979; Gochfeld 1979c; King and LeFevre 1979). Both adults and young frequently die from entanglements in fish line and plastic (Nickell 1964; Gochfeld 1973; Burger 1981a,e). Mortality from automobiles is a major problem for young terns when they are learning to fly. They seek open spaces such as parking lots and highways adjacent to the major beach colonies (Gochfeld 1978c).

Although deplorable and preventable, these impacts pale by comparison with the devastation wrought by commercial exploitation of terns in the nineteenth century which eliminated terns from all but the most remote

Table 8.5. Visits to Cedar Beach tern colony by nature photographers and bird watchers. Given are number of weekend and weekday days when data on intrusion were obtained and the average per 8-hour-day of people standing on the dune at the edge of the colony and actually entering the colony.

	Weekends			*Weekdays*		
		Average No. of Visitors			*Average No. of Visitors*	
Year	*No. of Days*	*Dune*	*Colony*	*No. of Days*	*Dune*	*Colony*
1975	9	1.9	0.22	15	0.4	0.07
1978	11	2.1	0.27	18	0.8	0.11
1985	10	0.5	0.10	12	0.3	0.08
1986	8	0.3	0	11	0.3	0.09
1987	6	0.5	0.33	8	0.1	0
1988	5	0.4	0.21	8	0	0.12
1989	6	0.2	0.03	4	0	0

areas (Nisbet 1973b). Contrary to perceptions in North America and Europe, commercial exploitation of seabirds persists in many parts of the world (e.g., egging of terns, leather from penguin flippers). Although we do not favor such exploitation, it is apparent that careful management of harvest in relation to bird productivity is an essential feature if extinction of populations is to be prevented.

Tern Response to Human Intruders

Common terns in all colonies in all years, in both salt marshes and sandy beaches, responded to the presence of people by mobbing and overt attacks (see fig. 6.5). The intensity of mobbing and the frequency of attack varied among the colonies (see chapter 6). Generally, however, when a person enters the colony, the edge birds fly up and over the intruder and begin to mob. If the intruder continues to approach, center birds fly up from their nests and join the mobbing group. In most cases, the loud cries, swirling masses of birds, and their overt attacks unnerve the intruder, and most people flee from the colony. The Alfred Hitchcock motion picture, *The Birds*, has contributed markedly to peoples' apprehensions and to tern protection. Over the years, we have repeatedly heard people allude to this movie as they covered their eyes and backed away from the mobbing terns, or waved sticks to ward off their attacks. Periodic reruns of this movie would contribute substantially to protecting tern colonies from human intrusion.

8.4 CONCLUSIONS

In addition to the piracy and predation threats described in chapters 6 and 7, common terns are exposed to tidal flooding, heavy rains and inclement weather, and human disturbance. Salt marsh colonies are particularly vulnerable to flooding, and beach colonies to human disturbance. All common terns may suffer from prolonged inclement weather, which makes fishing more difficult. In terms of actual damage to the populations we studied, tidal flooding was clearly the most frequent and serious threat to the salt marsh colonies. Often entire colonies were flooded out, regardless of whether there were eggs or chicks. When washouts occur early in the breeding cycle, the terns normally re-lay, but if all nests are washed out, the breeding effort may be abandoned, resulting in no productivity for the season. For example, in 1982 flood tides eliminated nests in 17 of the 18 colonies in the middle and southern parts of Barnegat Bay, and these birds abandoned the colonies and apparently did not attempt to breed again that year.

Presumably, common terns evolved with the threat of tidal flooding and their behavior is adapted to mitigate that threat. Common terns will often re-lay when they have lost eggs or chicks, and they build up their nests to reduce the impact of rising waters. Should eggs be washed out of a nest, the terns will retrieve the eggs by extending the neck and bill, tucking the bill over the egg, and pulling the egg back toward the body.

If prolonged adverse weather conditions strike early in the nesting season, it will delay egg-laying or prolong incubation. Later in the season, during the chick phase, such weather conditions result in large-scale chick mortality. Since colonies are not perfectly synchronized, presumably only part of a colony's yield will perish under such conditions.

Human disturbance, in the form of egging, may also be a threat that terns have experienced for thousands of years in North America. The native Americans of the Atlantic Coast often moved to the shore during the spring, presumably to harvest marine life including birds. Mosquitoes, however, may have forced them to higher ground during much of the breeding season, as they did the early European settlers. The wanton destruction of eggs and chicks, and the disturbance caused by picnickers, sunbathers, bird watchers, photographers, and exploring children may well be new threats made possible by mosquito control and construction of a network of roads, public transportation, and recreational facilities—not to mention leisure time. Environmental pollution also engenders new threats, such as the reduced hatchability of eggs with only a small amount of oil (Butler and Lukasiewicz 1979). Such threats can only be eliminated by continuous direct protection and public education and an improved stewardship of the environment we share.

NINE

Reproductive Success and Colony Dynamics

An individual's lifetime reproductive success is one measure of its fitness (Williams 1966). Ideally, one would like to measure the recruitment of offspring as successful breeders in the population. Further, it is usually difficult to determine reproductive success of individual pairs over their lifetime, particularly if their offspring disperse and breed far from their natal colony.

Study of population dynamics is challenging indeed, and some important long-term studies are in progress (e.g., Nisbet et al. 1984) which will enhance our understanding of the demography and dispersal of common terns. Colony site use and turnover is a complex result of the suitability and availability of sites as well as of the birds' prior success at the site. In the previous chapters we have discussed the behavior of the terns during the breeding season and have considered the number of young reared per pair per season as an estimate of reproductive success or productivity. Previously, birds that have bred successfully at a site show a high tendency of returning to that site in subsequent years (Austin 1949; McNicholl 1975). This site tenacity probably plays an important role in reproductive success of colonial birds, or any bird for that matter. In this chapter, we examine reproductive success, causes of reproductive failure, population dynamics, and turnover rate as the end measures of the success of the behavior and ecology of common terns.

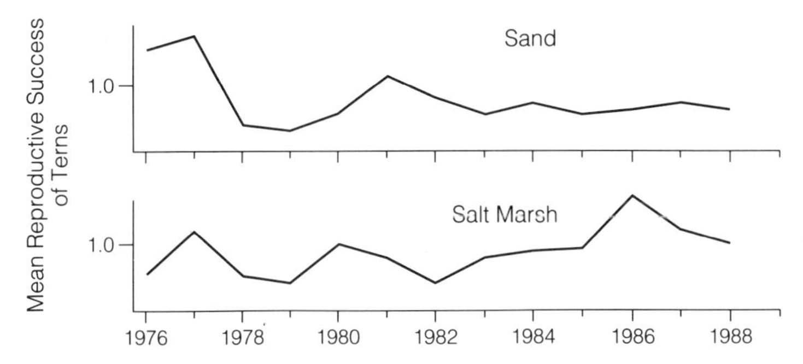

Figure 9.1. Reproductive success (mean number fledged/pair) variations for common terns in salt marsh and beach colonies in Barnegat Bay.

9.1 REPRODUCTIVE SUCCESS IN COMMON TERNS

Nisbet (1978) and DiCostanzo (1980) have provided estimates on common tern mortality, and on the recruitment of young birds to the breeding population. It is usually difficult to determine reproductive success of individual pairs over their lifetime. Even with extensive banding studies it is difficult to locate a large sample of young terns when they return to breed, although a few studies have accomplished this process successfully. Therefore, we estimated the number of young fledged, per pair, as a measure of the average breeding success of the group. In this chapter, we examine the reproductive success for common terns on a colony basis, and examine the cause of colony failures as a function of habitat, colony, and colony size. These data serve as a measure of the behavioral and ecological adaptations discussed throughout this book. Habitat choices, population dynamics, synchrony, territoriality, and antipredator behavior are of interest because of their effect on the reproductive success. In the beach colonies, productivity ranged from a low of 0 to about 1.40 young fledged per nest. Even within a single season (1972), productivity varied from 0 at Wantagh to nearly 1.40 at West End Beach (Gochfeld and Ford 1974).

Over the thirteen years that we measured productivity in New Jersey colonies (1976–1988), reproductive success varied from 0 to 2.10 young fledged per pair (table 9.1). At the Holgate beach colony, reproductive success varied from 0 to 1.26 young fledged/year (fig. 9.1). Overall, reproductive success at Barnegat Bay salt marsh colonies varied markedly from year to year, and there was no long term temporal trend in success (fig 9.1).

Table 9.1. Productivity of common terns colonies on salt marsh islands in Barnegat Bay, New Jersey (1976–1988) (best estimate of number of young fledged per pair in colony).

Colonies	*1976*	*1977*	*1978*	*1979*	*1980*	*1981*	*1982*	*1983*	*1984*	*1985*	*1986*	*1987*	*1988*	*Weighted*[a] *Mean ± SD*
N.W. Lavellette	1.30	1.44	0.56	0.06	0.58	1.40	1.07	0.60	1.33	1.08	1.10	1.25	1.50	1.14 ± .05
S.W. Lavellette	1.06	1.36	0.91	0.18	0.50	1.10	1.05	1.52	1.55	1.90	2.10	1.35	1.13	1.41 ± 0.5
N. Lavellette	1.33	1.40	0.93	0.67	0.51	1.46	1.14	1.56	1.48	1.17	2.05	1.87	1.33	1.27 ± 0.4
S. Lavellette	1.33	1.43	1.23	0.56	1.00	1.55	1.10	1.50	1.60	1.03	1.76	2.10	1.23	1.23 ± 0.3
Busters	0.00	0.00	0.00	0.00	0.00	0.00			0.21	0.41	0.65	1.00	2.00	0.32 ± 0.5
East Point			0.04	0.50	0.16				0.00					0.18 ± 0.2
Clam (small)	1.22	0.89	0.00							0.00				0.52 ± 0.6
High Bar (small)		1.13	0.00	0.88	0.90	0.00	0.00	0.00	0.00	0.40	0.88	1.28	0.10	0.37 ± 0.5
E. Vol	1.00	1.18	0.92	0.58	1.45	0.00	0.00	0.00		0.20	1.01	1.33	0.50	0.59 ± 0.6
W. Vol	1.36	1.04	0.02	0.08	1.40	0.51	0.00	0.00	0.66	0.39	1.50	1.60	0.60	0.65 ± 0.6
Gulf Point	0.93	1.31	0.00	0.00										0.56 ± 0.7
W. Sloop Sedge	0.00													0.00
E. Sloop Sedge	0.00													0.00
Sandy Island	0.00													0.00
Flat Creek	0.33	0.88	0.07	0.08	1.50		0.00		0.00					0.41 ± 0.6
W. Carvel	0.61	0.83	0.02	0.04	0.93	0.00	0.00			0.23				0.33 ± 0.4

E. Carvel	1.31	1.21	0.96	1.00	1.50	0.39	0.23	0.20	0.76	0.00	1.60	1.56	1.55	0.96 ± 0.8
W. Log Creek	1.23	1.22	0.00				0.00				1.10			0.67 ± 0.7
Log Creek	1.11	0.85	0.11	0.11	1.63		0.00		0.40	1.25	0.70	1.85	1.53	0.88 ± 0.6
Pettit	0.41	1.33	0.00	0.16	1.39	0.75	0.00	0.83	0.95	1.05	0.82	0.00	0.00	0.69 ± 0.5
Cedar Creek	0.13	0.83	0.14	0.06	1.14	0.76	0.00	0.29	0.37	0.27	0.51	0.00	0.00	0.41 ± 0.4
E. Cedar Bonnet	0.00													0.00
S.W. Cedar Bonnet	0.78	0.90	0.00	0.12	0.91	0.26	0.00	0.00				1.25	0.00	0.47 ± 0.4
Thorofare		1.21	0.00	0.00	0.30		0.00	0.34	0.40					0.32 ± 0.4
Egg (small)	0.00		0.00			0.00	0.00	0.00		0.40	0.20		0.00	0.07 ± 02
E. Ham	1.20	1.20	0.00	0.40	0.86	0.00	0.00	0.00	0.20	0.36			0.50	0.35 ± 0.4
W. Ham	0.44	1.40	0.00	0.10	0.19	0.32	0.00	0.32	0.23	0.45	0.50	1.25	1.00	0.55 ± 0.4
Marshelder				0.32								0.00	0.50	0.32
E. Long Point	0.00													0.00
W. Long Point	0.00								0.50	0.00	0.20	0.00	0.60	0.21 ± 0.3
Little	0.00	1.22	0.00	1.07	1.26	0.50	0.00	0.00	0.10	0.00		0.00		0.37 ± 0.6
Mordecai	0.00	0.93	0.00	0.20	1.26	0.00	0.00	0.00	0.48	0.00	0.26			0.29 ± 0.5
Hester Sedge	0.00	0.84	0.00	0.59	1.06	0.00	0.00	0.22	0.44	0.00				0.32 ± 0.4

[a] Weighted mean determined by multiplying annual colony size by annual productivity estimate.
(small) designates a small island in an island group with a single name.

Several conclusions can be drawn for salt marsh colonies (table 9.1): 1) reproductive success varied among colonies each year; 2) reproductive success varied among years for the same colony; 3) some colonies were consistently more successful than others; 4) every year, some colonies failed completely; and 5) in some years, there was little production in most colonies.

Overall, the means for the 33 colonies ranged from 0 to 1.41, although only 6 unusually successful colonies averaged > 1.0 young/pair. In general, the Lavellettes had high reproductive success while Buster, East Point, High Bar, and Egg had consistently low reproductive success. Overall, only 75 colony-years resulted in fledging over one young per nest, and 29 (39 percent) of these were at the four Lavellette Islands. None of the Lavellettes had a year with zero production, whereas there were 73 complete failures in the 227 colony-years.

Since reproductive success varied markedly among colonies and years (table 9.1), we examined the variation in terms of environmental or social variables. Reproductive success clearly related to habitat types. Furthermore, reproductive success was frequently dissimilar in the two habitats. That is, when success was high in salt marshes it was frequently low on the barrier beach (fig. 9.1). In only one year (1977) was reproductive success high in both habitats, and in one year (1979) it was low in both.

Reproductive success of common terns nesting in Barnegat Bay also varied significantly by colony size (table 9.2, fig. 9.2). Larger colonies consistently raised more young than smaller colonies. Colonies with fewer than 33 pairs raised no young 41 percent of the time, whereas colonies with more than 100 pairs failed only 8 percent of the time and raised > 1.0 young/pair 56 percent of the time.

Surprisingly, although clutch size is often the primary determinant of productivity, in this sample, productivity was not related to initial clutch size ($F = 1.60$, df = 1,141, $p > 0.20$). This probably relates to the unpredictability of tidal flooding which can (and does) wipe out entire colonies regardless of clutch size.

Long Island

Table 9.3 provides data on common tern productivity at various Long Island colonies. Unlike the marsh colonies, there were no complete failures. The

Table 9.2. Reproductive success in common tern salt marsh colonies in Barnegat Bay, New Jersey, as a function of colony size (number of nests per colony). Data from 1976 to 1985. Given are the number of colonies in each category (percent in parentheses).

Number of Nests per Colony[a]	*Number of Colonies*[b] %	*Number of Young Raised per Nest* *None* # %	*0–0.25* # %	*0.26–0.99* # %	*Over 0.99* # %
1–33	87 (38)	41 (47)	11 (13)	23 (40)	11 (13)
34–66	50 (22)	16 (32)	9 (18)	11 (22)	14 (28)
67–100	32 (14)	10 (31)	3 (9)	12 (38)	7 (22)
>100	59 (26)	5 (8)	7 (12)	14 (24)	33 (56)
Total (%)	227 (100%)	72 (32)	30 (13)	60 (26)	65 (29)

[a] $X^2 = 43.7$, $df = 9$, $P < 0.01$.

[b] Each entry represents 1 colony-year (33 sites, 10 years, 227 occupied colony-years).

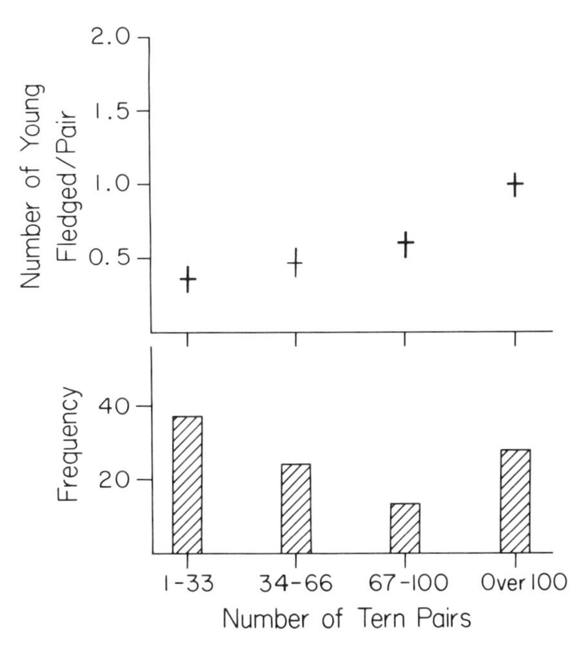

Figure 9.2. Reproductive success (mean number ± SD fledged/pair) for common terns as a function of colony size in Barnegat Bay.

Table 9.3. Productivity of terns in Long Island beach colonies. Shown are number of young fledged/nest. In some years, separate estimates were made for each area of the Cedar Beach colony. Other values indicate a range of estimated productivity for the colony as a whole.

	Cedar Beach			*West End*	*Breezy*	*Short*
	East	*Center*	*West*	*II*	*Point*	*Beach*
1972	——	1.08–1.25	——	1.14–1.40		0.94–1.21
1976	1.14	1.48	1.42			
1977	1.05	0.96	0.28–1.64			
1978	0.56	0.30	0.89			
1979	0.80	0.68	1.05–1.20	0.90	0.05	
1984	——	1.15–1.33	——			
1985	——	1.10–1.29	——			
1986	——	0.93–1.02	——			
1987	——	1.17–1.25	——			

beach colonies in New York have not suffered complete washouts due to flood tides, largely because the two large colonies (Cedar Beach, West End) are in interdune areas behind the outer dunes.

To examine the role of habitat in a different fashion, we followed the fate of common tern colonies nesting in different habitats on Long Island in the period from 1976 to 1980. There were 131 colony occupancies, 93 of them in salt marshes. There was only one colony on an outer beach and it failed completely. None of the spoil bank or salt marsh colonies produced more than 1.0 young/pair, while 40 percent of the interdune colony-years did so (table 9.4). There were fewer than 0.1 young/pair in 94 percent of the marsh colonies, a result substantially poorer than in Barnegat Bay (see table 9.2). 133 of 215 New Jersey marsh colonies-years had > 1.0 young compared with 0 of 93 in the New York marsh colonies ($p < 0.001$).

Discussion

Overall, it is not surprising that we found such variability in reproductive success, particularly for Barnegat Bay where we followed about 33 colonies over a 13-year period on quite diverse salt marsh islands. No similar data set is available for comparison. What is remarkable is the continued use of these islands despite a variety of predator, human, and tidal disruptions.

Table 9.4. Comparison of colony habitat types with respect to use and success of common tern colonies on Long Island.

Site Type	*Year*	*Total Occupied Sites*	*Number of Colonies Producing Young*			
			0	*<0.1*	*0.1–1.0*	*Over 1.0*
Interdune	1976	4	1	1		2
	1977	5	2		1	2
	1978	3		1	1	1
	1979	3	1		1	1
	1980	2		1		1
	Subtotal	17	24%	18%	18%	40%
Spoil island	1976	3	1	1	1	
	1977	2	1		1	
	1978	4	1	2	1	
	1979	6	2	1	1	
	1980	5	2	2		
	Subtotal	20	41%	24%	35%	
Salt Marsh	1976	7	5	2		
	1977	11	9	2		
	1978	23	6	7		
	1979	28	5	5	2	
	1980	24	2	7	2	
	Subtotal	93	51%	43%	6%	
Outer Beach	1980	1	1			
Summary of outcomes			38	32	12	7
% of known outcomes			42%	36%	13%	8%

The reproductive rates we report (0–2.10 young fledged per pair) are within the range reported in the literature (table 9.5). As is clear, most productivity studies cover only 1 or 2 years or only 1 or 2 colonies. It is not unusual for some colonies to fail completely while others are fairly successful. Since there is a worldwide pattern in variability, it suggests that the behavior and ecological adaptations of common terns accommodate this variable success. Common terns live for more than a decade, and may breed 10 or more seasons. Indeed, for such a long-lived species, it is not essential for an individual pair to be successful in any given season.

Table 9.5. Productivity of common terns from the literature.

Location	*Fledged/pair*	*Habitat*	*Source*
Great Britain	1.50–2.24	dry land	Langham 1972
Kustavi, SW Finland	1.46	dry land	Lemmetyinen 1973
New England (1970–71)	0.4–2.1 mean = 0.9	dry land	Nisbet & Drury 1972b
New England 1972	0.0–1.8 mean = 0.4	dry land	Nisbet & Drury 1972b
New England	0.0–2.0	dry land	Nisbet 1973
New England	1.1–1.5	dry land	Austin studies from Nisbet 1973
Massachusetts[a]	2.5	dry land	Nisbet et al. 1984
Great Gull Island (N.Y.)	0.4–0.7	dry land	Lecroy & Collins 1972
Great Gull Island (N.Y.)	0.8–1.0	dry land	Hays 1978
Cedar Beach (N.Y.) 1972	1.08–1.25	dry land	Gochfeld & Ford 1974
West End Beach (N.Y.) 1972	1.14–1.40	dry land	Gochfeld & Ford 1974
Wantagh (N.Y.) 1972	0.0	dry land	Gochfeld & Ford 1974
Short Beach (N.Y.) 1972	0.94–1.21	dry land	Gochfeld & Ford 1974
New York–New Jersey	0.0–1.90 mean = 0.54	salt marsh	This study
New York–New Jersey	0.0–.45 mean = 0.96	sandy beach	This study
Lake Ontario, Canada	0.49–1.16	dry land	Morris et al. 1980
Lake Erie, Canada	0.62–1.60	dry land	Richards & Morris 1984
Minnesota	0.15	dry land	McKearnan & Cuthbert 1989

[a] Early laid clutches only.

9.2 ADVERSITIES AFFECTING REPRODUCTIVE SUCCESS

The major causes of reproductive failure for the terns were starvation of chicks (particularly 2d and 3rd chicks), exposure to cold wet weather, predation, flooding and human disturbance. Minor causes of failure included attacks on wandering chicks by neighboring terns, defective eggs (fig. 9.3 top), disease, and congenital and developmental defects (fig. 9.3 middle and bottom).

Starvation and exposure were often linked. Several days of cold wet weather interfered with fishing (Dunn 1973), causing the foraging adult to be absent for longer periods and perhaps causing both adults to be absent. If chicks got wet and chilled during this period, they were likely to die.

Infectious disease is a major theoretical problem associated with coloniality, yet evidence for this occurring in common tern colonies is sparse. We have documented three episodes (twice on Long Island and once in Barnegat Bay) when dozens of fledgling terns and adults died. These were characterized by greenish fecal staining of the cloacal area. The cause was not identified, but we suspected that it was infectious.

However, Nisbet (1983c) has described paralytic shellfish poisoning in Massachusetts common terns, and this condition also resembles that observed in our colonies. It is due to a neurotoxin produced by the red tide dinoflagellate *Gonyaulax excavata*.

We examined reproductive failure by comparing mean number of young fledged, per pair, among colonies and years. In salt marshes, flooding was the major cause of failure, followed by predation (table 9.6, fig 9.4). Human disturbance was very slight as few people ever entered salt marsh colonies. For terns nesting on beaches at Holgate, Corson's Inlet, and Stone Harbor Point (New Jersey), the causes of colony failure were floods, human disturbance, and predators in about equal proportions. In some years, floods completely destroyed beach colonies; in other years predators (rats, fox, owls) completely eliminated reproduction; and, in others, human intruders caused destruction. In general, floods frequently destroyed a colony completely while predators and human disturbance reduced productivity, but usually not to zero. In New York beach colonies, floods and predators took a lower toll of eggs and chicks, and starvation (of the youngest chicks, fig 9.5) was of relatively greater importance. In 1971, an unknown nocturnal predator took most of the common tern chicks hatched at West End.

Discussion

The overall factors affecting reproductive success in terns were flood tides and predation. Human disturbance was a factor in some habitats but not others (see below), while prolonged cold wet rains accompanied by wind and fog were critical in some years. Floods were the primary cause of reproductive failures in salt marshes, whereas human disturbance and predation were the causes in beach colonies.

Figure 9.3. (top) Thin-shelled egg, possibly associated with contamination by chlorinated hydrocarbon pesticides, which that broke spontaneously in the nest. (bottom) A congenital defect: downlessness and phocomelia (lack of development of the wings and legs). (right) A fledgling common tern showing feather loss. All of the wing and tail feathers have broken off at the base due to a developmental defect probably associated with high levels of mercury or PCBs. Defects such as these were more common in the early 1970s and have declined in recent years.

Flood Tides

Flood tides were the primary cause of mortality in salt marshes and occasionally affected beach colonies. In some years, in New Jersey, few terns and no skimmers fledged because of high tides. When flood tides occurred early in the season, terns usually relaid and began a new reproductive cycle.

In large colonies of marsh or beach-nesting terns, floods seldom washed out all the nests because some tern nests were higher than others. Whenever a large segment of a tern colony was not washed out, the pairs with intact nests served as a nucleus of social activity. Pairs that had lost their nests merely relaid, sometimes in their previous nests. In smaller colonies where all nests were washed out, the terns were likely to abandon and move to a new colony site. Flood damage can destroy nests, wash away eggs or young chicks, or cause cold stress and death even in older chicks. Severe floods can wash out 200–300 nests at a time, completely destroying a colony.

The potential for flood damage may be difficult to assess early in the season, because so many factors affect the actual height of the highest tides. Storms and high winds determine which colonies will be flooded in any given year.

Flood tides are an important cause of mortality in a number of coastal species, including laughing gull (Bongiorno 1970; Montevecchi 1978b), clapper rails (Mangold 1974), herring gulls (Burger 1979a; Ward and Burger

Table 9.6. Causes of colony failure in salt marsh-nesting common terns (1976–1988) in Barnegat Bay, New Jersey. Given are the number of colonies in which an event occurred.

	1976	*1977*	*1978*	*1979*	*1980*	*1981*	*1982*	*1983*	*1984*	*1985*	*1986*	*1987*	*1988*	*Total*	*Percent*
Colonies that failed (<0.25 young/pair)	12	1	19	15	2	8	18	10	7	9	2	5	5	113	37
Failures due to:															
floods	6	1	12	7		6	18	9	7	0	1	3	4	74	25
predators	5		4	4	2	2		1	0	8				26	4
human disturbance (followed by predation)	1			1										2	<1
floods and predators (each at least 25%)			3	2						1	1	2	1	10	3
disease[a]				1										1	<1
Colonies that raised 0.26–0.99 young/pair	6	8	5	7	9	7	0	4	9	6	8	1	6	76	27
Failures due to:															
floods	4	6	5	2	6	5		4	8	1	7	1	6	55	19
predators	2	2		2	3	2			1	5	1			18	3
disease[a]				3										3	<1
Colonies that raised over 1 young/pair	11	15	2	3	11	5	4	4	4	6	8	12	8	103	36

[a] Dead young had diarrhea and were emaciated.

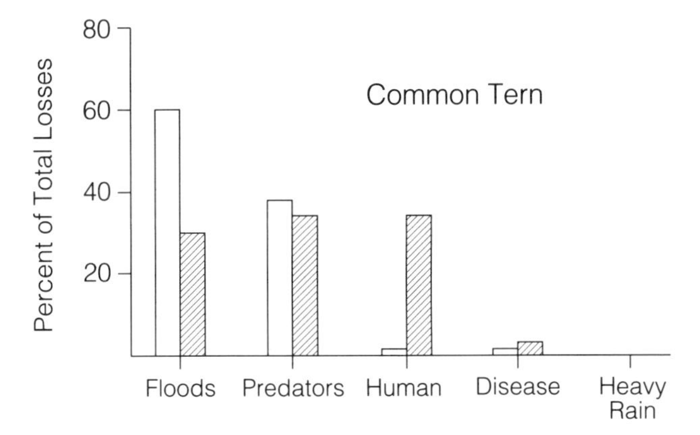

Figure 9.4. Causes of colony failures for common terns nesting on salt marsh (open bar) and beach colonies (hatched bar) in New Jersey colonies. Human refers to direct human disturbance.

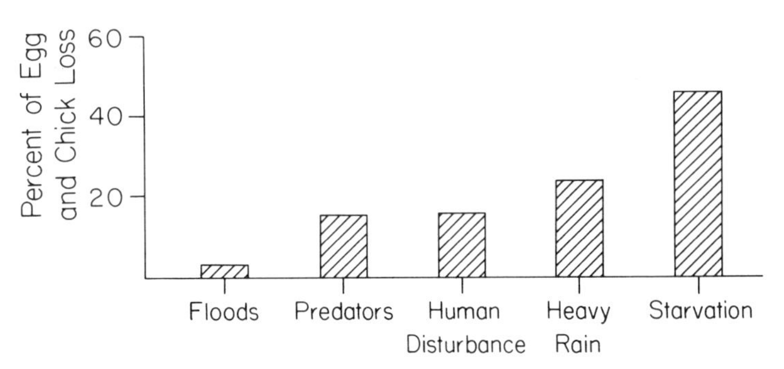

Figure 9.5. Causes of egg and chick loss of common terns in New York beach colonies.

1980) and least tern (Burger 1984a). The vulnerability of common terns to flood tides has frequently been noted, both for beach and marsh nesting colonies (Austin 1929, 1933; Greenhalgh 1974; Gochfeld and Ford 1974; Burger and Lesser 1978, 1979; Erwin et al. 1981). Several species of larids regularly nest in marshy habitats—where they are vulnerable to flooding during rain storms—including black terns (Baggerman et al. 1956), Franklin's gull (Burger 1974a), brown-hooded gull (Burger 1974b), laughing gulls (Bongiorno 1970), and Forster's terns (McNicholl 1982). Exposure to regular tidal inundation allows birds to adapt by building higher nests or selecting high nest sites (Burger 1979b). However, the unpredictability of occasional flood tides makes long-term adaptation difficult. Occasionally rain and wind storms coincide with high tides to produce massive flooding.

Predators

Predators are a major source of mortality to most birds. Predation generally accounted for 20 to 40 percent of the mortality over the 13 years of productivity data for New Jersey salt marshes and beach colonies and 15 percent of the mortality for the New York beach colonies. Predators usually take a small but steady toll of eggs and young, day after day. In some cases, however, the entire colony may be destroyed and subsequently abandoned because of predators. This is particularly true with rats (Austin 1929, 1948). Which outcome prevails depends upon the number and type of predator as well as the colony size and antipredator behavior of colony members.

The importance of predators to colonial-nesting species has been extensively reviewed (Kruuk 1964; Burger 1981d; Gochfeld 1985). Oystercatchers and gulls are diurnal avian predators that can destroy a large number of eggs. Unless suitable preadaptations exist, it is difficult for terns to deal effectively with man-influenced predators such as dogs, cats, and—particularly—rats. Further, man has influenced the quantity of predators by changing the habitat on barrier islands and providing a constant source of food, allowing species such as fox and raccoon, normally rare on barrier islands, to maintain themselves.

Human Disturbance

Human disturbance includes destruction of eggs or chicks, and indirect forms of disturbance such as walking through the colony, loud noises, or habitat destruction (Burger 1981a, f). Human disturbance at salt marsh colonies was minimal and was limited to a few occasions when a boater with a dog landed on the islands. Many people consider salt marshes unappealing or repugnant because they are wet, muddy, smelly, and full of biting insects. By contrast, however, at beach colonies human disturbance was often severe, because the nesting colonies are on beaches preferred for bathing. People walking through a colony keep adults off nests, exposing them to increased threat of predation and thermal stresses. Both eggs and young chicks are exceedingly vulnerable to heat stress (Austin 1933; Gunter 1982) as well as cold stress.

Deliberate egging has rarely been a problem in the beach colonies on Long Island, and we found no evidence of egging elsewhere. The effect of egging is potentially severe, and terns respond by re-laying if it is early in the season or abandoning their nest sites if it is late in the season. We believe the occurrence of egging in the Long Island colonies represents a tradition brought over by recent immigrants from Europe and the Orient. Since there are continual waves of human immigrants to the New York City area, the tradition persists.

9.3 COLONY DYNAMICS

Most of the New York colonies were small (fewer than 250 pairs), but most of the terns nest in the few large colonies (> 1000 pairs). This was also true for common terns in Poland (Wesolowski et al. 1985). The relationship was less apparent in New Jersey, where there were many medium-sized colonies and few large ones. The mean size of British common tern colonies was 263 pairs, while the mean of 31 Virginia colonies was 95 pairs (Erwin 1978). In Barnegat Bay, where common terns nest mainly on salt marsh islands, there were 34 (32 marsh, 1 small beach on a marsh, 1 sandy island) colony sites that were used for nesting, although in any given year terns nested at only 18–29 different sites (table 9.7). From 1976 to 1990, the number of occupied colony sites decreased steadily, while the number of nesting terns varied widely from year to year (table 9.7, fig. 9.6). Colony size generally ranged from 1 to 800 pairs, and most colonies were under 400 pairs, although in 1990 Pettit had over 1000 nesting pairs. In most years, 25 percent or more of the total tern population was concentrated on the Lavallette Islands (north section of the Bay). Other areas of high concentrations were on the Vol Sedges (East and West Vol), Carvel, and Pettit Islands (table 9.7).

The tern population figures for the Barnegat Bay salt marshes show a great deal of variation among years and colonies, typical of a mobile population. The beach colonies, in contrast, proved more stable both in New Jersey and on Long Island. The population of common terns on western Long Island has increased steadily, yet birds remain concentrated mainly at the few large beach colonies (fig. 9.7). The West End colony has been in existence since about 1950, although the actual nesting area has shifted somewhat as beach facility construction has usurped tern nesting areas. The Cedar Beach colony has existed only since 1970, but has grown to a very large size for this species

Table 9.7. Maximum number of nesting pairs of common tern in Barnegat Bay, New Jersey (1976–1990). Estimates are ± 5 percent.

	1976	*1977*	*1978*	*1979*	*1980*	*1981*	*1982*	*1983*	*1984*	*1985*	*1986*	*1987*	*1988*	*1989*	*1990*
N.W. Lavallette	115	125	180	250	320	30	75	50	75	750	450	350	300	300	350
S.W. Lavallette	104	125	110	105	56	250	200	250	200	210	300	510	440	350	250
N. Lavalette	113	250	350	450	75	120	123	225	250	350	800	750	750	600	200
S. Lavalette	554	350	300	225	150	200	192	100	175	300	600	550	500	550	350
Pelican[a]	0	0	2	40	0	0	0	0	0	0	0	0	0	0	0
Buster	826	10	38	45	10	5	0	0	15	18	65	170	125	75	0
East Point	0	0	0	52	60	76	0	0	18	0	0	0	0	0	0
Clam (small)	9	9	75	0	0	0	0	0	0	45	0	0	0	0	0
High Bar (small)	0	16	20	8	10	15	12	10	20	25	28	36	29	30	0
East Vol	1	38	252	300	58	79	2	5	0	25	25	25	75	80	150
West Vol	55	50	169	180	200	93	61	20	47	64	50	40	125	150	100
Gulf Point	83	70	42	110	0	0	0	0	0	0	0	0	0	0	0
West Sloop Sedge	87	0	0	0	0	0	0	0	0	0	0	0	0	0	0
East Sloop Sedge	12	0	0	0	0	0	0	0	0	0	0	0	0	0	0
Sandy Island	1	0	0	0	0	0	0	0	0	0	0	0	0	0	0
Flat Creek	33	8	14	13	10	0	13	0	5	0	0	0	0	0	0
West Carvel	46	36	58	47	44	42	19	0	0	26	0	0	0	0	0
East Carvel	45	87	265	210	156	166	300	50	88	90	5	260	320	350	0
West Log Creek	66	50	43	0	0	0	14	0	0	0	5	16	0	0	10

Log Creek	28	20	27	18	19	0	26	0	5	20	50	60	65	50	25
Pettit	29	42	94	90	98	127	353	326	409	505	12	36	25	200	1050
Cedar Creek	230	12	35	50	36	46	99	83	27	15	20	40	50	25	0
E. Cedar Bonnet	37	0	0	0	0	0	0	0	0	0	0	0	0	0	60
S.W. Cedar Bonnet	9	31	24	26	11	31	11	1	0	0	0	120	60	50	10
Thorofare	0	80	18	77	10	0	33	58	20	0	0	0	0	0	0
Egg (small)	13	0	3	0	0	6	2	11	0	45	110	0	25	20	20
East Ham	0	10	20	90	70	25	20	42	10	83	0	0	80	60	40
West Ham	16	40	70	49	16	78	51	38	52	55	300	150	30	20	300
Marshelder	0	0	0	250	0	0	0	0	0	0	0	4	220	200	8
E. Long Point	32	0	0	0	0	0	0	0	0	0	0	0	0	0	0
W. Long Point	43	0	0	0	0	0	0	0	20	8	6	18	28	20	20
Little	235	55	42	30	72	70	85	36	20	25	10	8	0	0	25
Mordecai	2	30	42	10	50	18	2	2	100	65	5	0	0	0	0
Hester Sedge	6	80	80	110	121	119	118	90	80	35	0	0	0	0	20
Total Number of Nests	2840	1624	2373	2835	1652	1596	1811	1397	1636	2759	2841	3113	3345	3130	2988
Number of Active Colonies	29	24	26	25	22	20	22	18	20	21	18	18	19	18	18

[a] Pelican Island is the only non-salt marsh colony. It is a dredge spoil island.

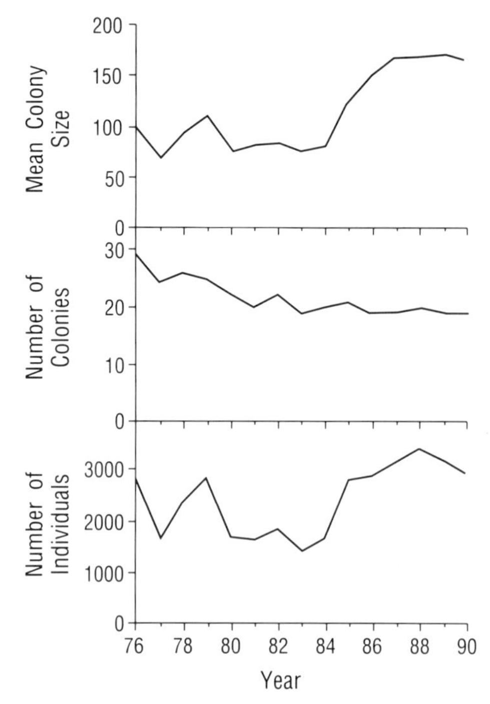

Figure 9.6. Total population size of common terns nesting on salt marshes in Barnegat Bay.

(over 6,000 pairs). Three beach colonies (Wantagh, Breezy Point, and West End II) disappeared during our studies (fig 9.7), although Breezy Point was recolonized after four years and the West End colony relocated about 1.5 km away.

In New Jersey, common terns nested on salt marshes and sandy beach colonies. The major beach tern colony, Holgate, was extremely stable and was active for the fifteen years of the study. In some cases, parts of the beach colonies were eventually washed out by high tides and the birds abandoned these sites, but the colony was always occupied early in the season.

Terns nesting in salt marsh colonies, however, were less stable. Only 8 marsh colonies were occupied in every year of the study, and several sites were occupied only once or twice (table 9.7). Similarly, the mean number of nests per colony varied and showed no clear pattern over the fifteen year study. The year to year turnover rate (Erwin et al. 1981) of salt marsh tern

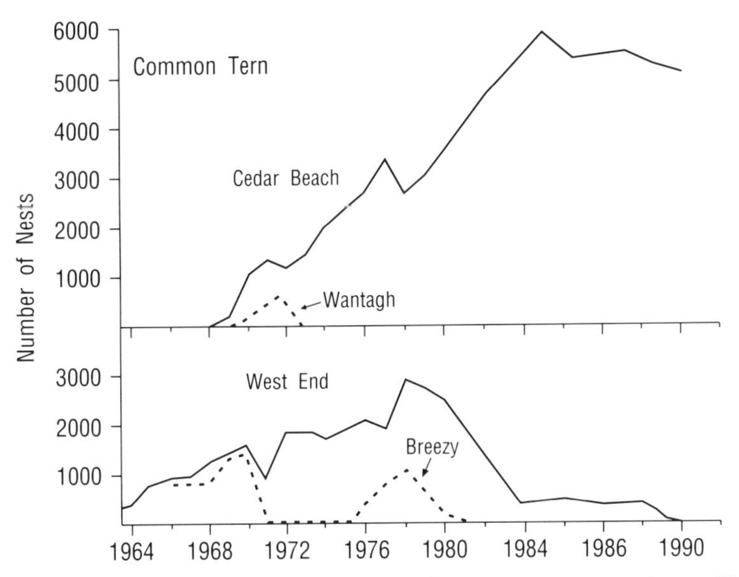

Figure 9.7. Total population size of common terns nesting at West End and Cedar Beach and nearby sand colonies (number of nests).

colonies ranged from 0.0 to 0.22 (mean = 0.10). Thus, in most years, tern colonies in salt marshes also were relatively stable (table 9.8).

Discussion

The total population of common terns in Barnegat Bay has fluctuated markedly in the fifteen years of the study, but has remained relatively high since 1985. This stability largely results from the lack of flooding in the period 1985–1990, when very few colonies suffered complete washouts. Since nonbreeding terns do not linger at nesting colonies, the counts represent actively breeding pairs. Even when pairs fail, they usually either re-lay in the same colony or abandon the colony. Thus, the variation in population levels reported fairly well represents changes in population size for Barnegat Bay.

Population levels, however, should be viewed on a more regional scale, and our banding results show that some terns that fail in a breeding attempt in the Barnegat Bay salt marshes may move to one of the large beach colonies in southern New Jersey or Long Island to nest later in the same season.

Colonial birds often exhibit a high degree of site tenacity, returning to the

Table 9.8. Colony size and turnover rates in common tern colonies in Barnegat Bay, New Jersey.

	1976[a]	*1977*	*1978*	*1979*	*1980*	*1981*	*1982*	*1983*	*1984*	*1985*	*1986*	*1987*	*1988*	*1989*	*1990*
Total maximum number of nests	2830	1624	2373	2835	1652	1596	1811	1397	1636	2759	2841	3113	3345	3130	2988
Total number of colonies	29	24	26	25	22	20	22	18	20	21	18	18	19	18	18
Range in number nests/colony	1–826	8–350	2–350	8–450	10–320	5–250	2–353	1–326	5–250	8–750	5–800	4–750	10–750	20–600	8–1050
Number colonies reused (from previous year)	24	21	24	23	22	19	18	18	15	16	18	17	16	18	14
Number colonies lost from previous year	1	7	0	3	3	3	2	4	3	3	3	2	2	0	4
Number of new colonies	4	3	2	2	0	1	4	0	5	3	0	1	3	0	4
Turnover rate	0.08[a]	0.18	0.04	0.10	0.06	0.09	0.01	0.09	0.21	0.14	0.07	0.08	0.13	0.00	0.22

[a] 30 colonies were present in 1975, but detailed nest counts were not made.

same colony year after year (Austin 1949; McNicholl 1975; Southern and Southern 1980). In a stable environment, it is advantageous to return to the same site year after year, particularly if reproductive success has been high. Over a longer time frame, however, tern numbers fluctuate and colony sites may shift (Nisbet 1983).

In ephemeral or unstable habitats, it is advantageous to be able to shift sites if a site proves unsuitable. Indeed, species such as black-billed gull (Beer 1966), Franklin's gull (Burger 1974a), brown-hooded gull (Burger 1974b), and laughing gull (Bongiorno 1970; Montevecchi 1978) that nest in marshes or on riverine sand bars will shift nests and colony sites depending on the condition they find at the start of the nesting season. Common terns nesting in salt marshes might also be expected to shift sites (Burger and Lesser 1978; Storey 1987a,b) when tidal conditions warrant.

Turnover rate in the New York common tern colonies was very low, and the two primary beach colonies were used almost every year (Cedar Beach from 1970 to 1990 and West End II from 1969 to 1987). In New Jersey, however, turnover varied from year to year, but was still relatively low. The common terns appear to have selected islands that are relatively free from predators and that more often than not avoid tidal flooding. The turnover rates for salt marsh common terns we found are low compared to those reported by Erwin et al. (1981) for marsh-nesting herring gulls (0.18–0.32), black skimmers (0.36–0.49) and common terns (0.44). Erwin et al's data however, were based on only two years. In contrast, we found that turnover rates for salt-marsh-nesting terns were 0–0.22 while for skimmers they were 0.07–0.41, based on 15 years of data, suggesting that—over a long period—nesting in salt marshes may be more stable than initially postulated.

We also found that large tern colonies (> 250 pairs) tended to have lower turnover than smaller colonies, and that the very large beach colonies were the most stable of all. Stability or persistence and colony size seem closely related in common terns, as we found for black skimmers (Burger and Gochfeld 1990a). Wesolowski et al. (1985) also reported that large common tern colonies in Poland had low turnover.

9.4 CONCLUSION

Reproductive success in the salt marsh and beach colonies varied from 0 to 2.10 young fledged/pair. Success varied within and among years and colonies. Major causes of reproductive failures were floods, predation, and human disturbance. Floods could completely wash out some colonies, while predators and people usually only lowered reproductive success. Salt marsh colonies were the most vulnerable to flood tides, while the beach colonies were most vulnerable to predators and human disturbance. Thus, by nesting on a variety of habitats in many colonies, the tern population is buffered against one or another adversity. Terns produced young somewhere in all years. Successful colonies were generally more stable, providing suitable colony sites for terns that abandoned sites that had suffered flooding or predation. Terns nesting in salt marshes generally nest in smaller colonies than those on beaches. Consequently, whenever there was a disturbance in salt marshes, the entire colony usually responded. This was not true in the large beach colonies. Moreover, when predation or flood tides struck marsh colonies they often completely wiped out the colony for that year; whereas beach colonies seldom failed completely.

The more observations and data one assembles on the breeding of colonial birds, the more complex the picture and the more unanswered questions arise. Why, for example, don't all the terns breed in very large colonies at the few sites which show the greatest success? Why should a handful of terns pioneer at some tiny salt marsh islet, where they are subsequently flooded out? Are such "mistakes" due to inexperience or are they costs of the flexibility necessary to achieve a multihabitat flexible evolutionary strategy?

Historical accounts of huge numbers of terns rising from the beaches like swirling snow, suggests that in precolonial times, colonies in excess of tens of thousands may have typified the Atlantic coast, while the island of Griend in the Netherlands had about 25,000 pairs until the mid 1950s (Roselaar 1985). These pale besides estimates that the common tern population of Sable Island, Nova Scotia, may have been about one million in historic times (Anthony Locke, personal communication). Why then don't we see a few very large colonies thriving on a few well-protected refuges? The Cedar Beach site, for example, currently among the largest common tern colonies in the world, has a relatively low density of nesting birds. It could probably accommodate all the common terns for 100 km around if they were to nest at maximum density. Is there selection for not putting all of one's eggs in one basket? Conversely, perhaps it is the low density nesting at Cedar Beach which allows it to be so successful and thereby to attract new recruits.

In addition to contributing to our understanding of the biology of the common tern and colonial seabirds in general, answers to some of these questions have important conservation and management implications. The breeding population of common terns in the northeastern United States seems to be stable on the order of about 75,000 pairs (Kress et al. 1983). However, this follows a very stormy history. The millions of terns of the precolonial era were reduced to a remnant by constant exploitation. The recent history reflects the near extinction of terns in the late nineteenth century and their gradual recovery under protection beginning about 1913 and continuing, with marked ups and downs (Nisbet 1973b), to the present time. Nisbet (1978) calculated that the decline in New England's common terns during the 1960s and 1970s was not due to lowered reproductive success or recruitment, but probably to adult mortality, and he inferred that this must be occurring on the wintering grounds. Common terns and other coastal birds are harvested for food in northern South America (Trull 1983) and the magnitude, geographic extent, and impact of this harvest remain to be determined.

Part of the difficulty we have had in studying and portraying the breeding biology of the terns is the interrelatedness of all phenomena. Colony site selection is intimately related to feeding and predation. Nest site selection is intimately related to protection from aggression, from weather, and from predators. Good feeding success frees adults to guard their chicks. Gaston and Nettleship (1981) conclude their monograph on the thick-billed murre by noting that all these interrelated phenomena pivot on the availability of food, which is the one variable that we have, until recently, had the greatest difficulty measuring.

We are convinced that there are limitations to trying to predict where terns will nest and what habitats they will prefer. Many terns carry with them an individual history of perhaps a dozen prior breeding seasons. Moreover, they are highly social, moving in groups, some being attracted to large aggregations, others remaining in small flocks. These factors must influence the movement and settling behavior of terns in ways that have thus far eluded our inquiry, and they pose abundant challenges (both for the terns and for ourselves) in the years that lie ahead.

TEN

Conclusions and Summary

Common terns nest in a variety of habitats worldwide, ranging from sandy beaches, spits, and dunes, to rocky or shingle beaches, to vegetated rocky islands and salt marshes. In general, however, these are all flat relatively open habitats, with good visibility and sparse vegetation, allowing, for example, early detection of approaching predators. In the colonies we studied, common terns nested in both salt marshes and sandy beaches, and in both habitats experienced predation and flooding to varying degrees. Although visibility was higher on sandy beach colonies, on salt marsh islands the terns nesting on mat had open visibility in all directions. In both habitats, some terns had sufficient visibility to warn conspecifics of mammalian or avian predators approaching at a distance. Although risks of flooding were higher on the marshes, the risk of predation was lower. Terns nesting on sandy beaches could merely walk between territories or nests to defend territories, court mates, or interact with chicks while those nesting among salt marsh grasses usually flew to encounter neighbors or intruders. Thus, the habitat differences resulted in some differences in reproductive behavior regarding nest defense and courtship behavior, as well as differences in vulnerability to tidal flooding and visibility of predators.

Having selected a marsh or a sandy beach for nesting, common terns then select territories which vary in size, degree of vegetative cover, and location within the colony. In salt marshes, the decision involves nesting on wrack or on the grass itself, whereas on beaches it involves nesting on open sand or nearer vegetation. In both habitats, terns encounter black skimmers, and the larger-bodied skimmers usually can displace common terns from the skimmers' preferred locations. In both habitats, skimmers often nested in the center of open spaces, either mat on the marshes, or open sand surrounded by vegetation on the sandy beach. Even though terns prefer edges with some vegetation, in the absence of skimmers some terns settle in the open habitats otherwise used by nesting skimmers.

Common terns establish territories that they use for courtship, mate acquisition, copulation, egg-laying, incubation, and chick care. When undisturbed, the young remain on the territory until after they can fly. However, our periodic intrusions often disrupted this tendency. Territory size varied among colonies and years, with larger colonies usually being denser. Birds nested closer together on salt marsh mat, at intermediate distances on sandy beaches, and farthest apart in *Spartina* grass. The small internest distances on mat are a result of intense competition, since on most salt marsh islands there is only a small amount of mat. Both salt marsh grass and sandy beaches provide abundant habitat, but even within these relatively homogeneous habitats slight differences in elevation or vegetative cover make particular spots preferred. Although individual terns may develop a preference for either salt marsh or sandy beach, some individuals do move between habitats, even in the same nesting season, if they lose a clutch.

Aggressive displays as well as overt attacks and chases are used to defend territories, and territorial aggression varies seasonally depending on the stage in reproduction and the vulnerability of the reproductive unit. There are high levels of aggression early in the season during the territory acquisition stage. Aggression is particularly high toward skimmers during the terns' egg-laying period when skimmers are establishing territories and may be usurping space. Aggression levels then decrease during incubation, when merely covering the eggs protects them from territorial disputes and predation. Aggression still occurs during incubation because some pairs without territories are still trying to obtain them, and some unmated birds land in an attempt to acquire mates. Aggression increases markedly during hatching, when small chicks are vulnerable to attacks from territorial neighbors (either terns or skimmers), from intruders seeking space, or from potential predators.

In sandy beach colonies, there is a secondary peak in aggression when chicks are two to three weeks old and begin to wander into neighboring territories. They are still being guarded most of the time by one parent, who defends them against the attack of neighbors. As chicks get older and larger, parental aggression levels again decrease because chicks can defend themselves and even to some extent their territory, and the risk of adopting a neighbor's chick is decreased.

Not only the amount of aggression, but the type of aggression varies seasonally. The terns performed more high-intensity displays (ground and aerial fights) during the early chick stage than during incubation. Again, this reflects the vulnerability of small chicks that can be easily pecked to death or even eaten by skimmers.

Because of differences in habitat, terns nesting in open areas of sandy beaches use walking displays, whereas terns nesting in vegetation in the same colonies use flying displays. Similar differences occur in salt marshes where terns nesting in *Spartina* are unable to walk between nests.

Aggression is not only a result of internal factors and reproductive stage, but of intruder pressure. If there are no intruders, aggression levels reflect only encounters with neighbors. Intruders can be seeking space for territories, seeking mates, or attempting to steal fish from other terns. Early in the season, territory-seeking birds are usually unmated males or pairs, and, if they are persistent, they usually succeed in insinuating themselves between existing territories.

During the courtship feeding stage, some intruders may attempt to steal fish, while others land with fish to present to an already paired female, apparently to secure a mate. While a female's mate is out searching for fish to courtship feed her, other males may attempt to win her by presenting fish themselves. Nesting females almost invariably resist such offerings. Before and during the egg-laying period, intruders often interrupt copulation, and this cost can be quite high for pairs nesting in optimal habitats. During the egg-laying period 11 percent of the copulations were interrupted. Common terns nesting on mat in salt marshes were interrupted at the highest levels, and those nesting in the *Spartina* grass itself were seldom interrupted. These interruption levels corresponded to territory size differences; terns nesting in dense, highly preferred habitats were interrupted more often.

Following the peak of egg-laying most pairs have mates or territories, and the number of intruders is low and limited to late-nesting pairs seeking space. Later in the season, potential pirates intrude in an attempt to steal fish from parents feeding chicks, or from young unable to quickly swallow a fish.

In beach colonies, terns were exposed to more tern intruders; whereas in salt marsh colonies, terns were exposed to more skimmer intruders. This difference reflects relative species composition. In salt marshes, skimmers often dominate the subcolonies on mats and there may be equal numbers of terns and skimmers. However, in sandy beaches when both species occurred, terns were almost always more numerous.

The aggression costs, in terms of both time and potential injury, are clearly higher when the terns are nesting in colonies compared to solitarily. Where common terns nest solitarily or in groups of two or three (on some salt marsh islands), the birds engage in almost no aggression, and there are almost no intruders. As colony size and nesting density increase, the number of intruders and the amount of aggression increase. The highest levels of aggression occur in the densest colonies with the highest numbers of breeding pairs. Further, the costs of copulation interruption are also highest in large, dense colonies or subcolonies. In addition to the time costs involved in aggression, territorial clashes can lead to injury, particularly for chicks that wander into adjacent territories. Sometimes, bloody chicks are found dead; at other times, we have observed skimmers eating tern chicks that wandered into their territories. Thus, the aggression levels are a clear cost of nesting in colonies, and that cost increases in dense subcolonies.

Predation is a primary selective factor on reproductive success for all birds, regardless of whether or not they nest in colonies. Nonetheless, the location of colonies is often obvious to predators because of the conspicuousness of the assemblage. Birds nesting in colonies often nest in remote and/or inaccessible locations, where they avoid mammalian and even avian predators. Predators can either take eggs or chicks, or they can take adults, and the latter clearly presents a greater threat to overall fitness. Thus terns vigorously mob egg and chick predators, but are more likely to flee from a species that takes adults.

Whereas many seabirds nest on inaccessible, oceanic islands devoid of indigenous mammalian predators, common terns nest on coastal habitats (as well as interiorly) where mammals are at least potential predators. Certainly, the common tern colonies on barrier beaches are accessible to predators, and even most salt marsh colonies are near enough to land to allow predators to swim or to walk across winter ice to reach them. In most cases, the salt marsh islands are sufficiently small and low so that there is not enough food to maintain predators throughout the year (e.g., fox, raccoon) and high winter tides overwash them killing any small mammalian predators (e.g., rat). Thus, mammalian predation is a much greater threat on the barrier beach sand colonies than the salt marsh colonies.

Potential mammalian predators at the beach colonies include fox, raccoon, skunk, weasel, mink, grey squirrel, dog, cat, and Norway rat. All these species were historically rare on barrier beaches (Van Gelder 1984), but have increased in abundance with increased human populations on the barrier beaches. We observed fox, mink, dog, and rat on salt marsh islands, and squirrel, raccoon, dog, cat, and rat on the beach colonies. The human commensals (rats, dogs, and cats) were the most frequent predators in tern colonies which seems to be a general phenomenon around the world. Raccoon, squirrel, and rats ate mainly eggs; mink ate only chicks; dogs ate eggs and chicks; cats ate chicks and adults; and foxes ate all three.

Avian predators, unlike mammals, have easy access to all the tern colonies we studied. Potential avian predators include gulls, owls, hawks, turnstones, oystercatchers, crows, grackles, and blackbirds. Except for harriers, a species endangered in New Jersey, all other avian predators noted above occurred in the New Jersey salt marsh colonies. Except for Boat-tailed Grackle, a relative newcomer to New York, all the avian predators were present in the sandy beach colonies. Gulls were the most common predators in both areas. Except for hawks and owls, all the avian predators ate eggs. Chicks were eaten only by the gulls, hawks, and owls, and adults were eaten only by the hawks and owls.

Alexander (1974) pointed out that the costs of coloniality—conspicuousness, competition, and interference are assured; whereas the putative benefits—food exploitation and information transfer, vigilance, antipredator defense, and predator swamping—are not. Birkhead (1985) suggests that the benefits of coloniality in terms of increased reproductive success are more easily studied than are the costs. In this study, we have attempted to address some of these issues, recognizing that the costs and benefits may operate differently in different colonies and at different times.

Nesting in colonies increases the risk of predation because of conspicuousness and the constancy of the food resource, but colony members benefit from the enhanced opportunity for early warning and for active antipredator defense. Both activities require some investment in time, and the risk of potential injury from predators; but the advantages of early warning and group defense are greater. Terns can passively avoid predation by being a member of a large group through "predator swamping" (Hamilton 1971) whereby the individual's chance of becoming a victim is reduced. The spatial pattern of nests is another passive antipredator mechanism because tern eggs and chicks are cryptic, and sparse nesting reduces their chances of being discovered (Tinbergen et al. 1967). Thus the optimal nesting pattern reflects the balance between close nesting to enhance defense and spacing out to enhance the benefits of crypsis.

Active avoidance includes vigilance and early warning, and mobbing and overt attack. Terns nesting in colonies derive advantages through increased vigilance of colony members and decreased time each individual need devote to vigilance. Further, the cost of one individual not being vigilant when a predator happens by is low because some of the other colony members no doubt will be watching.

Group defense, another advantage of nesting in tern colonies, involves mobbing and overt attack on potential predators. Mobbing is clearly effective in deterring predators (see summary in Andersson 1976), and is effective in common terns. Common terns chase, mob, attack, and strike predators, and their response differs depending on the intruder. Terns are more vulnerable when they attack or strike a predator, and less vulnerable when they chase or mob a predator. They can be more easily captured when they are closer to the predator. They dive and strike predators that don't normally take adults (but only take eggs and chicks), and mob those that could take them, maintaining a safe avoidance distance. Thus, common terns chase most birds and squirrels that won't kill them, but they don't chase or strike foxes, dogs, and cats that could snatch them from the air. Although they attack many species of birds, they do not strike hawks that might catch them.

Observations of predator intrusions and tern responses give a picture of group defense, but don't allow a quantitative comparison of the effects of habitat, density, and stage in the breeding cycle on group defense. Thus, we used a live hawk and gull as models to examine antipredator behavior. In general, more terns mobbed the hawk than the gull, but far fewer dove at or struck the hawk, and terns stayed farther from the hawk than the gull. This is consistent with the hypothesis that they do not expose themselves unduly to predators (such as the hawk) that could grab them from the air.

In general, more terns responded to the hawk when it was in a dense nesting area compared to a low density area. Further, fewer terns mobbed the hawk when it was in dense vegetation cover compared to open habitats, and they mobbed it more when it was moving rather than stationary. We used the hawk only at the beach to avoid undue stress to it.

Terns showed habitat differences in their response to the herring gull. In the marsh, nearly twice as many terns flew over, dove, and struck the gull as in the beach, although the salt marsh terns remained higher above the gull. The rate of dives was similar between the two habitats. As with the hawk, more terns mobbed the gull in dense nesting compared to sparse-nesting parts of the colony. Nest density was the most important factor affecting defense. Defensive behavior increased when terns had chicks compared to when they were only incubating eggs.

Habituation to both models occurred, but it was slower to the hawk. The number of terns overhead decreased faster than the number of birds diving, suggesting that the divers are the nearest territory holders, and they do not as easily habituate to the predator.

The experiments indicated that terns nesting in salt marshes responded more vigorously to the gull model than those in sandy beaches (note that the hawk was only used at the beach), defensive behavior increased with nesting density, and habituation occurred gradually. Moreover, on the beach, defensive behavior was greater to the hawk than the gull; the terns stayed farther away from the hawk than from the gull. These observations corroborate our naturalistic observations but only by such experimentation can we control and evaluate the factors affecting defense. Our natural and experimental observations clearly show that there is a time and energy cost associated with predator defense, but that the presence of other conspecifics lowers the individual cost as well as decreasing the likelihood of falling a victim while engaging in defense.

Piracy is another cost of colonial-nesting for common terns because the availability of large numbers of terns flying with visible fish in their beaks, and the presence of young with large fish protruding from their mouths, makes piracy a viable method of food acquisition for opportunistic individuals. Fish are potentially available to pirates during the courtship period, when males bring back comparatively large fish for females, and during the chick-rearing phase.

Piracy does not relate linearly to fish availability, either daily or seasonally. For example, fish delivery (to the colony) is highest in the very early morning, midmorning, and in the early afternoon, whereas piracy rates are low before noon and increase in the afternoon. However, when chicks are ten days and older, piracy rates increase throughout the day, perhaps reflecting that pirates are also feeding large chicks that require increasing amounts of food.

Piracy in the air is similar regardless of habitat, but piracy rates are higher on chicks standing in the open compared to those standing near vegetation. This provides another advantage (in addition to lowered aggression rates and shelter for chicks) for nesting near vegetation rather than in the open.

Piracy rates are higher on larger fish because they are easier to see, harder for parents to transport, and harder for chicks to swallow rapidly. Younger chicks, however, lose more fish to pirates than older chicks that could swallow larger fish faster, and could hang on more tightly. Thus, parents can reduce piracy attempts by feeding chicks fish small enough for them to easily swallow. Older chicks lose fish mainly during a struggle with a sibling in which each pulls on the fish, preventing the other from swallowing it and rendering the fish conspicuous and vulnerable to a potential pirate.

Parents can reduce piracy by feeding their chicks when they are near vegetation, and thus less visible to potential pirates. Parents can also avoid landing if they see a potential pirate, and circle until it is safe to land. Terns flying low over the colony with fish are seldom pirate victims. Indeed, we found that terns flying toward the colony with large fish decrease their flight height as they enter the colony, whereas terns with small fish (that would normally not be pirated) do not.

Like all birds nesting coastally, common terns are faced with inclement weather, tidal flooding, and human disturbance. Inclement weather can include heavy and prolonged rain, fog, strong winds, and excessive heat or cold. Human disturbance is a cost for most birds, except those nesting on remote or inaccessible islands.

Common terns can avoid tidal flooding by nesting on higher islands, by selecting high nest sites, by building tall nests, or by placing nests on mats that float. Thus, colony and nest site selection are particularly critical for the terns nesting on lowlying salt marsh islands. Tidal flooding is not limited to salt marsh islands, since lowlying portions of beach colonies such as at Holgate and Wantagh sometimes experience tidal losses.

We found that, although particularly low sites on salt marsh islands are always vulnerable to tidal flooding, it is more difficult to predict when intermediate or high sites will be flooded. Birds nesting on wrack usually suffer fewer losses than those nesting in *Spartina* grass. The relative height of the tide is affected not only by lunar forces, but by wind direction and velocity. Thus, in some years, intermediate height sites are flooded in northern Barnegat Bay; in other years, such sites are flooded in the southern part of the bay. Islands that are closer to inlets are also more vulnerable. Thus, if the birds at a particular colony site have been successful for several years, and are then completely washed out in one year, they are likely to try that site again in future years.

Human disturbance varies markedly among habitats and colonies, but it generally is a byproduct of where the terns nest, rather than that people are attracted to the colony itself. Salt marshes are muddy and wet, with mosquitos and biting flies; whereas barrier beaches offer attractive swimming beaches, as well as fishing, sunbathing, and other recreation. In the fifteen years of this study in New Jersey, only two salt marsh colonies suffered severe losses due to people. On the other hand, the Cedar Beach and West End colonies were adjacent to beaches used by thousands of people on sunny days, and frequently experienced intrusion. Although visitors cause a disturbance, it is usually temporary, and the birds quickly settle down. Wanton destruction of eggs or chicks is potentially more critical, and unfortunately it sometimes occurs. Such wanton destruction or deliberate egging is a greater threat in colonial birds because people can find large numbers of nests. Mobbing by the birds is usually enough to warn off the casual intruder, but the persistent vandal ignores the circling and diving terns.

The final measure of the adaptive significance of the social behavior and colonial nesting of the common terns is their reproductive success. In the beach colonies, reproductive success varied from 0 to 1.40 young fledged per pair. Even within a single season productivity can vary.

Over the thirteen years of the salt marsh productivity study, reproductive success varied from 0 to 2.10 young fledged per pair. Reproductive success varied among colonies each year and among years for the same colony, although some colonies were consistently more successful than others. In general, reproductive success was higher in northern Barnegat Bay than in the southern part, mainly because there is no inlet there; and, as a result, the tides are moderated (it takes longer for tidal waters to reach there). Moreover, larger colonies usually raised more young per pair than smaller colonies, probably due to increased social facilitation and effective antipredator behavior.

Predation, flooding, and—to a lesser extent—human disturbances, were the causes of lowered reproductive success. In salt marshes, flooding was the major cause; whereas, in the beach colonies, all three were important causes of lowered reproductive success in some years, in some colonies. Predation losses could be minimized by early warning and group defense, whereas flooding could be reduced only by the colony initially forming in areas where the threat of tidal flooding was lower. Still, individuals select territories and nest sites, and they can select and compete for the higher sites. Thus, the selection factors of flooding and predation affect the social behavior of the terns. Common terns that nested in dense colonies with many breeding pairs had higher reproductive success than those nesting in less dense, smaller colonies. Solitary pairs were few in number, and almost never produced young. Thus, although abundant, suitable habitat is available in New Jersey salt marshes, the birds nest together on one part of an island, incurring the added costs of resource competition (nest and mate), piracy, and aggression to reap the benefits of coloniality, such as social facilitation, increased information transfer concerning food sources and safe-nesting sites, early warning of predators, and group defense.

Suitable undeveloped beach nesting sites are much fewer than marsh sites. However, some of the available sites are very spacious, which allows the formation of very large colonies, such as the 6000 pairs that nested at Cedar Beach in the mid-1980s.

When we first began this study, we were persuaded by the then prevalent attitude that salt marshes provided only marginal habitat for terns that had been displaced from traditional and more suitable beach habitat. The evidence, however, shows that common terns are adapted to exploit both of these quite different habitats and can reproduce quite successfully in both (table 10.1). Although in both habitats they confront serious threats, the distribution of threats differs by habitat as well as by year and place. Far from finding it a marginal habitat, in some years terns on salt marsh fare better than those on nearby beaches. Utilization of both habitats increases the likelihood that some birds will breed successfully in every year. Preservation of both habitat types and protection of the nesting colonies is necessary for preservation of the terns and their associated species, although many of the threats facing the population occur on their tropical wintering ground where there has been little study and negligible protection.

Overall, it seems that the terns are exploiting two different habitat types with different selection pressures. Dry land colonies suffer mammalian predation while colonies in salt marsh are often devastated by flood tides. Some individuals nest in both habitats during their reproductive lifetime, although others may show preferences for one or the other. Recurrent reproductive failures presage a relocation to a new site, and that site need not be in the same habitat as the original colony. On balance, both habitats support reproduction in most years, although in some years birds in only one of the

Table 10.1. Comparison of nesting ecology and behavior of common terns in our study.

	Salt Marsh	*Sandy Beach*
Physical features		
Colony size (number of pairs)	1–1025	500–6000
Physiognomy	uniform, flat	flat with dunes
Substrate	marsh grass or mat	sand and shell
Vegetation height	0–0.5 m	0–1 m
Vegetation cover	mostly vegetated >60%	mostly open <40%
Visibility	low to high	high
Predator barriers	surrounded by water	none
Behavioral features		
Nearest neighbor distance	87–514 cm	80–310 cm
Aggression levels (max) (per pair/hour)	2.2	1.7
Aggression levels during		
territory acquisition	high	high
egg-laying	high	high
chick phase	low	high
Predominant display type	aerial in grass ground on mats	ground
Copulation interruption frequency	very low in grass medium on mats	medium
Risks		
Flooding	high	low
Avian predation	moderate	moderate
Mammalian predation	low-moderate	moderate-high
Human disturbance	low	high
Annual Reproductive Success	very variable 0 to very high	less variable low to high

habitats raise young. In the final analysis, their habitat choices and the adversities they face are influenced by what is available as well as by their previous experience. Habitat availability is closely tied to human activity and demands, and the study of breeding biology and habitat selection must play a role in the conservation and management of habitats and populations.

APPENDIX A

Scientific Names of Organisms Referred to in Text

The following species have been referred to in the text or tables. Widely used alternative common names are given. Names in { } indicate recently revised scientific names, when the older names may be more familiar to most readers.

Plant Names	
Bayberry	*Myrica pensylvanica*
Beach Grass	*Ammophila breviligulata*
bushes in salt marsh	*Baccharis halimifolia*
bushes in salt marsh	*Iva frutescens*
Common Reed or Phragmites	*Phragmites communis*
cordgrasses	*Spartina* spp.
Japanese Black Pine	*Pinus nigra*
Salt Hay	*Spartina patens*
Salt Marsh Cordgrass	*Spartina alterniflora*
Sea Rocket	*Cakile edentula*
Seaside Goldenrod	*Solidago sempervirens*
Wild Rose	*Rosa rugosa*
Insects	
digger wasps	*Sphex ichneumoneus*

17 year cicada	*Magicicada* spp.
Urania moth	*Urania fulgens*
Fish	
Bay Anchovy	*Anchoa mitchelli*
Bluefish	*Pomatomus saltatrix*
Butterfish	*Peprilus triacanthus*
killifish	*Fundulus majalis* & *F. heteroclitus*
pipefish (= Northern pipefish)	*Syngnathus fuscsu*
Sand Lance or Sand Eel	*Ammodytes americanus*
shark	*Lamna nasus*
Silversides or menidia	*Menidia menidia* & *M. americanum*
Reptiles	
Diamond-backed Terrapin	*Malaclemys terrapene*
Garter Snake	*Thamnophis sirtalis*
Rat Snake	*Elaphe obsoleta*
Birds	
Rockhopper Penguin	*Eudyptes crestatus*
Black-footed (= Jackass) Penguin	*Spheniscus demersus*
Adelie Penguin	*Pygoscelis adeliae*
Magellanic Penguin	*Spheniscus magellanicus*
Common Loon	*Gavia immer*
Horned Grebe	*Podiceps auritus*
Rolland's Grebe	*Podiceps rolland*
Silvery Grebe	*Podiceps occipitalis*
Laysan Albatross	*Diomedea immutabilis*
Northern Fulmars	*Fulmarus glacialis*
Band-Rumped Storm-Petrel	*Oceanodroma castro*
White Pelican	*Pelecanus erythrorhynchos*
Northern Gannet	*Sula bassana*
Double-crested Cormorant	*Phalacrocorax auritus*
Pelagic Cormorant	*Phalacrocorax pelagicus*
Frigatebirds	*Fregata* spp
Green-backed [= Green] Heron	*Butorides striatus*
formerly known as Green Heron,	*Butorides virescens*
Little Blue Heron	*Egretta [= Florida] caerulea*
Cattle Egret	*Bubulcus [= Ardeola] ibis*
Black-crowned Night Heron	*Nycticorax nycticorax*
White Ibis	*Eudocimus albus*
Glossy Ibis	*Plegadis falcinellus*

Roseate Spoonbill	*Ajaja ajaia*
Whooper Swan	*Cygnus cygnus*
Snow Goose	*Anser caerulescens*
mergansers	*Mergus* spp.
Lesser Scaup	*Aythya affinus*
Common Eider	*Somateria mollisima*
Shelducks	*Tadorna* spp.
Northern Harrier or Marsh Hawk	*Circus cyaneus*
Peregrine	*Falco peregrinus*
American Kestrel	*Falco sparverius*
Forest falcons	*Micrastur* spp
Crested Caracara	*Polyborus plancus*
Domestic chicken	*Gallus gallus*
American coot	*Fulica americana*
Clapper Rail	*Rallus longirostris*
American oystercatcher	*Haematopus palliatus*
African black oystercatcher	*Haematopus moquini*
Eurasian oystercatcher	*Haematopus ostralegus*
Eurasian lapwing	*Vanellus vanellus*
Killdeer	*Charadrius vociferus*
Piping Plover	*Charadrius melodus*
Ruddy Turnstone	*Arenaria interpres*
Dunlin	*Calidris alpinus*
Red Knot	*Calidris canutus*
Willet	*Catoptrophorus semipalmatus*
Lesser Sheathbill	*Chionus minor*
Parasitic Jaeger or Arctic Skua	*Stercorarius parasiticus*
Laughing Gull	*L. atricilla*
Franklin's Gull	*L. pipixcan*
Black-headed Gull	*Larus ridibundus*
Common Black-headed Gull	
Brown-hooded Gull	*Larus maculipennis*
Black-billed Gull	*L. bulleri*
Heermann's Gull	*Larus heermanni*
Gray Gull	*Larus modestus*
Mew or Common Gull	*Larus canus*
Black-tailed Gull	*Larus crassirostris*
Ring-billed Gull	*Larus delawarensis*
Herring Gull	*Larus argentatus*
Lesser Black-backed Gull	*Larus fuscus*
Western Gull	*Larus occidentalis*

Glaucous-winged Gull	*Larus glaucescens*
Kelp Gull or Southern Black-backed Gull	*Larus dominicanus*
Great Black-backed Gull	*Larus marinus*
Black-legged Kittiwake	*Rissa tridactyla*
Ivory Gull	*Pagophila eburnea*
Terns	
Gull-billed Tern	*Gelochelidon nilotica*
Caspian tern	*Sterna [=Hydroprogne] caspia*
Royal Tern	*Sterna maxima*
Crested Tern	*Sterna bergii*
Lesser-crested Tern	*Sterna bengalensis*
Chinese Crested Tern	*Sterna zimmermanni*
Elegant Tern	*Sterna elegans*
Sandwich tern	*Sterna sandvicensis*
Cayenne Tern	*Sterna eurygnatha* or *Sterna sandvicensis eurygnatha*
River Tern	*Sterna aurantia*
Roseate Tern	*Sterna dougallii*
Common Tern	*Sterna hirundo*
South American Tern	*Sterna hirundinacea*
Arctic Tern	*Sterna paradisaea*
Antarctic Tern	*Sterna vittata*
Kerguelen Tern	*Sterna virgata*
Black-naped Tern	*Sterna sumatrana*
Forster's Tern	*Sterna forsteri*
Trudeau's Tern	*Sterna trudeaui*
White-cheeked Tern	*Sterna repressa*
Black-bellied Tern	*Sterna melanogaster*
White-fronted Tern	*Sterna striata*
Black-fronted Tern	*Sterna albostriata*
Little Tern	*Sterna albifrons*
Least Tern	*Sterna antillarum*
Yellow-billed Tern	*Sterna superciliaris*
Peruvian Tern	*Sterna lorata*
Saunder's Tern	*Sterna saundersi*
Damara Tern	*Sterna balaenarum*
Fairy Tern	*Sterna nereis*
Aleutian Tern	*Sterna aleutica*
Gray-backed Tern	*Sterna lunata*

Bridled Tern	*Sterna anaethetus*
Sooty Tern	*Sterna fuscata*
Large-billed Tern	*Phaetusa simplex*
Whiskered Tern	*Chlidonias hybridus*
White-winged Black Tern	*Chlidonias leucopterus*
Black Tern	*Chlidonias nigra*
Blue-grey Noddy or Ternlet	*Procelsterna cerulea*
Brown Noddy	*Anous stolidus*
Black (White-capped) Noddy	*Anous minutus*
Lesser Noddy	*Anous tenuirostris*
White (Fairy) Tern or Noddy	*Gygis alba*
Inca Tern	*Larosterna inca*
Black Skimmer	*Rynchops niger*
Great Auk	*Pinguinis impennis*
Common Murre or Guillemot	*Uria aalge*
Thick-billed Murre	*Uria lomvia*
Atlantic Puffin	*Fratercula arctica*
Black Guillemot	*Cepphus grylle*
Passenger Pigeon	*Ectopistes migratorius*
Mourning Dove	*Zenaida macroura*
Short-eared Owl	*Asio flammeus*
Great-horned Owl	*Bubo virginianus*
Acorn Woodpecker	*Melanerpes formicivorus*
Eastern Kingbird	*Tyrannus tyrannus*
Barn Swallow	*Hirundo rustica*
Bank Swallow	*Riparia riparia*
Cliff Swallow	*Petrochelidon pyrrhonota*
Lesser Skylark	*Alauda gulgula*
Horned Lark	*Eremophila alpestris*
crows	*Corvus* (Family Corvidae)
Fish Crow	*Corvus ossifragus*
American Crow	*Corvus brachyrhynchos*
Carrion Crow	*Corvus corone*
Bluejay	*Cyanocitta cristata*
Piñon Jay	*Gymnorhinus cyanocephala*
Magpie or Black-billed Magpie	*Pica pica*
Yellow Warbler	*Dendroica petechia*
Pampas Red-breasted Meadowlark	*Sturnella defilippii*
Red-winged blackbird	*Agelaius phoeniceus*
Brown-headed Cowbird	*Molothrus ater*
Common Grackle	*Quiscalus quiscala*

Boat-tailed Grackle	*Quiscalus [Cassidix] major*
Eastern Meadowlark	*Sturnella magna*
Yellow-headed Blackbird	*Xanthocephalus xanthocephalus*
Tricolored Blackbird	*Agelaius tricolor*
Brewer's Blackbirds	*Euphagus cyanocephalus*
Common, Eurasian, or European Starling	*Sturnus vulgaris*
Fieldfare	*Turdus pilaris*
American Robin	*Turdus migratorius*
Mountain Bluebird	*Sialia currucoides*
flycatchers	*Ficedula hypoleucus*
reed warblers	*Acrocephalus* spp.
chickadees	*Parus atricapillus*
Bishop	*Euplectes orix*
weaverbirds	*Ploceus* spp.
Village Weaver	*Ploceus cucullatus*
Quelea	*Quelea quelea*
Rosy Finch	*Leucosticte tephrocotis*
Mammals	
Hedgehog	*Erinaceus europaeus*
Coyote	*Canis latrans*
Red Fox	*Vulpes fulva*
Domestic Dog	*Canis familiaris*
Domestic Cat	*Felis cattus* or *Felis sylvestris cattus*
Striped Skunk	*Mephitis mephitis*
Raccoon	*Procyon lotor*
Mink	*Mustela vison*
Long-tailed Weasel	*Mustela frenata*
Stoat	*Mustela erminea*
River Otter	*Lutra canadensis*
Meerkat	*Suricata suricatta*
seals	*Ommatiphoca rossi*
Grey Squirrel	*Sciurus carolinensis*
Cottontail Rabbit	*Sylvilagus floridanus*
Norway Rat	*Rattus norvegicus*
Polynesian Rat	*Rattus exulans*
Prairie Dog	*Cynomys* sp.
Deer Mouse	*Peromyscus maniculatus*
Bison	*Bison bison*

APPENDIX B

Terns of the World

To facilitate comparison of the common tern with its close relatives, we provide a synopsis of the species, distribution, and habitats of the 44 species of terns (see appendix A for scientific names). A similar listing for the 44 species of gulls is given in Burger (1974). The species order in appendix B follows that of *The Checklist of Birds of the World* (Peters 1934), with the following exception. The Caspian Tern is placed close to the other crested terns. The Saunder's and Least Terns are considered distinct species. To give some idea of the diversity among the terns, the number of subspecies recognized by Peters (1934) is indicated in parentheses.

The noddies and the dark backed terns (including the sooty, bridled, and gray-backed terns) are pantropical in distribution. In addition, the crested terns (with the notable exception of Caspian) have a mainly tropical breeding distribution. The groups of small and medium-sized capped terns are mainly either North or South Temperate.

Habitats indicated include breeding substrates and habitats. Almost all species are aquatic feeders, the gull-billed tern being the notable exception. We have provided more habitat detail for the less well-known species. We have had the good fortune to observe the breeding of most of the tern species, and have indicated our personal observations with the designation *PO* under citation. Many fine habitat photographs are included in some of the references such as Marples and Marples 1934, Frith 1983, and Robertson 1984.

Species	*Breeding Range*	*Nesting Habitat*	*Races*	*Reference*
Whiskered Tern	Palearctic, Ethiopian Oriental, Australia	floating nests or among reeds inland lakes or marshes small to medium, sparse colonies usually not with other terns	7	Frith 1983, Fasola 1986 Ali & Ripley 1969 Cramp 1985
White-winged Black Tern	Palearctic, Ethiopian Oriental, Australia	floating nests in lakes muddy islets, flooded grassland small sparse inland colonies	1	Urban et al. 1986 Fasola 1986 Cramp 1985
Black Tern	Cosmopolitan	floating nests in fresh or brackish lakes or marshes; vegetated islets small sparse colonies	2	Bent 1921, PO Dunn 1979, Fasola 1986 Cramp 1985
Large-billed Tern	South America	riverine sandbars lagoons	2	Teague 1955 Hartert & Venturi 1909
Gull-billed Tern	Cosmopolitan	inland lakes, beaches, marshes salt marsh mats, salt pans, grass sand banks, shell & rock bars, riverine sand bars solitary to large colonies	7	Bent 1921, PO Frith 1983 Fasola 1986 Cramp 1985
River Tern	India	riverine sandbars & beaches colonial or widely scattered often with other tern species	1	Ali & Ripley 1969
South American Tern	southern South America	beaches, sand bars, shingle sand amid vegetation small to large colonies usually coastal islands or spits	1	Humphrey et al. 1970 Johnson & Goodall 1967 PO

Common Tern	Cosmopolitan	beaches, shingle, shell bars fresh & salt marshes muskrat houses, rocky islets piers, boats, lake shore subArctic to tropics coastal to inland small to large colonies, subdense	4	Jones 1906, Bent 1921 Marples & Marples 1934 Palmer 1941, PO Cramp et al. 1974 Burger & Lesser 1978, 1979 Cramp 1985 Fasola 1986
Arctic Tern	Holarctic	beaches, rocky islets, shingle grassy islands marshes & bogs Arctic to temperate coastal to inland usually small colonies, not dense	1	Bent 1921, Austin 1933 Lemmetyinen 1971 Cramp et al. 1974 Marples & Marples 1934 Cramp 1985 PO
Antarctic Tern	Subantarctic Islands and Antarctica	rocky bar and beaches, ground & cliffs moss-covered slopes amid rocks and/or vegetation small to medium, often dense colonies	3	Bailey & Sorensen 1962 Falla et al. 1966 Parmalee & Maxson 1974 Murphy 1936
Kerguelen Tern	Subantarctic islands: Marion, Crozets, and Kerguelen Islands	ground with vegetation shell or pebble loosely colonial, small groups	1	Derenne et al. 1974 Watson 1975
Forster's Tern	North America	floating vegetation or wood on inland & coastal lakes, marshes, estuaries; salt marsh mats or mud, sand & gravel usually small sparse colonies	1	McNicholl 1982, PO Bent 1921 Cramp 1985

Species	*Breeding Range*	*Nesting Habitat*	*Races*	*Reference*
Trudeau's Tern	Temperate South America	floating vegetation on marshes or among reeds small colonies	1	Holland 1890 Gibson 1920
Roseate Tern	Worldwide but not in South America	beaches in open, under bushes grassy tunnels, sand dunes, spits coral islets in open or vegetation coastal, usually inshore crevices in coral, offshore stacks small to large colonies sparse to dense	5	Nisbet 1981, Feare 1979 Gochfeld & Burger 1987 Burger & Gochfeld 1988 Britton & Brown 1971 Wetmore & Swales 1931 Cramp et al. 1974 Marples & Marples 1934 Cramp 1985
White-fronted Tern	New Zealand	estuaries, sandy beaches & dunes shell, shingle, rocky islets offshore stacks steep cliffs & cliff ledges ground amid tussock grass small to large, dense colonies	3	Robertson 1984 Frith 1983 Falla et al. 1966, PO
White-cheeked Tern	Red Sea & Indian Ocean	sand, rocky or coral coastal islets amid sparse vegetation large, often dense colonies	1	Britton & Brown 1971 Etchecopar & Hue 1967 Cramp 1985
Black-naped Tern	Indian Ocean SW Pacific Islands Australia	sand near sparse vegetation hollows in rocks sandbars & pebble beach few pairs/small colonies	2	Gillham 1977, PO Amerson 1969 Frith 1983

Black-bellied Tern	India to se Asia	riverine sandbars & spits small to moderate sparse colonies often with other tern species	1	Cuthbert 1985 Ali & Ripley 1969
Black-fronted Tern	New Zealand (South Island)	shingle bars on braided rivers small loose colonies, usually with gulls	1	Robertson 1984, PO Falla et al 1966
Aleutian Tern	Aleutians & eastern Siberia	grassy islands sand amid vegetation	1	Bent 1921, PO Cramp 1985
Gray-backed Tern	Central Pacific Easter Island	islets, usually sparse vegetation, sometimes in or under bushes among coral, on cliff crevices, rocky slopes, piers; small groups to large colonies	1	Woodward 1972, Ely & Clapp 1973 Clapp et al. 1977 Amerson et al. 1974, PO Clapp & Wirtz 1975 Clapp & Kridler 1977
Bridled Tern	Pantropical	hidden under bushes, rocks, small caves, burrows, rock crevices on sand or gravel small sparse marine colonies often solitary	5	Hartert 1893, PO Frith 1983, Cramp 1985 Murphy 1936 Nicholls 1977
Sooty Tern	Pantropical	barren sand or with vegetation among bushes, pavement occasionally in dense bushes or grass large dense marine colonies	6	Chapin 1954, Ashmole 1963, Cramp 1985 Woodward 1972, Feare 1976 Firth 1983 Robertson 1964 Burger & Gochfeld 1986 Clapp & Wirtz 1975 Saliva & Burger, 1989

Species	*Breeding Range*	*Nesting Habitat*	*Races*	*Reference*
Fairy Tern	Australia, New Zealand New Caledonia	barren sandhills, coral beaches river beds beaches, non-colonial	4	Falla et al. 1966 Robertson 1984
Yellow-billed Tern	northern South America	riverine sandbars small groups	1	Murphy 1936
Damara Tern	southern Africa	sandy coastal flats with scattered rocks widely scattered nests	1	Frost & Shaughnessy 1976 PO
Peruvian Tern	Ecuador to Chile	beach & gravel flats, coastal dunes, up to 2 km inland widely scattered nests	1	Murphy 1936, Johnson & Goodall 1967 PO
Least Tern	North America Caribbean	beaches, sand, gravel, coral riverine sand bars, gravel roofs little or no vegetation coastal and inland small to medium sparse colonies	2	Bent 1921, PO Gochfeld 1983b Massey 1974 Burger 1989
Little Tern	Old World	barren beach, shingle, coral sand lagoons, river banks muddy islands	6	Austin 1972, Fasola 1986 Etchecopar & Hue 1967 Cramp et al. 1974 Marples & Marples 1934 Cramp 1985
Saunder's Tern	Red Sea & Indian Ocean	sand or stony desert sparse vegetation sometimes solitary, usually small sparse groups	1	Urban et al. 1986 Cramp 1985

Crested or Swift Tern	South Africa Australia Indian Ocean	bare gravel, sand, coral rock marine or littoral islands dense colonies	5	Frith 1983, PO Ali & Ripley 1969 Cramp 1985
Royal Tern	New World Africa, Galapagos	sandy beaches & spits, shingle coral or rocky islets, shell bars warm water coasts dense colonies	2	Bent 1921, PO Buckley & Buckley 1972 Cramp 1985
Lesser Crested Tern	e Africa, Indian Ocean to Australia	sandy coral cays, spits, beaches amid sparse vegetation dense colonies on islands	2	Frith 1983, PO Cramp 1985
Chinese Crested Tern	Coastal China Southeast Asia	islands (unknown)	1	King 1981
Cayenne Tern	Caribbean to South America	rocky islets, rock or sand with or without vegetation	1	Voous 1963, Sick 1965 PO
Elegant Tern	Baja California to southern California	sandy or stony ground & on salt marshes large dense colonies	1	Boswall & Barrett 1978 PO
Sandwich Tern	Atlantic North America Europe, Caribbean	sandy or stony islets, spits, dunes beaches, saline waters coastal or inland seas little or no vegetation bare areas amid vegetation dense colonies	2	Veen 1977, Bent 1921, PO Etchecopar & Hue 1967 Fasola 1986, Cramp 1985 Cramp et al. 1974 Marples & Marples 1934

Species	*Breeding Range*	*Nesting Habitat*	*Races*	*Reference*
Caspian Tern	Worldwide except South America	sand or gravel, pebble banks coastal dunes, estuaries inland pans occasionally solitary	2	Bent 1921, PO Frith 1983 Urban et al. 1986 Cramp 1985
Inca Tern	Ecuador to Chile	ledges or crevices in sea cliffs abandoned barges small groups, often solitary	1	Murphy 1936, PO Johnson & Goodall 1967
Blue-gray Noddy or Ternlet	Central & South Pacific Ocean Easter Island	crevices in rocky cliffs, gravel bar coral rubble, under tussock grass small sparse, marine colonies	7	Rauzon et al. 1984 Schreiber & Ashmole 1970 Soper 1969, Amerson 1969
Brown Noddy	Pantropical	sandy, shell or coral substrates on ground or low bushes, bare rock rarely in trees also cliffs and rock crevices tropical marine islands small to medium colonies with other marine terns	5	Frith 1983, Bent 1921 Cramp 1985, PO Clapp 1977
Lesser Noddy	Seychelles to Western Australia	tree branches, rarely on ground nest of algae & excreta often in mangroves large, often dense colonies	2	Frith 1983

White-capped or Black Noddy	tropical oceans SE Asia	tree branches, or cliff ledges rarely on ground nest of algae & excreta, coastal islands or reefs large, often dense colonies	7	Frith 1983, PO Falla et al. 1986 Clapp & Wirtz 1975 Clapp & Kridler 1977
White or Fairy Tern	Central & South Pacific South Atlantic Islands Indian Ocean	tree branches, no nest bare rock surface or crevice solitary or small sparse colonies	6	Frith 1983, PO Johnson & Goodall 1967

APPENDIX C

Physical and vegetational characteristics and species composition of primary colonies at which we studied nesting common terns.

	Key to Map (Fig 2.4)	Size of Site (ha)	Distance to Nearest Land (m)	Over 95% Grass[a]	Over 15% Bush/ Reeds[b]	Common Terns (pairs)	Black Skimmers (pairs)	Herring Gulls (pairs)
Beach Colonies								
New York								
West End II		7.9	0			500–2500	1–200	0
Cedar Beach		10.5	0			1100–6200	50–225	1
Breezy Point		5.5	0			100–2000	20–40	0
New Jersey								
Holgate	35	5.0	0			100–200	43–400	0
Pelican		0.9	130		x	2–40	0	0
Salt Marsh Colonies								
New York								
Lane's	4	4.5	100	x		100–1000	25–35	0
Seganus	5	2.8	70	x	x	250–400	0	0
New Jersey								
N.W. Lavallette		45.0	130		x	30–750	0	2–80
S.W. Lavallette		17.9	130	x		56–510	0	2–200
North Lavallette		6.3	30	x		75–800	0	1–20
South Lavallette		7.1	30	x		100–600	0	1–5
Buster		6.2	30	x		5–826	10–11	2–4
East Point	7	0.3	40	x		18–76	1	0
Clam (small)	8	0.6	30	x		9–75	1	600–800

High Bar (small)	9	1.9	5	x		8–36	0	2–25
East Vol	10	7.3	80			1–300	2	5–45
West Vol	11	13.0	80			20–200	15–36	40–150
Gulf Point	12	0.4	110	x		42–110	1	2
West Sloop Sedge	13	5.0	100			87	13	100–200
East Sloop Sedge	14	1.0	100			12	0	2–50
Sandy Island	15	27.5	400			1	0	50–60
Flat Creek	16	17.9	5	x		5–33	2	0
West Carvel	17	4.4	280			19–58	2–20	0
East Carvel	18	5.4	280			5–350	1–45	0
West Log Creek	19	0.6	8	x		6–66	8–21	0
Log Creek	20	2.0	3	x		5–65	3–14	0
Pettit	21	1.9	680	x		12–1050	1–20	1–2
Cedar Creek	22	0.5	90	x		12–230	1–8	0
E. Cedar Bonnet	23	0.1	12	x		37	0	0
S.W. Cedar Bonnet	24	9.2	8	x		1–120	1–13	0
Thorofare	25	13.8	13	x		10–80	2	1–3
Egg	26	1.7	1.2	x		2–110	1–8	0
East Ham	27	0.5	112	x		10–90	3–55	0
West Ham	28	10.0	112			16–300	4–43	20–250
Marshelder	29	18.0	122		x	250	82	1–2
East Long Point	30	0.4	30	x		32	0	0
West Long Point	31	0.4	30	x		6–43	0	0
Little	32	2.6	739	x		8–235	0–2	0
Mordecai	33	25.0	80		x	2–100	1–71	1–2
Hester Sedge	34	0.8	20	x		6–121	3–4	0
Tow Island[c]	36	2.0	250	x		2–30	325	1

[a]More than 95% of area covered with *Spartina*.

[b]More than 15% of area covered by bushes *Iva frutescens* and *Baccharis halimifolia* or *Phragmites communis*.

[c]Tow Island is located just south of Holgate and Beach Haven Inlet in Atlantic County. This island was only visited from 1985 onward, when it was used by nesting terns and skimmers.

APPENDIX D

Clutch Size Distribution for Common Terns

Raw data on clutch size distribution for common terns in Barnegat Bay salt marsh colonies (Part A) (1976–1985) and in Long Island marsh and beach colonies (Part B) (1970–1985). Colonies listed alphabetically (c/l = 1 egg clutches).

Colony	*Year*	*c/1*	*c/2*	*c/3*	*c/4*	*Nests*	*Mean*	*%C/3*
Part A. New Jersey								
Buster	1976	19	92	713	1	825	2.84	86%
Cedar Creek	1976	4	28	198	0	230	2.84	86%
Clam (small)	1976	0	2	7	0	9	2.78	78%
Egg (small)	1976	2	3	8	0	13	2.46	62%
E. Carvel	1976	2	11	31	0	44	2.66	70%
E. Cedar Bonnet	1976	10	13	14	0	37	2.11	38%
E. Long Point	1976	1	9	22	0	32	2.66	69%
E. Sloop Sedge	1976	1	1	10	0	12	2.75	83%
Hester Sedge	1976	0	2	4	0	6	2.67	67%
Little	1976	0	5	77	1	83	2.95	93%
Log Creek	1976	1	7	20	0	28	2.68	71%
N.W. Lavallette	1976	5	20	90	0	115	2.74	78%
N. Lavallette	1976	4	14	94	0	112	2.80	84%
Pettit	1976	4	12	13	0	29	2.31	45%
S.W. Cedar Bonnet	1976	2	1	1	0	4	1.75	25%

Colony	*Year*	*c/1*	*c/2*	*c/3*	*c/4*	*Nests*	*Mean*	*%C/3*
S.W. Lavallette	1976	9	24	71	0	104	2.60	68%
S. Lavallette	1976	27	46	479	0	552	2.82	87%
West Ham	1976	3	6	7	0	16	2.25	44%
W. Carvel	1976	0	5	41	0	46	2.89	89%
W. Log Creek	1976	6	21	39	0	66	2.50	59%
W. Long Point	1976	4	5	33	0	42	2.69	79%
W. Sloop Sedge	1976	2	21	64	0	87	2.71	74%
W. Vol	1976	5	11	27	1	44	2.55	61%
Buster	1977	2	6	2	0	10	2.00	20%
Cedar Creek	1977	2	8	0	0	10	1.80	0%
East Ham	1977	2	8	0	0	10	1.80	0%
E. Carvel	1977	10	50	17	0	77	2.09	22%
E. Vol	1977	10	18	10	0	38	2.00	26%
Flat Creek	1977	2	5	1	0	8	1.88	13%
Gulf Point	1977	0	50	20	0	70	2.29	29%
Hester Sedge	1977	12	16	52	0	80	2.50	65%
High Bar (small)	1977	2	10	4	0	16	2.13	25%
Little	1977	20	30	5	0	55	1.73	9%
Log Creek	1977	10	10	0	0	20	1.50	0%
Mordecai	1977	8	12	10	0	30	2.07	33%
N.W. Lavallette	1977	18	41	46	0	105	2.27	44%
N. Lavallette	1977	40	110	100	0	250	2.24	40%
Pettit	1977	10	21	9	0	40	1.98	23%
S.W. Cedar Bonnet	1977	10	21	0	0	31	1.68	0%
S.W. Lavallette	1977	18	41	46	0	105	2.27	44%
S. Lavallette	1977	40	110	200	0	350	2.46	57%
Thorofare	1977	8	46	26	0	80	2.23	33%
West Ham	1977	2	11	27	0	40	2.63	68%
W. Carvel	1977	10	18	9	0	37	1.97	24%
W. Log Creek	1977	10	25	15	0	50	2.10	30%
W. Vol	1977	10	18	22	0	50	2.24	44%
Buster	1978	8	14	16	0	38	2.21	42%
E. Carvel	1978	30	38	17	0	85	1.85	20%
E. Vol	1978	19	22	153	0	194	2.69	79%
Flat Creek	1978	1	2	0	0	3	1.67	0%
Little	1978	5	16	14	1	36	2.31	39%
N.W. Lavallette	1978	10	100	70	0	180	2.33	39%
N. Lavallette	1978	20	70	250	0	340	2.68	74%
S.W. Lavallette	1978	10	39	60	1	110	2.47	55%
S. Lavallette	1978	5	45	249	0	299	2.82	83%
W. Carvel	1978	10	34	8	0	52	1.96	15%
W. Vol	1978	28	20	80	0	128	2.41	63%

Colony	*Year*	*c/1*	*c/2*	*c/3*	*c/4*	*Nests*	*Mean*	*%C/3*
Buster	1979	6	12	3	0	21	1.86	14%
Cedar Creek	1979	1	3	2	0	6	2.17	33%
East Ham	1979	1	10	12	0	23	2.48	52%
E. Carvel	1979	10	80	120	0	210	2.52	57%
E. Vol	1979	5	125	200	0	330	2.59	61%
Gulf Point	1979	8	12	17	0	37	2.24	46%
Little	1979	10	31	5	0	46	1.89	11%
Log Creek	1979	1	2	1	0	4	2.00	25%
Marshelder	1979	12	31	66	0	109	2.50	61%
Mordecai	1979	1	5	2	0	8	2.13	25%
N.W. Lavallette	1979	16	46	46	0	108	2.28	43%
N. Lavallette	1979	12	23	12	0	47	2.00	26%
S.W. Cedar Bonnet	1979	3	8	15	0	26	2.46	58%
S.W. Lavallette	1979	10	60	35	0	105	2.24	33%
S. Lavallette	1979	15	60	150	0	225	2.60	67%
Thorofare	1979	4	8	9	0	21	2.24	43%
West Ham	1979	0	2	7	0	9	2.78	78%
W. Carvel	1979	1	7	3	0	11	2.18	27%
Cedar Creek	1980	2	10	24	0	36	2.61	67%
East Ham	1980	16	23	24	0	63	2.13	38%
E. Carvel	1980	25	26	32	0	83	2.08	39%
Flat Creek	1980	0	2	3	0	5	2.60	60%
Little	1980	10	27	32	0	69	2.32	46%
Log Creek	1980	3	6	10	0	19	2.37	53%
Mordecai	1980	9	13	20	0	42	2.26	48%
N.W. Lavallette	1980	28	66	40	1	135	2.10	30%
N. Lavallette	1980	5	25	45	0	75	2.53	60%
Pettit	1980	27	38	28	0	93	2.01	30%
S.W. Lavallette	1980	4	5	4	0	13	2.00	31%
West Ham	1980	5	4	7	0	16	2.13	44%
W. Carvel	1980	8	23	13	0	44	2.11	30%
W. Vol	1980	31	28	22	0	81	1.89	27%
East Ham	1981	1	13	6	0	20	2.25	30%
E. Carvel	1981	2	7	32	0	41	2.73	78%
E. Vol	1981	8	12	58	1	79	2.66	73%
Mordecai	1981	1	5	3	0	9	2.22	33%
N.W. Lavallette	1981	10	18	2	0	30	1.73	7%
N. Lavallette	1981	25	75	20	0	120	1.96	17%
Pettit	1981	7	25	46	0	78	2.50	59%
S.W. Cedar Bonnet	1981	4	9	17	0	30	2.43	57%
S.W. Lavallette	1981	25	70	155	0	250	2.52	62%
S. Lavallette	1981	25	75	100	0	200	2.38	50%

Colony	*Year*	*c/1*	*c/2*	*c/3*	*c/4*	*Nests*	*Mean*	*%C/3*
West Ham	1981	7	16	48	0	71	2.58	68%
W. Vol	1981	3	30	35	1	69	2.49	51%
E. Carvel	1982	28	56	52	0	136	2.18	38%
Little	1982	16	34	35	0	85	2.22	41%
N.W. Lavallette	1982	5	40	30	0	75	2.33	40%
N. Lavallette	1982	10	32	79	2	123	2.59	64%
Pettit	1982	16	83	95	0	194	2.41	49%
S.W. Lavallette	1982	20	70	100	2	192	2.44	52%
S. Lavallette	1982	2	20	75	0	97	2.75	77%
W. Carvel	1982	2	7	10	0	19	2.42	53%
W. Vol	1982	14	23	19	0	56	2.09	34%
Cedar Creek	1983	2	9	44	0	55	2.76	80%
East Ham	1983	2	15	25	0	42	2.55	60%
Egg (small)	1983	0	1	9	0	10	2.90	90%
E. Carvel	1983	4	8	35	0	47	2.66	74%
E. Vol	1983	1	0	4	0	5	2.60	80%
Little	1983	2	2	32	0	36	2.83	89%
N. Lavallette	1983	25	75	125	0	225	2.44	56%
Pettit	1983	5	55	150	0	210	2.69	71%
S.W. Lavallette	1983	20	125	155	0	300	2.45	52%
Thorofare	1983	2	12	21	0	35	2.54	60%
West Ham	1983	1	10	17	0	28	2.57	61%
W. Vol	1983	3	2	6	0	11	2.27	55%
Cedar Creek	1984	2	11	10	0	23	2.35	43%
E. Carvel	1984	12	20	41	0	73	2.40	56%
Mordecai	1984	9	18	6	0	33	1.91	18%
N.W. Lavallette	1984	10	35	25	0	70	2.21	36%
N. Lavallette	1984	20	125	100	0	245	2.33	41%
Pettit	1984	18	22	35	0	75	2.23	47%
S.W. Lavallette	1984	15	100	85	0	200	2.35	43%
S. Lavallette	1984	25	75	75	0	175	2.29	43%
West Ham	1984	12	15	25	0	52	2.25	48%
W. Vol	1984	7	6	23	0	36	2.44	64%
East Ham	1985	14	39	15	0	68	2.01	22%
Egg (small)	1985	2	14	3	0	19	2.05	16%
E. Vol	1985	3	8	14	0	25	2.44	56%
N.W. Lavallette	1985	26	72	66	0	164	2.24	40%
Pettit	1985	43	65	39	0	147	1.97	27%
S. Lavallette	1985	30	128	115	0	273	2.31	42%
West Ham	1985	7	27	10	0	44	2.07	23%
W. Carvel	1985	2	3	11	0	16	2.56	69%
W. Vol	1985	7	18	32	2	59	2.49	54%

Colony	*Year*	*c/1*	*c/2*	*c/3*	*Nests*	*Mean*	*%C/3*
Part B. New York							
West End	1970	628	1169	368	2168	1.88	17%
Short Beach	1971	8	90	31	129	2.17	24%
West End	1971	160	1616	467	2243	2.14	21%
Cedar Beach	1972	1	5	9	15	2.53	60%
Meadowbrook	1972	2	5	9	16	2.44	56%
Short Beach	1972	26	67	31	124	2.23	38%
Wantagh	1972	13	60	68	141	2.39	48%
West End	1972	37	200	153	390	2.29	39%
Cedar Beach	1974	5	25	42	72	2.51	58%
Short Beach	1974	7	39	74	121	2.57	62%
West End	1974	20	144	412	512	2.68	72%
Cedar Beach	1975	6	47	163	216	2.72	75%
Loop	1975	5	5	49	59	2.76	83%
West End	1975	23	128	363	515	2.66	71%
West Moriches	1975	3	9	50	62	2.76	81%
Cartwright	1976	10	16	51	77	2.53	66%
Cedar Beach	1976	9	20	13	42	2.09	31%
Lanes Island	1976	17	46	82	145	2.45	57%
North Line	1976	5	7	9	21	2.19	43%
Shinnecock Bay	1976	9	13	21	43	2.28	49%
Shinnecock Is.	1976	6	10	21	37	2.41	57%
West Moriches	1976	16	54	107	177	2.52	60%
Cedar Beach	1977	27	67	308	402	2.69	76%
Islet Z	1977	9	49	39	97	2.30	40%
Lanes Island	1977	0	9	11	20	2.55	55%
West End	1977	6	14	56	76	2.65	74%
West Moriches	1977	4	21	19	45	2.38	43%
Bostwick	1978	7	22	71	100	2.64	71%
Cedar Beach	1978	42	102	346	492	2.62	68%
Cobb Is.	1978	9	8	9	26	2.00	35%
Hicks Is.	1978	0	1	8	9	2.89	89%
Lane's Island	1978	8	22	52	82	2.54	63%
Seganus	1978	11	14	13	39	2.10	34%
South Line	1978	2	3	4	9	2.22	44%
West End	1978	29	138	159	326	2.40	49%
Cedar Beach	1979	31	121	227	214	2.51	57%
Cedar Beach	1980	23	139	438	600	2.69	73%
Cedar Beach	1982	39	79	67	186	2.15	36%
West End	1982	5	13	9	27	2.15	33%
West End	1983	1	11	49	61	2.78	80%
Cedar Beach	1984	25	116	229	372	2.55	62%
Cedar Beach	1985	64	218	109	391	2.11	27%

Bibliography

Abramson, M. 1979. Vigilance as a factor influencing flock formation among curlews *(Numenius arquata)*. *Ibis* 121:213–216.

Alados, C. 1985. An analysis of vigilance in the Spanish ibex *(Capra pyrenaica)*. *Z. Tierpsychol.* 68:58–64.

Alatalo, R. V., A. Lundberg, and K. Stahlbrandt. 1984. Female mate choice in the pied flycatcher *(Ficedula hypoleuca)*. *Behav. Ecol. Sociobiol.* 14:253–261.

Alexander, R. D. 1974. The evolution of social behavior. *Ann. Rev. Ecol. Syst.* 5:325–383.

Ali, S. and S. D. Ripley. 1969. *Handbook of the Birds of India and Pakistan*. vol. 3. London: Oxford University Press.

Altmann, S. A. 1956. Avian mobbing behavior and predator recognition. *Condor* 58:241–253.

A.O.U. (American Ornithologists' Union). 1983. *Check-List of North American Birds*. 6th ed. Lawrence, Kansas: Allen Press.

Amerson, A. B. Jr. 1969. Ornithology of the Marshall and Gilbert Islands. *Atoll Res. Bull.* 127:1–348.

Amerson, A. B. Jr, R. C. Clapp, and W. O. Wirtz II. 1974. The natural history of Pearl and Hermes Reef, northwestern Hawaiian Islands. *Atoll Res. Bull.* 174:1–306.

Amlaner, C. J. and J. F. Stout. 1978. Aggressive communications by *Larus glaucescens*. VI: interactions of territory residents with a remotely controlled, locomotory model. *Behaviour* 66:223–249.

Anderson, S. S., R. W. Burton, and C. F. Summers. 1975. Behavior of grey seals *(Halichoerus grypus)* during a breeding season at North Rona. *J. Zoologyy* (London). 177:179–195.

Andersson, M. 1976. Predation and kleptoparasitism by Skuas in a Shetland seabird colony. *Ibis* 118:208–217.

Andersson, M. and F. Gotmark. 1980. Social organization and foraging ecology in the arctic skua *Stercorarius parasiticus:* A test of the food defendability hypotheses. *Oikos* 35:63–71.

Andersson, M., F. Gotmark, and C. G. Wiklund. 1981. Food information in the black-headed gull, *Larus ridibundus*. *Behav. Ecol. Sociobiol.* 9:199–202.

Andersson, M., C. G. Wiklund, and H. Rundgren. 1980. Parental defence of offspring: a model and an example. *Animal Behav.* 28:536–542.

Andrew, R. J. 1961. The motivational organization controlling the mobbing calls of the blackbird *(Turdus merula)*. I. Effects of flights on mobbing calls. *Behaviour* 17:224–246.

Annett, C. and R. Pierotti. 1989. Chick hatching as a trigger for dietary switching in the Western gull. *Colonial Waterbirds* 12:4–11.

Arnason, E. and P. R. Grant. 1978. The significance of kleptoparasitism during the breeding season in a colony of Arctic skuas *Stercorarius parasiticus* in Iceland. *Ibis* 120:38–54.

Ashcroft, R. E. 1978. Survival rates and breeding biology of puffins on Skomer Island, Wales. *Ornis Scan.* 10:100–110.

Ashmole, N. P. 1962. The black noddy *Anous tenuirostris* on Ascension Island. 1: General Biology. *Ibis* 103b:297–364.

Ashmole, N. P. 1963. The biology of the Wide-awake or sooty tern *Sterna fuscata* on Ascension Island. *Ibis* 103b:297–364.

Ashmole, N. P. 1971. Sea bird ecology and the marine environment. In D. S. Farner and J. R. King, eds., *Avian Biology*. 1:224–286. New York: Academic.

Ashmole, N. P. and M. J. Ashmole. 1967. Comparative feeding ecology of sea birds of a tropical oceanic island. New Haven, Connecticut: Peabody Mus. Nat. Hist.

Atkinson, I. A. E. 1985. The spread of commensal species of Rattus to oceanic islands and their effects on island avifaunas. In P. J. Moors, ed., *Conservation of Island Birds*: 35–81. Cambridge, England: International Council for Bird Preservation. Tech. Publ. 3.

Austin, O. L., Jr. 1929. Contributions to the knowledge of the Cape Cod Sterninae. *Bull. NE Bird Band. Assoc.* 5:123–140.

Austin, O. L., Jr. 1932. Further contributions to the knowledge of the Cape Cod Sterninae. *Bird Banding* 3:123–139.

Austin, O. L., Jr. 1933. The status of Cape Cod terns in 1933. *Bird Banding* 4:190–198.

Austin, O.L. 1940. Some aspects of individual distribution in the Cape Cod tern colonies. *Bird Banding* 11:155–169.

Austin, O. L. 1942. The life span of the common tern *(Sterna hirundo)*. *Bird Banding* 13:159–176.

Austin, O. L. 1944. The status of Tern Island and the Cape Cod Terns in 1943. *Bird Banding* 15:10–27.

Austin, O. L. 1946. The status of Cape Cod terns in 1944: a behavior study. *Bird Banding* 17:10–27.

Austin, O. L. 1948. Predation by the common rat *Rattus norvegicus* in the Cape Cod colonies of nesting terns. *Bird Banding* 19:60–65.

Austin, O. L. 1949. Site tenacity, a behaviour trait of the common tern (*Sterna hirundo* Linn.). *Bird Banding* 20:1–39.

Austin, O. L. 1951. Group adherence in the common tern. *Bird Banding* 22:1–10.

Austin, O. L., Jr. 1972. The birds of Korea. *Mus. Comp. Zoo. Bull.* 101:11–301.

Austin, O. L. and O. L. Austin, Jr. 1956. Some demographic aspects of the Cape Cod population of common terns (*Sterna hirundo*). *Bird Banding* 27:55–66.

Austin, O. L. Jr., and N. Kuroda. 1953. The birds of Japan: their status and distribution. *Bull. Mus. Comp. Zool.* 109:280–637.

Baerrends, G. P., and R. H. Drent. 1970. The herring gull and its egg. Leiden: E. J. Brill.

Baggerman, B., G. P. Baerrends, H. S. Heikens, and J. H. Mook. 1956. Observations on the behaviour of the black tern *Chlidonias n. niger* L. in the breeding area. *Ardea* 44:1–71.

Bailey, A. M. and J. H. Sorensen. 1962. Subantarctic Campbell Island. *Denver Mus. Nat. Hist. Proc.* 10:1–303.

Baird, P. A. and R. A. Moe. 1978. The breeding biology and feeding ecology of marine birds in the Sitkalidak Strait area, Kodiak Island, 1977. U.S. Dept. of Commerce and U.S. Dept. Interior Environ. *Assess. Alaskan Continental Shelf. Ann. Rept. Principal Invest.* 3:313–524.

Barash, D. P. 1976. Mobbing behavior by crows: the effects of the "crows in distress" model. *Condor* 78:120.

Barker, R. J. and R. M. Hand. 1981. Mobbing response in adélie penguins. *Emu* 81:169.

Barnard, C. J. 1980a. Flock feeding and time budgets in the house sparrow (*Passer domesticus* L.). *Anim. Behav.* 28:295–309.

Barnard, C. J. 1980b. Factors affecting flock size mean and variance in a winter population of house sparrows (*Passer domesticus* L.) *Behaviour* 74:114–127.

Barnard, C. J. and H. Stephens. 1981. Prey size selection by lapwings in lapwing/gull associations. *Behaviour* 77:1–22.

Barnard, C. J., D. B. A. Thompson, and H. Stephens. 1982. Time budgets, feeding efficiency, and flock dynamics in mixed species flocks of lapwings, golden plovers, and gulls. *Behaviour.* 80:43–69.

Barrie, N. 1975. When push comes to shove. *Internatl. Wildlife* 5(1):18–19.

Bartlett, L. M. 1957. Ring-billed gull steals food from coot. *Wilson Bull.* 69:182.

Bayer, R. D. 1982. How important are bird colonies as information centers? *Auk* 99:31–40.

Becker, P. H. 1984. Wie richtet eine Flusseeschwalbenkolonie (*Sterna hirundo*) ihr Abwehrverhalten auf den Feinddruck durch Silbermowen (*Larus argentatus*) *Zeits. Tierpsychologie* 66:265–288.

Becker, P. H. and M. Erdelen. 1982. Windrichtung und Vegetationsdeckung am Nest der Silbermöwen *(Larus argentatus)*. *J. Ornithol.* 123:117–130.

Becker, P. H., P. Finck, and A. Anlauf. 1985. Rainfall preceding egg-laying—a factor of breeding success in common terns *(Sterna hirundo)*. *Oecologia* 65:431–436.

Beer, C. G. 1961. Incubation and nest-building behaviour of black-headed gulls. I: Incubation behaviour in the incubation period. *Behaviour* 18:62–106.

Beer, C. G. 1962. Incubation and nest-building behaviour of black-headed gulls. II: Incubation behaviour in the laying period. *Behaviour* 19:283–304.

Beer, C. G. 1966. Adaptations to nesting habitat in the reproductive behaviour of the black-billed gull, *Larus bulleri*. *Ibis* 103:394–410.

Belopol'skii, L. O. 1957. Ecology of sea colony birds of the Barents Sea. Moscow, Izdatel'stvo Akademia Nauk. (Translated from Russian by Israel Program for Scientific Translations, Jerusalem 1961.)

Bent, A. C. 1921. Life histories of North American Gulls and Terns. *Smithsonian Institute Bull.* 113:1–337.

Bergman, G. 1960. Uber neue Futtergewohnheiten der Mowen an den Kusten Finnlands. *Ornis Fennica* 37:11–28.

Bergman, G. 1980. Single-breeding versus colonial breeding in the Caspian tern *Hydroprogne caspia*, the common tern *Sterna hirundo*, and the Arctic tern *Sterna paradisea*. *Ornis Fennica* 57:141–152.

Bernstein, N. P. and S. J. Maxson. 1984. Sexually distinct daily activity patterns of blue-eyed shags in Antarctica. *Condor* 86:151–156.

Bertram, B. C. R. 1980. Vigilance and group size in ostriches. *Anim. Behav.* 28:278–286.

Bianki, V. V. 1967. Gulls, shorebirds, and alcids of Kandalaksha Bay. Murmanskoe Knizhnoe Izdatel'stvo, Murmansk. (Translated from Russian by Israel Program for scientific translations, Jerusalem 1977.)

Bildstein, K. L. 1982. Responses of northern harriers to mobbing passerines. *J. Field Ornithol.* 53:7–14.

Birkhead, T. R. 1979. Mate guarding in the magpie *(Pica pica)*. *Anim. Behav.* 27:866–874.

Birkhead, T. R. 1982. Timing and duration of mate guarding in magpies, *Pica pica*. *Anim. Behav.* 30:277–283.

Birkhead, T. R. 1985. Coloniality and social behaviour in the Atlantic Alcidae. In D. N. Nettleship and T. R. Birkhead, eds., *The Atlantic Alcidae*, 355–383. New York: Academic Press.

Birt, V. L., T. P. Birt, D. Goulet, D. K. Cairns, and W. A. Montevecchi. 1987. Ashmole's halo: direct evidence for prey depletion by a seabird. *Marine Ecology* 40:205–208.

Black, B. B. and L. D. Harris. 1981. Winter foraging pattern of Gulf coast black skimmers. *Col. Waterbirds* 4:187–193.

Blancher, P. J. and R. J. Robertson. 1982. Kingbird aggression: does it deter predation? *Anim. Behav.* 30:929–945.

Blokpoel, H., R. D. Morris, and P. Trull. 1982. Winter observations of common terns in Trinidad, Guyana, and Suriname. *Colonial Waterbirds* 5:144–147.

Blokpoel, H., R. D. Morris, and G. D. Tessier. 1984. Field investigations of the biology of common terns wintering in Trinidad. *J. Field. Ornithol.* 55:424–434.

Blus, L. J. and C. J. Stafford. 1980. Breeding biology and relation of pollutants to black skimmers and gull-billed terns in South Carolina. Fish and Wildlife Service Special Scientific Report—Wildlife no. 230.

Bongiorno, S. F. 1970. Nest-site selection by adult laughing gulls (*Larus atricilla*). *Anim. Behav.* 18:434–444.

Boshoff, A. F. 1980. Mobbing by kelp gulls *Larus dominicanus* as a possible cause of cape vulture *Gyps coprotheres* mortality. *Cormorant* 8:15–16.

Boswall, J. and M. Barrett. 1978. Notes on the breeding birds of Isla Raza, Baja California. *Western Birds* 9:93–99.

Brearey, D. and O. Hildon. 1985. Nesting and egg-predation by turnstones *Arenaria interpres* in larid colonies. *Ornis Scand.* 16:283–292.

Britton, P. L. and L. H. Brown. 1971. Breeding sea-birds at the Kiunga Islands, Kenya. *Ibis* 113:364–366.

Brockman, H. J. 1980. House sparrows kleptoparasitize digger wasps. *Wilson Bull.* 92:394–398.

Brockman, H. J. and C. J. Barnard. 1979. Kleptoparasitism in birds. *Anim. Behav.* 27:487–514.

Brown, C. R. 1984. Laying eggs in a neighbor's nest: benefit and cost of colonial nesting in swallows. *Science* 224:518–519.

Brown, C. R. and M. B. Brown. 1987. Group-living in cliff swallows as an advantage in avoiding predators. *Behav. Ecol. Sociobiol.* 21:97–107.

Brown, J. L. 1964. The evolution of diversity in avian territorial systems. *Wilson Bull.* 76:160–169.

Brown, R. G. B. 1967a. Breeding success and population growth in a colony of herring and lesser black-backed gulls, *Larus argentatus* and *Larus fuscus*. *Ibis* 109:502–515.

Brown, R. G. B. 1967b. Courtship behaviour in the lesser black-backed gull, *Larus fuscus*. *Behaviour* 29:122–153.

Brun, E. 1974. Breeding success of gannets *Sula bassana* at Noardmjele, Andoya, North Norway. *Astarte* 7:77–82.

Buck, J. and E. Buck. 1978. Toward a functional interpretation of synchronous flashing by fireflies. *Am. Nat.* 112:471–492.

Buckley, F. G. and P. A. Buckley. 1972. The breeding biology of royal terns *Sterna (Thalasseus) maxima maxima*. *Ibis* 114:344–359.

Buckley, F. G. and P. A. Buckley. 1982. Microenvironmental determinants of survival in saltmarsh-nesting common terns. *Colonial Waterbirds* 5:39–48.

Buckley, P. A., M. Gochfeld, and F. G. Buckley. 1978. Efficacy and timing of helicopter censuses of black skimmers and common terns on Long Island, N.Y.: a preliminary analysis. *Proc. Colonial Waterbird Group* 1:48–61.

Buitron, D. 1983. Variability in the responses of black-billed magpies to natural predators. *Behaviour* 87:209–234.

Bullough, W. S. 1942. Observations on the colonies of the Arctic tern *(Sterna macrura* Naumann) on the Farne Islands. *Proc. Zool. Soc. London* 112A:1–12.

Burger, A. E. 1981. Time budgets, energy needs and kleptoparasitism in breeding lesser sheathbills *(Chionis minor)*. *Condor* 83:106–112.

Burger, J. 1974a. Breeding adaptations of Franklin's gull *(Larus pipixcan)* to a marsh habitat. *Anim. Behav.* 22:521–567.

Burger, J. 1974b. Breeding biology and ecology of the brown-hooded gull in Argentina. *Auk* 91:601–613.

Burger, J. 1974c. Determinants of colony and nest-site selection in the silver grebe *(Podiceps occipitalis)* and Rolland's grebe *(Rollandia rolland)*. *Condor* 76:301–306.

Burger, J. 1976. Daily and seasonal activity patterns in breeding laughing gulls. *Auk* 93:308–323.

Burger, J. 1977. Role of visibility in nesting behavior of *Larus* gulls. *J. Comp. Physiol. Psych.* 91:1347–1358.

Burger, J. 1978a. Competition between cattle egrets and native North American herons, egrets, and ibises. *Condor* 80:15–23.

Burger, J. 1978b. The pattern and mechanism of nesting in mixed-species heronries. In A. Sprunt, J. Ogden, and S. Wickler, eds., *Wading Birds*. Nat. Aud. Soc. Res. Rept. No. 7, pp. 45–60. New York: National Audubon Society.

Burger, J. 1979a. Colony size: A test for breeding synchrony in herring gull *Larus argentatus* colonies. *Auk* 96:694–703.

Burger, J. 1979b. Nest repair behavior in birds nesting in salt marshes. *J. Comp. Physiol. Psychol.* 11:189–199.

Burger, J. 1979c. Competition and predation: herring gulls versus laughing gulls. *Condor* 81:269–277.

Burger, J. 1980a. The transition from dependence to independence and post-fledging parental care in marine birds. In J. Burger, B. L. Olla, and H. E. Winn, eds., *Behavior of Marine Organisms: Perspectives in Research*. Vol. 4: *Marine Birds*, pp. 45–60. New York: Plenum Press.

Burger, J. 1980b. Nesting adaptations of herring gulls *Larus argentatus* to salt marshes and storm tides. *Biol. Behav.* 5:147–162.

Burger, J. 1981a. Effects of human disturbance on colonial species, particularly gulls. *Colonial Waterbirds* 4:28–36.

Burger, J. 1981b. On becoming independent in herring gulls: parent-young conflict. *Am. Nat.* 117:444–456.

Burger, J. 1981c. Sexual differences in parental activities of breeding black skimmers. *Am. Nat.* 117:975–984.

Burger, J. 1981d. A model for the evolution of mixed species colonies of Ciconiiformes. *Quart. Rev. of Biology* 56:143–167.

Burger, J. 1981e. The effect of human activity on birds at a coastal bay. *Biol. Cons.* 21:231–241.

Burger, J. 1981f. Behavioral responses of herring gulls *(Larus argentatus)* to aircraft noise. *Env. Pollut.* 24:177–184.

Burger, J. 1982a. The role of reproductive success in colony site selection and abandonment in black skimmers *(Rynchops niger)*. *Auk* 99:109–115.

Burger, J. 1982b. Jamaica Bay studies: I. Environmental determinants of abundance and distribution of common terns *(Sterna hirundo)* and black skimmers *(Rynchops niger)* at an east coast estuary. *Col. Waterbirds* 5:148–160.

Burger, J. 1983. Competition between two species of nesting gulls: On the importance of timing. *Behavioral Neuroscience* 97:492–501.

Burger, J. 1984a. Colony stability in least terns. *Condor* 86:61–67.

Burger, J. 1984b. Abiotic factors affecting migrant shorebirds. In J. Burger and B. Olla, eds., *Behavior of Marine Animals*. Vol. 6: *Shorebirds: Migrating and Foraging Behavior*, 1–73. New York: Plenum Press.

Burger, J. 1984c. Pattern, mechanism, and adaptive significance of territoriality in herring gulls *(Larus argentatus)*. *Ornithological Monographs* 34:1–92.

Burger, J. 1984d. Grebes nesting in gull colonies: protective associations and early warning. *Am. Nat.* 123:327–337.

Burger, J. 1985a. Advantages and disadvantages of mixed species colonies of seabirds, 905–918. Proc. 18th Intern. Orn. Congress, Moscow.

Burger, J. 1985b. Habitat selection in temperate marsh-nesting birds. In M. L. Cody, ed., *Habitat Selection in Birds*, 243–281. New York: Academic Press.

Burger, J. 1989. Least tern populations in coastal New Jersey: monitoring and management of a regionally-endangered species. *J. Coastal Research* 5:801–811.

Burger, J. and C. G. Beer. 1975. Territoriality in the laughing gull (L. *atricilla)*. *Behaviour* 55:301–320.

Burger, J. and M. Gochfeld. 1979. Age differences in ring-billied gull kleptoparasitism on starlings. *Auk* 96:806–808.

Burger, J. and M. Gochfeld. 1981a. Age-related differences in piracy behaviour of four species of gulls, *Larus*. *Behaviour* 77:242–267.

Burger, J. and M. Gochfeld. 1981b. Discrimination of the threat of direct versus tangential approach to the nest by incubating herring and great black-backed gulls. *J. Comp. Physiol. and Psychol.* 95:676–684.

Burger, J. and M. Gochfeld. 1983. Behavioural responses to human intruders of herring gulls *(Larus argentatus)* and great black-backed gulls *(L. marinus)* with varying exposure to human disturbance. *Behav. Processes* 8:327–344.

Burger, J. and M. Gochfeld. 1984. Great Black-backed gull predation on kittiwake fledglings in Norway. *Bird Study* 31:149–151.

Burger, J. and M. Gochfeld. 1985. Early postnatal lead exposure: behavioral effects in common tern chicks *(Sterna hirundo)*. *J. Tox. and Env. Health* 16:869–886.

Burger, J. and M. Gochfeld. 1986. Nest site selection in Sooty Terns *(Sterna fuscata)* in Puerto Rico and Hawaii. *Colonial Waterbirds* 9:31–45.

Burger, J. and M. Gochfeld. 1988. Nest-site selection and temporal patterns in habitat use of roseate and common terns. *Auk* 105:433–438.

Burger, J. and M. Gochfeld. 1990a. *Black Skimmers: Social Dynamics of a Colonial Species*. New York: Columbia University Press.

Burger, J. and M. Gochfeld. 1990b. Nest site selection in least terns *(Sterna antillarum)* in New Jersey and New York, *Colonial Waterbirds* 13:31–40.

Burger, J. and M. Gochfeld. 1990c. Early experience and vegetation preferences in common tern chicks. *Wilson Bulletin* 102:328–333.

Burger, J. and C. Hahn. 1977. Crow predation on black-crowned night heron eggs. *Wilson Bull.* 89:350–351.

Burger, J. and F. Lesser. 1978. Selection of colony sites and nest sites by common terns *Sterna hirundo* in Ocean County, New Jersey. *Ibis* 120:433–449.

Burger, J. and F. Lesser. 1979. Breeding behavior and success in salt marsh common tern colonies. *Bird Banding* 50:322–337.

Burger, J. and J. Shisler. 1978. Nest site and competitive interactions of herring and laughing gulls in New Jersey. *Auk* 95:252–266.

Burger, J. and J. Shisler. 1980. Colony and nest site selection in laughing gulls in response to tidal flooding. *Condor* 82:251–258.

Burley, N. 1981. Mate choice by multiple criteria in a monogamous species. *Am. Nat.* 117:515–528.

Butler, R. G. and P. Lukasiewicz. 1979. A field study of the effect of crude oil on herring gull *(Larus argentatus)* chick growth. *Auk* 96:809–812.

Butler, R. G., and W. Trivelpiece. 1981. Nest spacing, reproductive success, and behavior of the great black-backed gulls *(Larus marinus)*. *Auk* 98:99–107.

Butler, R. W. 1982. Wing fluttering by mud-gathering cliff swallows: avoidance of "rape" attempts? *Auk* 95:758–761.

Cade, T. J. 1967. Ecological and biological aspects of predation by the northern shrike. *Living Bird* 6:43–86.

Cain, S. A., and G. M. de Oliverira Castro. 1971. *Manual of Vegetation Analysis*. New York: Hofner.

Cairns, D. K. 1987. The ecology and energetics of chick provisioning by black guillemots. *Condor* 89:627–635.

Caraco, T. 1982. Flock size and the organization of behavioral sequences in juncos. *Condor* 84:101–105.

Carter, L. R. and L. B. Spear. 1986. Costs of adoption in western gulls. *Condor* 88:253–256.

Case, J. F. and J. Buck. 1963. Control of flashing in fireflies. II: Role of central nervous system. *Biol. Bull.* 125:234–250.

Case, N. A. and O. H. Hewitt. 1963. Nesting and productivity of the red-winged blackbird in relation to habitat. *Living Bird* 2:7–20.

Chapin, J. P. 1954. The calendar of Wideawake Fair. *Auk* 71:1–15.

Clapp, R. B. 1972. The natural history of Gardner Pinnacles, Northwestern Hawaiian Islands. *Atoll Res. Bull.* 163:1–25.

Clapp, R. B. 1977. Notes on the vertebrate fauna of Tongareva Atoll. *Atoll Res. Bull.* 198:1–7.

Clapp, R. B. and E. Kridler. 1977. The natural history of Necker Island, northwestern Hawaiian Island. *Atoll Res. Bull.* 206:1–102.

Clapp, R. B., E. Kridler, and R. B. Fleet. 1977. The natural history of Nihoa Island, northwestern Hawaiian Islands. *Atoll Res. Bull.* 207:1–147.

Clapp, R. B. and W. O. Wirtz, II. 1975. The natural history of Lisianski Island, northwestern Hawaiian Islands. *Atoll. Res. Bull.* 186:1–196.

Clark, P. J. and F. C. Evans. 1954. Distance to nearest neighbor as a measure of spatial relationships in populations. *Ecology* 35:445–453.

Clark, K. L. and R. J. Robertson. 1978. Spatial and temporal multispecies nesting aggregations in birds as antiparasite and antipredator defenses. *Behav. Ecol. Sociobiol.* 5:359–371.

Cody, M. L. 1968. On the methods of resource division in grassland bird communities. *Am. Nat.* 102:107–147.

Cody, M. L. 1973. Coexistence, coevolution, and convergent evolution in seabird communities. *Ecology.* 54:31–44.

Cody, R. P. and J. K. Smith. 1985. *Applied Statistics and the SAS Programming Language.* New York: North-Holland.

Collias, N. E. and E. C. Collias. 1969. Size of breeding colony related to attraction of mates in a tropical passerine bird. *Ecology* 50:483–488.

Collias, N. E. and E. C. Collias. 1977. Weaverbird nest aggregation and evolution of the compound nest. *Auk* 94:50–64.

Collias, N. E. and E. C. Collias. 1980. Behavior of the grey-capped social weaver *Pseudonigrita arnaudi* in Kenya. *Auk* 97:213–226.

Collias, N. E., and J. K. Victoria. 1978. Nest and mate selection in the village weaverbird *(Ploceus cucullatus)*. *Anim. Behav.* 26:470–479.

Collias, N. E., M. Brandman, J. K. Victoria, L. F. Kiff, and C. E. Rischer. 1971. Social facilitation in Weaverbirds: Effects of varying the sex ratio. *Ecology* 52:829–836.

Collins, C. T. 1970. The black-crowned night heron as a predator of tern chicks. *Auk* 87:584–586.

Connor, P. F. 1971. The mammals of Long Island, New York. *New York State Mus. Bull.* 416:1–78.

Connors, P. G., V. C. Anderlini, R. W. Risebrough, M. Gilbertson, and H. Hays. 1975. Investigations of heavy metals in common tern populations. *Field-Nat.* 89:157–162.

Conover, M. R. 1984. Occurrence of supernormal clutches in the Laridae. *Wilson Bulletin* 96:249–267.

Conover, M. R. 1987. Acquisition of predator information by active and passive mobbers in ring-billed gull colonies. *Behaviour* 102:41–58.

Conover, M. R., and D. E. Miller. 1979. Reaction of ring-billed gulls to predators and human disturbance at their breeding colonies. *Proc. Colonial Waterbird Group* 2:41–47.

Conover, M. R. and D. E. Miller. 1980. Daily activity patterns of breeding ring-billed and California gulls. *J. Field Ornithol.* 51:329–339.

Cooper, D. M., H. Hays, and C. Pessino. 1970. Breeding of the common and

and roseate terns on Great Gull Island. *Proc. Linn. Soc. New York* 71:83–104.

Corkhill, P. 1973. Manx shearwaters on Skomer: population and mortality due to gull predation. *Brit. Birds* 66:136–143.

Coulson, J. C. 1966. The influence of pair-bond and age on the breeding biology of the kittiwake gull, *Rissa tridactyla*. *J. Anim. Ecol.* 35:269–279.

Coulson, J. C. 1968. Differences in the quality of birds nesting in the centre and on the edges of a colony. *Nature* 217:478–479.

Coulter, M. C. 1986. Assortative mating and sexual dimorphism in the common tern. *Wilson Bull* 98:93–100.

Courtney, P. A. and H. Blokpoel. 1980. Food indicators of food availability for common terns on the lower Great Lakes. *Can. J. Zool.* 58:1318–1323.

Craig, A. J. K. 1974. Reproductive behaviour of the male red bishop bird. *Ostrich* 45:149–160.

Cramp, S. ed. 1985. *Handbook of the Birds of Europe, the Middle East, and North Africa*. vol. 4. Oxford: Oxford University Press.

Cramp, S., W. R. P. Bourne, and D. Saunders. 1974. *The Seabirds of Britain and Ireland*. New York: Taplinger.

Crawford, R. D. 1977. Breeding biology of year-old and older female red-winged and yellow-headed blackbirds. *Wilson Bull.* 89:73–80.

Crook, J. H. 1964. The evolution of social organization and visual communication in the weaver birds *(Ploceinae)*. *Behaviour Suppl.* 10:1–178.

Crook, J. H. and P. A. Butterfield. 1968. Effects of testosterone propionate and luteinizing hormone on agonistic and nest building behavior of *Quelea quelea*. *Anim. Behav.* 19:370–384.

Crossin, R. S. and L. N. Huber. 1970. Sooty tern egg predation by ruddy turnstones. *Condor* 72:372–373.

Crowell, E. M. and S. Crowell. 1946. The displacement of terns by herring gulls at the Weepecket Islands. *Bird Banding* 17:1–9.

Crowley, M. and J. Bovet. 1980. Social synchronizastion of circadian rhythms in deer mice *(Peromyscus maniculatus)*. *Behav. Ecol. Sociobiol.* 7:99–105.

Croxall, J. P. ed. 1987. *Seabirds: Feeding Ecology and Role in Marine Ecosystems*. Cambridge, England: Cambridge University Press.

Croze, H. 1970. *Searching Image in Carrion Crows*. Berlin: Paul Parey.

Cullen, E. 1957. Adaptations in the kittiwake to cliff-nesting. *Ibis* 99:275–303.

Cullen, J. M. 1954. The diurnal rhythm of birds in the arctic summer. *Ibis* 96:31–46.

Cullen, J. M. 1960. Some adaptations in the nesting behavior of terns. *Proc. 12th Int. Ornith. Congr.* 12:153–157.

Curio, E. 1967. Die Adaptation einer Handlung ohne den zugehorigen Bewegungsablauf. *Verh. Dtsch. Zool. Ges.* 13:153–163.

Curio, E. 1969. Funktionsweise und Stammesgeschichte des Flug feinderkennens einiger Darwinfinken *(Geospizinae)* *Z. Tierpsychol.* 26:394–487.

Curio, E. 1975. The functional organization of antipredator behavior in the pied flycatcher: a study of avian visual perception. *Anim. Behav.* 23:1–115.

Curio, E. 1978. The adaptive significance of avian mobbing. I: Telenomic hypotheses and predictions. *Z. Tierpsychol.* 47:175–183.

Curtis, P. E. 1986. Mobbing behaviour by adult breeding chickens. *Vet. Rec.* 119:273–274.

Daan, S., and J. Tinbergen. 1979. Young guillemots *(Uria lomvia)* leaving their Arctic breeding cliffs: a daily rhythm in numbers and risk. *Ardea* 67:96–100.

Darling, F. F. 1938. *Bird Flocks and the Breeding Cycle.* London: Cambridge University Press.

Davis, D. E. 1963. The hormonal control of aggressive behavior. *Proc. 13th Internat. Ornithol. Congress.* 13:994–1003.

Davis, J. W. F. 1973. Aspects of the breeding ecology and feeding of certain gulls. Ph.D. Thesis, Oxford University.

Dawkins, R. and T. R. Carlisle. 1976. Parental investment, mate desertion and a fallacy. *Nature* 262:131–133.

Delius, J. D. 1970. The effects of daytime, tides and other factors on some activities of lesser black-backed gulls, *Larus fuscus. Rev. Comp. Anim.* 4:3–11.

Denson, R. D. 1979. Owl predation on a mobbing crow. *Wilson Bull.* 91:133.

Derenne, P., G. X. Lufbery, and B. Tollu. 1974. L'avifaune de l'Archipel Kerguelen. *Recherches Ecol. et Physiol. Faune Terres Australes et Antarctique Française.* Paris: Museum National d'Hist. Naturelle.

Dexheimer, M., and W. E. Southern. 1975. Breeding success relative to nest location and density in ring-billed gull colonies. *Wilson Bull.* 85:288–290.

Dicostanzo, J. 1980. Population dynamics of a common tern colony. *J. Field Ornithol.* 51:229–243.

Drent, R. H. 1967. Functional aspects of incubation in the herring gull *(Larus argentatus Pont). Behaviour Supp.*

Duffy, D.C. 1977. Incidence of oil contamination on breeding common terns. *Bird Banding* 48:370–371.

Duffy, D. C. 1983. Competition for nesting space among Peruvian guano birds. *Auk* 100:680–688.

Duffy, D. C. 1986. Foraging at patches: interactions between common and roseate terns. *Ornis Scand.* 17:47–52.

Dugan, P. J. 1981. The importance of nocturnal foraging in shorebirds: A consequence of increased invertebrate prey activity. In N. V. Jones and W. J. Wolff, eds., *Feeding and Survival Strategies of Estuarine Organisms.* pp. 251–260. New York: Plenum Press.

Dulude, A.-M., G. Baron and R. McNeil. 1987. Role of male and female ring-billed gulls in the care of young and territorial defense. *Can. J. Zool.* 65:1535–1540.

Dummigan, K. A. 1977. Food-piracy by Iceland gull on oystercatchers. *Brit. Birds* 70:392.

Dunn, E. H. 1979. Nesting biology and development of young in Ontario black terns. *Canadian Field-Naturalists* 93:276–281.

Dunn, E. K. 1973. Robbing behaviour of roseate terns. *Auk* 90:641–651.

Dwernychuk, L. W., and D. A. Boag. 1972. Ducks nesting in association with gulls —an ecological trap. *Can. J. Zool.* 50:559–563.

Elliot, R. D. 1985. The exclusion of avian predators from aggregations of nesting lapwings *(Vanellus vanellus)*. *Anim. Behav.* 33:308–314.

Elston, S. F. and W. E. Southern. 1983. Effects of intraspecific piracy on breeding ring-billed gulls. *Auk* 100:217–220.

Elston, S. F., C. D. Rymal, and W. E. Southern. 1978. Intraspecific kleptoparasitism in breeding ring-billed gulls. *Proc. Col. Waterbird Group* 1:102–109.

Ely, C. A. and R. B. Clapp. 1973. The natural history of Laysan Island, northwestern Hawaiian Islands. *Atoll Res. Bull.* 171:1–361.

Emlen, J. T., Jr. 1952. Social behavior in nesting cliff swallows. *Condor* 54:177–199.

Emlen, J. T., Jr., D. E. Miller, R. M. Evans, and D. H. Thompson. 1966. Predator induced parental neglect in a ring-billed gull colony. *Auk* 83:677–679.

Emlen, S. T. 1971. Adaptive aspects of coloniality in the bank swallow. *Am. Zool.* 11:47.

Emlen, S. T. and N. J. Demong. 1975. Adaptive significance of synchronized breeding in a colonial bird: A new hypothesis. *Science* 188:1029–1031.

England. M. E. 1986. Harrier kills mobbing willet. *Raptor Res.* 20:78–79.

Erwin, R. M., 1977a. Black Skimmer breeding ecology and behavior. *Auk* 94:709–717.

Erwin, R. M. 1977b. Foraging and breeding adaptations to different food regimes in three seabirds: the common tern, *(Sterna hirundo)*, royal tern, *(Sterna maxima)* and black skimmer, *(Rynchops niger)*. *Ecology* 58:389–397.

Erwin, R. M. 1978, Coloniality in terns: the role of social feeding. *Condor* 80:211–215.

Erwin, R. M. 1979. Species interactions in a mixed colony of common terns *(Sterna hirundo)* and black skimmers *(Rynchops nigra)*. *Anim. Behav.* 27:1054–1062.

Erwin, R. M. 1988. Correlates of nest-defense behavior of common terns. *J. Field Ornithol.* 59:135–142.

Erwin, R. M., J. Galli, and J. Burger. 1981. Colony site dynamics and habitat use in Atlantic coast seabirds. *Auk* 98:550–556.

Erwin, R. M. and D. C. Smith. 1985. Habitat comparisons and productivity in nesting common terns on the mid-Atlantic coast. *Colonial Waterbirds* 8:155–165.

Etchecopar, R. D. and F. Hue 1967. *The Birds of North Africa*. Edinburgh: Oliver & Boyd.

Evans, P. R. 1976. Energy balance and optimal foraging strategies in shorebirds: Some implications for their distribution and movement in the non-breeding season. *Ardea* 64:117–139.

Evans, R. M. 1982a. Colony desertion and reproductive synchrony of black-billed gulls *Larus bulleri*. *Ibis* 124:491–501.

Evans, R. M. 1982b. Foraging flock recruitment at a black-billed gull colony: implications for the information center hypothesis. *Auk* 99:24–30.

Ewald, P. W., G. L Hunt, Jr., and M. Warner. 1980. Teritory size in western gulls:

Importance of intrusion pressure, defense, investments, and vegetation structure. *Ecolony* 61:80–87.

Falla, R. A., R. R. Sjkibson, and E. G. Tjurbott. 1966. *A Field Guide to the Birds of New Zealand and Outlying Islands*. Boston: Houghton Mifflin.

Farraway, A., J. Thomas and H. Blokpoel. 1986. Common tern egg predation by ruddy turnstones. *Condor* 88:521–522.

Fasola, M. 1986. *Distribuzione e popolazione del Laridi e Sternidi nidificanti in Italia*. vol. 11. Bologna: Instituto Nazion. Biol. Dell Selvaggina.

Fautin, R. W. 1941. Incubation studies of the yellow-headed blackbird. *Wilson Bull.* 53:107–122.

Feare, C. J. 1976. The breeding of the sooty tern *Sterna fuscata* in the Seychelles and the effects of experimental removal of its eggs. *J. Zool. Lond.* 179:317–360.

Feare, C. J. 1979. Ecology of Bird Island, Seychelles. *Atoll Res. Bull.* 226:1–36.

Feekes, F. 1981. Biology and colonial organization of two sympatric caciques *Cacicus cela cela* and *Cacicus haemorrhous haemorrhous*, Icteridae Aves, in Surinam. *Ardea* 69:83–107.

Ferguson, R. S. 1981. Territorial attachment and mate fidelity by horned grebes. *Wilson Bull.* 93:560–561.

Findley, C. S. and F. Cooke. 1982. Breeding synchrony in the lesser snow goose *(Anser caerulescens caerulescens)* I: Genetic and environmental components of hatch date variability and their effects on hatch synchrony. Evolution 36:342–351.

Fisher, H. I. 1971. The Laysan albatross: Its incubation, hatching, and associated behaviors. *Living Bird* 10:19–78.

Fisher, J. 1952. *The Fulmar*. London: Collin's New Naturalist.

Fisher, J. and R. M. Lockley. 1954. *Sea-Birds*. Boston: Houghton-Mifflin.

Fitch, M. A. 1979. Monogamy, polygamy, and female-female pairs in Herring Gulls. *Proc. Colonial Waterbird Group* 3:44–48.

Floyd, C. B. 1928. Notes on banding terns at Chatham, Massachusetts, for 1928. *Bull. Northeastern Bird Band. Assoc.* 4:125–140.

Floyd, C. B. 1929. Notes on banding terns at Chatham, Massachusetts, for 1929. *Bull. Northeastern Bird Band. Assoc.* 5:144–148.

Fordham, R. A. 1964a. Breeding biology of the southern Black-backed Gull. I: Pre-egg and egg stage. *Notornis* 11:3–34.

Fordham, R. A. 1964b. Breeding biology of the southern Back-backed Gull. II: Incubation and chick stage. *Notornis* 11:110–126.

Frankenberg, E. 1981. The adaptive significance of avian mobbing IV. "alerting others" and "perception advertisement" in blackbirds facing an owl. *Z. Tierpsychol.* 55:97–118.

Freer, V. M. 1979. Factors affecting site tenacity in New York bank swallows. *Bird Banding* 50:349–357.

French, N. R. 1959. Life history of the black rosy finch. *Auk* 75:159–180.

French, T. W. 1982. Piracy by a great black-backed gull on a shark. *Wilson Bull.* 94:96.

Friedmann, H. 1960. The parasitic weaverbirds. U.S. Natl. Mus. Bull. 223.
Frith, H. J., ed. 1983. *Reader's Digest Complete Book of Australian Birds*, 615 pp. Sydney: Reader's Digest Services.
Frost, P. G. H. and G. Shaughnessey. 1976. Breeding adaptations of the Damara Tern *Sterna balaenarum*. Madoqua 9:33–39.
Fry, D. M. and C. K. Toone. 1981. DDT-induced feminization of gull embryos. *Science* 213:922–924.
Fuchs, E. 1977a. Kleptoparasitism of sandwich terns *(Sterna sandvicensis)* by black-headed gulls *(Larus ridibundus)*. *Ibis* 119:183–190.
Fuchs, E. 1977b. Predation and anti-predator behaviour in a mixed colony of terns *Sterna* sp. and black-headed gulls *(Larus ridibundus)* with special reference to the sandwich tern *(Sterna sandvicensis)*. *Ornis. Scand.* 8:17–32.
Furness, B. L. and R. W. Furness. 1980. Apostatic selection and kleptoparasitism in the parasitic jaeger: A comment. *Auk* 97:832–836.
Furness, R. W. 1977. Effects of great skuas on Artic skuas in Shetland. *Brit. Birds* 70:96–107.
Furness, R. W. 1978. Kleptoparasitism by great skuas (*Catharacta skua* Brunn.) and Arctic skuas (*Stercorarius parasiticus* L.) at a Shetland seabird colony. *Anim. Behav.* 26:1167–1177.
Furness, R. W. 1987. Kleptoparasitism in seabirds. In J. P. Croxall, ed., *Seabirds: Feeding Ecology and Role in Marine Ecosystems*, pp. 77–100. New York: Cambridge University Press.
Galli, J. and R. Kane. 1979. Colonial waterbird populations of New Jersey. *Records of N. J. Birds* 7:38–42.
Galusha, J. G., Jr., and C. J. Amlaner, Jr. 1978. The effects of diurnal and tidal periodicities in the numbers and activities of herring gulls *Larus argentatus* in a colony. *Ibis* 120:322–328.
Garson, P. J. 1980. Male behaviour and female choice: Mate selection in the wren? *Anim. Behav.* 28:491–502.
Gaston, A. J. and D. N. Nettleship. 1981. The Thick-billed Murres of Prince Leopold Sound. Ottawa: Canadian Wildlife Service Monograph no. 6.
Gauzer, M. E. and M. T. Ter-Mikhaelyan. 1987. [Spatial and temporal population structure of common terns on islands of Krasnovodsk Bay.] *Zool Zh.* 66:110–118.
Gemperle, M. E. and F. W. Preston. 1955. Variation of shape in the eggs of the common tern in the clutch-sequence. *Auk* 72:184–198.
Gibson, E. 1920. Further ornithological notes from the neighbourhood of Cape San Antonio, Province of Buenos Aires, part 3. *Ibis* 1920:1–97.
Gillham, M. E. 1977. Vegetation of sea and shore-bird colonies on Aldabra Atoll. *Atoll. Res. Bull.* 200:1–19.
Gladstone, D. E. 1979. Promiscuity in monogamous colonial birds. *Am. Nat.* 114:545–557.
Gochfeld, M. 1971. Premature feather loss: A new disease of terns on Long Island, New York. *Kingbird* 21:206–211.
Gochfeld, M. 1973. Effect of artefact pollution on the viability of seabird colonies on Long Island, New York. *Env. Pollut.* 4:1–6.

Gochfeld, M. 1974. An incipient distraction display of the lest tern. *Proc. Linn. Soc.* (N. Y.) 73:80–82.

Gochfeld, M. 1975a. Comparative ecology and behavior of the red-breasted meadowlarks and their interactions in sympatry. Unpubl. Ph.D. thesis, City University of New York.

Gochfeld, M. 1975b. Developmental defects in common terns of western Long Island, New York. *Proc. Linn. Soc.* (N.Y.) 72:63–76.

Gochfeld, M. 1975c. Hazards of tail-first fish-swallowing in the common tern. *Condor* 77:345–346

Gochfeld, M. 1976. Waterbird colonies of Long Island, New York. 3: Cedar Beach ternery. *Kingbird* 26:62–80.

Gochfeld, M. 1977a. Intraclutch egg variation: The uniqueness of the common tern's third egg. *Bird Banding* 48:325–332.

Gochfeld, M. 1977b. Social system and possible lek behavior in brown-headed cowbirds. Abstract 1977 Annual Meeting, Amer. Ornithol. Union.

Gochfeld, M. 1978a. Colony and nest site selection by black skimmers. *Proc. Colonial Waterbird Group* 1:78–90.

Gochfeld, M. 1978b. Social facilitation of singing: group size and flight song rates in the pampas meadowlark *Sturnella defilippii*. *Ibis* 120:338–339.

Gochfeld, M. 1978c. Terns in traffic. *Nat. Hist.* 87:54–61.

Gochfeld, M. 1978d. Observations on feeding ecology and behavior of common terns. *Kingbird* 28:84–89.

Gochfeld, M. 1978e. Incubation behaviour in common terns: influence of wind speed and direction on orientation of incubating adults. *Animal Behav.* 26:848–851.

Gochfeld, M. 1979a. Group adherence in emigration in common terns. *Bird Banding* 50:365–366.

Gochfeld, M. 1979b. Breeding synchrony in the black skimmer: Colony vs. subcolonies. *Proc. Colonial Waterbird Group* 2:171–177.

Gochfeld, M. 1979c. Prevalence of oiled plumage of terns and skimmers on western Long Island, New York: Baseline data prior to petroleum exploration. *Env. Pollut.* 20:123–130.

Gochfeld, M. 1980a. Timing of breeding and chick mortality in central and peripheral nests of Magellanic penguins. *Auk* 97:191–193.

Gochfeld, M. 1980b. Mechanisms and adaptive value of reproductive synchrony in colonial seabirds. In J. Burger, B. Olla, and H. Winn, eds., *Behavior of Marine Animals*, vol. 4: Marine Birds, 207–270. New York: Plenum.

Gochfeld, M. 1980c. Tissue distribution of mercury in normal and abnormal young common terns. Marine Pollution Bull. 11:362–377.

Gochfeld, M. 1980d. Learning to eat by young common terns: consistency of presentation as an early cue. *Colonial Waterbirds* 3:108–118.

Gochfeld, M. 1981. Human influences on common tern *(Sterna hirundo)* reproduction in New York: ecological and chemical. In J. Cooper, ed., *Proceedings of the Symposium on Birds of the Sea and Shore, 1979*: 303–314. Cape Town: African Seabird Group.

Gochfeld, M. 1983a. The roseate tern: World distribution and status of a threatened species. *Biol. Conserv.* 25:103–125.

Gochfeld, M. 1983b. Colony site selection by Least Terns: Physical attributes of sites. *Colonial Waterbirds* 6:205–213.

Gochfeld, M. 1984. Antipredator behavior: aggressive and distraction displays of shorebirds. In J. Burger and B. L. Olla, eds., *Shorebirds: Breeding Behavior and Populations*, 289–377. New York: Plenum.

Gochfeld, M. 1985. Predation and coloniality in seabirds. *Proc. XVIII Int. Ornithol. Congr.* vol. 2 (V. D. Ilyichev & V. M. Gavrilov, eds), 882–891. Moscow: Academy of Sciences of the USSR.

Gochfeld, M., and Burger, J. 1981a. Boat-tailed grackle in Hewlitt Bay, Long Island. *Kingbird* 31:214.

Gochfeld, M., and J. Burger. 1981b. Age-related differences in piracy of frigatebirds from laughing gulls. *Condor* 83:79–82.

Gochfeld, M., and J. Burger. 1982. Feeding enhancement by social attraction in the sandwich tern. *Behav. Ecol. and Sociobiol.* 10:15–17.

Gochfeld, M. and J. Burger. 1987. Nest site selection: comparison of roseate and common terns *(Sterna dougallii* and *S. hirundo)* in a Long Island, New York colony. *Bird Behav.* 7:58–66.

Gochfeld, M., J. Burger, and J. R. Jehl, 1984. The classification of the shorebirds of the world. In J. Burger and B. Olla, eds., *Behavior of Marine Animals*, vol. 5: *Shorebirds: Breeding Behavior and Populations*, 1–16. New York: Plenum.

Gochfeld, M. and D. B. Ford. 1974. Reproductive success in common tern colonies near Jones Beach, Long Island, New York in 1972: a hurricane year. *Proc. Linn. Soc. New York* 72:63–76.

Goss-Custard, J. D., S. E. A. Durell, and Ens, B. J. 1982. Individual differences in aggressiveness and food stealing among wintering oystercatchers, *Haematopus ostralegus* L. *Anim. Behav.* 30:917–928.

Gotmark, F. 1982. Coloniality in five larus gulls: A comparative study. *Ornis Scandinavica* 13:211–234.

Gowaty, P. A. 1982. Sexual terms in sociobiology: Emotionally evocative and, paradoxically, jargon. *Anim. Behav.* 30:630–631.

Grant, P. R. 1971. Interactive behaviour of puffins (*Fratercula arctica* L.) and skuas (*Stercorarius parasiticus* L.). *Behaviour* 40:263–281.

Graves, J. S. and A. Whitten. 1980. Adoption of strange chicks by herring gulls. *Z. Tierpsychol.* 54:267–278.

Greenhalgh, M. E. 1974. Population, growth, and breeding success in a salt marsh common tern colony. *Yorkshire Naturalist* 932:121–127.

Greig-Smith, P. W. 1981a. The role of alarm responses in the formation of mixed-species flocks of heathland birds. *Behav. Ecol. Sociobiol.* 8:7–10.

Greig-Smith, P. W. 1981b. Responses to disturbance in relation to flock size in foraging groups of barred ground doves *Geopelia striata*. *Ibis* 123:103–106.

Greig-Smith, P. W. 1982. Dispersal between nest-sites by stonechats *Saxicola torquata* in relation to previous breeding success. *Ornis Scandinavica* 13:232–238.

Gunter, G. 1982. Heat death of least tern chicks on the Gulfport, Mississippi, beach in 1980. *Gulf Research Reports* 7(2):163–166.

Hailman, J. P. 1964. Breeding synchrony in the equatorial swallow-tailed gull. *Am. Nat.* 98:79–83.

Halkin, S. L. 1983. Resting birds tuck bills toward outside of group. *Auk* 100:997–998.

Hall, J. R. 1970. Synchrony and social stimulation in colonies of the black-headed weaver *Ploceus cucullatus* and Vieillot's black weaver *Melanopteryx nigerrimus*. *Ibis* 112:93–104.

Hamilton, W. D. 1971. Geometry for the selfish herd. *J. Theor. Biol.* 31:295–311.

Hammerstein, P. 1981. The role of asymmetries in animal contests. *Anim. Behav.* 29:193–205.

Hand, J. L., G. L. Hunt, Jr., and M. Warner. 1981. Thermal stress and predation: influences on the structure of a gull colony and possibly on breeding distributions. *Condor* 83:193–203.

Harding, C. F., and Follett, B. K. 1979. Hormone changes triggered by aggression in a natural population of blackbirds. *Science* 203:918–920.

Harlow, H. 1932. Social facilitation of feeding in the albino rat. *J. Gen. Psych.* 43:211.

Harris, M. P. 1969. The biology of storm petrels in the Galapagos Islands. *Proc. Calif. Acad. Sci.* 4th ser. 37:95–165.

Harris, M. P. 1973. The biology of the waved albatross *Diomedea irrorata* of Hood Island, Galapagos. *Ibis* 115:483–510.

Harris, M. P. 1978. Supplementary feeding of young puffins, *Fratercula arctica*. *J. Anim. Ecol.* 47:15–23.

Harrison, P. L., R. C. Babcock, G. D. Bull, J. K. Oliver, C. C. Wallace, and B. L. Willis. 1984. Mass spawning on tropical reef, corals. *Science* 223:1184–1189.

Hartert, E. 1893. On the birds of the islands of Aruba, Curaçao, and Bonaire. *Ibis* 1893:289–338.

Hartert, E. and Venturi S., 1909. Notes sur les oiseaux de la République Argentine. *Novit. Zool.* 16:159–267.

Hatch, J. J. 1970. Predation and piracy by gulls at a ternery in Maine. *Auk* 87:244–254.

Hatch, J. J. 1975. Piracy by laughing gulls *Larus atricilla*: an example of the selfish group. *Ibis* 117:357–365.

Hays, H. 1970. Common terns pirating fish on Great Gull Island. *Wilson Bull.* 82:99–100.

Hays, H. 1984. Common terns raise young from successive broods. *Auk* 101:274–280.

Hays, H., and R. W. Risebrough. 1972. Pollutant concentration in abnormal young terns from Long Island sound. *Auk* 89:19–35.

Hayward, J. L., Jr., C. J. Amlaner, Jr., W. H. Gillett, and J. F. Stout. 1975. Predation on nesting gulls by a river otter in Washington State. *Murrelet* 56:9–10.

Hebert, P. N., and Barclay, R. M. R. 1986. Asynchronous and synchronous hatching: effect on early growth and survivorship of herring gull, *Larus argentatus*, chicks. *Can. J. Zool.* 64:2357–2362.

Heppleston P. B. 1971. The feeding ecology of oystercatchers *Haematopus ostralegus L.* in winter in northern Scotland. *J. Anim. Ecol.* 41:651–672.

Hewitt, O. H. 1941. Common terns nesting on muskrat houses. *Auk* 58:579–580.

Hinde, R. A. 1954. Factors governing the changes in strength of a partially inborn response as shown by the mobbing behaviour of the chaffinch *(Fringilla coelebs)*. I: The nature of the response, and an examination of its course. *Proc. Royal Society*. ser. B. 142:306–331.

Hockey, P. 1980. Kleptoparasitism by kelp gulls *Larus dominicanus* of African black oystercatchers *Haematopus moquini*. *Cormorant* 8:97.

Hoffman, W. D., D. Heinemann, and J. A. Wiens. 1981. The ecology of seabird feeding flocks in Alaska. *Auk* 98:437–456.

Hogstad, O. 1983. Is nest predation really selecting for colonial breeding among fieldfares *Turdus pilaris? Ibis* 125:366–369.

Holland, A. H. 1890. On some birds of the Argentine Republic. *Ibis* 1890:424–428.

Hoogland, J. L. 1979a. The effect of colony size on individual alertness of prairie dogs (Sciuridae: *Cynomys* spp.) *Animal Behaviour* 27:394–407.

Hoogland, J. L. 1979b. Agression, ectoparasitism, and other possible costs of prairie dog (*Sciuridae: Cynomys* spp.) coloniality. *Behaviour* 69:1–35.

Hoogland, J. L. and P. W. Sherman. 1976. Advantages and disadvantages of bank swallow *(Riparia riparia)* coloniality. *Ecol. Monographs*. 46:33–56.

Hopkins, C. D., and R. H. Wiley. 1972. Food parasitism and competition in two terns. *Auk* 89:583–594.

Hoppensteadt, F. C. and J. B. Keller. 1976. Synchronization of periodical cicada emergences. *Science* 194:335–337.

Horn, H. S. 1968. The adaptive significance of colonial nesting in the Brewer's blackbird *(Euphagus cyanocephalus)*. *Ecology* 49:682–694.

Houde, A. E. 1983. Nest density, habitat choice, and predation in a common tern colony. *Colonial Waterbirds* 6:178–184.

Howell, T. R., B. Araya, and W. R. Millie. 1974. Breeding biology of the gray gull, *Larus modestus. Univ. of California Publ. in Zool.* 104:1–57.

Hulsman, K. 1976. The robbing behavior of terns and gulls. *Emu* 76:143–149.

Hulsman, K. 1984. Selection of prey and success of silver gulls robbing crested terns. *Condor* 86:130–138.

Humphrey, P. S., D. Bridge, P. W. Reynolds, and R. T. Peterson. 1970. Birds of Isla Grande (Tierra del Fuego). Washington, D.C.: Smithsonian Institution.

Hunt, G. L., Jr. and M. W. Hunt. 1976. Gull chick survival: The significance of growth rates, timing of breeding, and territory size. *Ecology* 57:62–75.

Hunt, G. L., Jr. and M. W. Hunt. 1977. Female-female pairing in Western gulls *(Larus occidentalis)* in southern California. *Science* 196:1466–1467.

Hunter, R. A. and R. D. Morris. 1976. Nocturnal predation by a black-crowned night heron at a common tern colony. *Auk* 93:629–633.

Hussell, D. J. T. 1972. Factors affecting clutch size in Arctic passerines. *Ecological Monographs* 42:317–364.

Hutchinson, G. E. and R. MacArthur. 1959. Appendix: On the theoretical significance of aggressive neglect in interspecific competition. *Amer. Nat.* 93:133–134.

Hutson, G. D. 1977. Observations on the swoop and soar display of the black-headed gull *Larus ridibundus*. *Ibis* 119:67–73.

Ingolfsson, A. 1969. Behaviour of gulls robbing eiders. *Bird Study* 16:45–52.

Jackson, B. J. S. and J. A. Jackson. 1985. Killdeer nest abandonment possibly caused by ants. *Miss. Kite* 15:5–6.

Johnson, R. A. 1938. Predation of gulls in murre colonies. *Wilson Bull.* 50:161–170.

Johnson, A. W. and J. D. Goodall. 1967. The Birds of Chile and Adjacent Regions. Buenos Aires: Platt Establecimientos Graficos.

Jones, L. 1906. A contribution to the life history of the common *(Sterna hirundo)* and the roseate *(Sterna dougalli)* terns. *Wilson Bull.* 18:35–47.

Joyner, D. E. 1974. Duck nest predation by gulls in relation to water depth. *Condor* 76:339–340.

Kadlec, J. A. 1971. Effects of introducing foxes and raccoons on herring gull colonies. *J. Wild Mgmt.* 35:625–637.

Kallander, HJ. 1975. Gratrut *Larus argentatus* som narings—parasit pasangsvan *Cygnus cygnus*. *Anser* 14:29–38.

Kallander, H. 1977. Piracy by black-headed gulls on lapwings. *Bird study* 24:186–194.

Kallander, H. 1979. Skrattmasen *(Larus ridibundus)* som kleptoparasit pa tofsvipa *(Vanellus vanellus)*. *Fauna och flora* 74:200–207.

Kane, R. and R. B. Farrar. 1976. 1975 Coastal colonial bird survey of New Jersey. *N. J. Audubon* 2:8–14.

Kane, R. and R. B. Farrar. 1977. 1977 Coastal colonial bird survey of New Jersey. *N. J. Audubon* 3:188–194.

Kaverkina, N. P. 1986a. [Pairing behavior of terns.] *Bull. Moscow Society for Natural History Research (Biology)* 91:40–47.

Kaverkina, N. P. 1986b. [Nesting biology of the Kamchatka {Aleutian} tern]. In N. M. Litvinenko, ed. [Seabirds of the Far East]. Vladivostock: Academy of Sciences of the USSR.

Kaverkina, N. P. 1989. [Feeding and hunting behavior in terns]. *Bull. Zoolog. Herald* (Leningrad Univ) 1989+:50–53.

Kaverkina, N. P. and U. K. Roschevski. 1984. [Defensive behavior of the Arctic tern.] *Bull. Leningrad Univ.* 15:19–25.

Kepler, C. B. 1967. Polynesian rat predation on nesting Laysan albatrosses and other Pacific seabirds. *Auk* 84:426–30.

Kharitonov, S. P. and D. Siegel-Causey. 1988. Colony formation in seabirds. In R. F. Johnson, ed., *Current Ornithology* 5:223–272. New York: Plenum Press.

Khleboselov, E. I. 1986. [On possible mechanisms of regulating the number of birds in feeding flocks]. In V. D. Illitchev, ed., [Current Problems in Ornithology]. Moscow: Academy of Sciences of USSR.

Kilham, L. 1981. Courtship feeding and copulation of royal terns. *Wilson Bull.* 93:390–391.

Kilham, L. 1982. Common crows pulling the tail and stealing food from a river otter. *Fla. Field Nat.* 10:39–40.

King, B. 1977. "Alarms," "dreads," and "panics" at terneries. *British Birds* 70:81–82.

King, W. B. (compiler). 1981. *Endangered Birds of the World: The ICBP Bird Red Data Book.* Washington, D. C.: Smithsonian Inst. Press.

King, K. A. and C. A. Lefever. 1979. Effects of oil transferred from incubating gulls to their eggs. *Marine Pollut. Bull.* 10:319–321.

Klint, T. 1980. Influence of male nuptial plumage on mate selection in the female mallard *(Anas platyrhynchos). Anim. Behavior* 28:1230–1238.

Klomp, H. 1970. The determination of clutch size in birds. A Review. *Ardea* 58:1–124.

Knopf, F. L. 1979. Spatial and temporal aspects of colonial nesting of white pelicans. *Condor* 81:353–363.

Knowlton, N. 1979. Reproductive synchrony, parental investment, and the evolutionary dynamics of sexual selection. *Anim. Behav.* 27:1022–1033.

Koskimies, J. 1957. Terns and gulls as features of habitat recognition for birds nesting in their colonies. *Ornis Fenn.* 34:1–6.

Kramer, G., and U. von St. Paul. 1951. Uber angeborenes und erworbenes Feinderkennen beim Gimpel (*Pyrrhula pyrrhula* L.). *Behaviour* 3:243–251.

Krebs, J. R. 1973. Behavioral aspects of predation. In P. P. G. Bateson and P. H. Klopfer, eds., *Perspectives in Ethology*, 73–112. New York: Plenum Press.

Krebs, J. R. 1974. Colonial nesting and social feeding as strategies for exploiting food resources in the great blue heron *(Ardea herodias). Behaviour* 50:99–134.

Krebs, J. R. 1978. Colonial nesting in birds, with special reference to the Ciconiiformes. In A. Sprunt, J. C. Ogden, and S. Winkler, eds., *Wading Birds*, 299–314. New York: National Audubon Society.

Krebs, J. R., M. H. MacRoberts, and J. M. Cullen. 1972. Flocking and feeding in the great tit *Parus major*—an experimental study. *Ibis* 144:507–530.

Kress, S. W., E. H. Weinstein and I. C. T. Nisbet. 1983. The status of tern populations in Northeastern United States and adjacent Canada. *Colonial Waterbirds* 6:84–106.

Kruijt, J. P., I. Bossema, and G. J. Lammers. 1982. Effects of early experience and male activity on mate choice in mallard females *(Anas platyrhynchos). Behaviour* 79:32–42.

Kruuk, H. 1964. Predators and anti-predator behavior of the black-headed gull *(Larus ridibundus L.) Behaviour*, suppl. 11:1–129.

Kruuk, H. 1972. *The Spotted Hyena.* Chicago: University of Chicago Press.

Kruuk, H. 1976. The biological function of gulls' attraction towards predators. *Anim. Behav.* 24:146–153.

Kury, C. R., and M. Gochfeld. 1975. Human interference and gull predation in cormorant colonies. *Biol. Conservation* 8:23–34.

Kushlan, J. A. 1973. Promiscuous mating behavior in the whtie ibis. *Wilson Bull.* 85:331–332.

Kushlan, J. A. 1977. The significance of plumage colour in the formation of feeding aggregations in the Ciconiiforms. *Ibis* 119:361–364.

Lack, D. 1954. *The Natural Regulation of Animal Numbers*. Oxford: Clarendon Press.

Lack, D. 1968. *Ecological Adaptations for Breeding in Birds*. London: Methuen.

Lamore, D. 1953. Ring-billed gulls stealing fish from female American mergansers. *Wilson Bull.* 65:210–211.

Langham, N. P. E. 1972. Chick survival in terns with particular reference to the common tern. *J. Animal Ecol.* 41:385–395.

Langham, N. P. E. 1974. Comparative breeding biology of the sandwich tern. *Auk* 91:255–277.

Langham, N. P. and K. Hulsman. 1986. The breeding biology of the crested tern *Sterna bergii*. *Emu* 86:23–32.

Lauro, B. 1986. Habitat and nest site selection in American oystercatcher. Unpubl. m.s. thesis, Rutgers University, New Brunswick, N.J.

Lauro, B. and J. Burger 1989. Habitat and nest site selection in American Oystercatcher. *Auk* 106:185–192.

Lazarus, J. 1979a. The early warning function of flocking in birds: An experimental study with captive *Quelea*. *Anim. Behav.* 27:855–865.

Lazarus, J. 19979b. Flock size and behaviour in captive red-billed weaverbirds *(Quelea quelea)*: Implications for social facilitation and the functions of flocking. *Behaviour* 71:127–1145.

Lazarus, J., and I. R. Inglis. 1978. The breeding behaviour of the pinkfooted goose: Parental care and vigilant behaviour during the fledging period. *Behaviour* 65:62–88.

Lazell, J. D., and I. C. T. Nisbet. 1972. The tern-eating snakes of No Mans Land. *Man and Nature* 1972:27–29.

LeCroy, M. and S. LeCroy. 1974. Growth and fledging in the common tern *(Sterna hirundo)*. *Bird Banding* 45:327–340.

Lemmetyinen, R. 1971. Nest defence behaviour of common and Arctic terns and its effects on the success achieved by predators. *Ornis Fennica* 48:13–24.

Lemmetyinen, R. 1972. Growth and mortality in the chicks of Arctic terns in Kongsfjord area, Spitsbergen, in 1970. *Ornis Fennica* 49:45–53.

Lemmetyinen, R. 1973. Breeding success in *Sterna paradisaea Pontnopp.* and *S. hirundo L.* in Southern Finland. *Ann. Zool. Fennici* 10:526–535.

Lemmetyinen, R. 1974. Comparative breeding ecology of the Arctic and the common tern with special reference to feeding. Report no. 3. Univ. of Turku, Dept. of Zool.

Lendrem, D. W. 1983. Sleeping and vigilance in birds. I: Field observations of the mallard *(Anas platyrhynchos)*. *Anim. Behav.* 34:532–538.

Lenington, S. 1980. Female choice and polygyny in red-winged blackbirds. *Anim. Behav.* 28:347–361.

Lobkov, E. G. and N. M. Golovina. 1978. [Comparative analysis of the biology of Aleutian and common terns on Kamchatka.] *Bull. Society Nature Study (Biology)* [Moscow] 83:27–37.

Loftin, R. W. and S. Sutton. 1979. Ruddy turnstones destroy royal tern colony. *Wilson Bull.* 91:133–135.

Lowther, P. E., W. J. Olsen, R. T. Young, and M. J. Luttenton, 1984. Observations on the mobbing response. *Iowa Bird life* 57:72–75.

Lumpkin, S., K. Kessel, P. G. Zenone, and C. J. Erikson, 1982. Proximity between the sexes in ring doves: Social bonds or surveillance. *Anim. Behav.* 30:506–513.

Mackay, G. H. 1895. The terns of Muskeget Island, Massachusetts. *Auk* 12:32–48.

Mackay, G. H. 1898. The terns of Muskeget Island, Massachusetts, part 4. *Auk* 15:168–172.

MacRoberts, B. R. and M. H. MacRoberts. 1972. Social stimulation of reproduction in herring and lesser black-backed gulls. *Ibis.* 114:496–506.

MacRoberts, M. H. 1973. Extra marital courting in lesser black-backed gulls and herring gulls. *Z. Tierpsychol.* 32:62–74.

MacRoberts, M. H. and B. R. MacRoberts. 1980. Toward a minimal definition of animal communication. *Psychological Record* 30:387–396.

Magurran, A. E., W. J. Oulton, and T. J. Pitcher. 1985. Vigilant behavior and shoal size in minnows. *Z. Tierpsychol.* 67:167–178.

Mangold, R. E. 1974. Research on shore and upland migratory birds in New Jersey: Clapper rail studies. Report of Div. of Fish, Game and Wildlife, New Jersey Department of Environmental Protection. Trenton, N.J.

Manuwal, D. A. 1974. Effects of territoriality on breeding in a population of Cassin's auklet. *Ecology* 55:1399–1406.

Marples, G. and A. Marples. 1934. *Sea Terns or Sea Swallows.* London: Country Life Ltd.

Marshall, N. 1942. Night desertion by nesting common terns. *Wilson Bull.* 54:25–31.

Massey, B. W. 1974. Breeding biology of the California Least Tern. *Proc. Linnaean Soc.* (N.Y.) 72:1–24.

Massey, B. W. and J. L. Atwood. 1981. Second-wave nesting of the California least tern: age composition and reproductive success. *Auk* 98:596–605.

Maxson, S. J. and N. P. Bernstein. 1982. Kleptoparasitism by South Polar skuas on blue-eyed shags in Antarctica. *Wilson Bull.* 94:269–281.

Mayr, E. 1935. Bernard Altum and the territory theory. *Proc. Linn. Soc.* (N.Y.) 45-46:24–38.

McClintock, M. K. 1971. Menstrual synchrony and suppression. *Nature* 229:244–245.

McClintock, M. K. 1981. Social control of the ovarian cycle and the function of estrous synchrony. *Am. Zool.* 21:243–256.

McKearnan, J. E. and F. J. Cuthbert. 1989. Status and breeding success of common terns in Minnesota. *Colonial Waterbirds* 12:185–1990.

McNicholl, M. K. 1973. Habituation of aggressive responses to avian predators by terns. *Auk* 90:902–904.

McNicholl, M. K. 1975. Larid site-tenacity and group adherence in relation to habitat. *Auk* 92:98–104.

McNicholl, M. K. 1979. Territories of Foster's terns. *Proc. Col. Waterbird Group* 3:196–203.

McNicholl, M. K. 1982. Factors affecting reproductive success of Forster's terns at Delta Marsh, Manitoba. *Colonial Waterbirds* 5:32–39.

Meanley, B. 1955. A nesting study of the little blue heron in eastern Arkansas. *Wilson Bull.* 67:84–99.

Meinertzhagen, R. 1959. *Pirates and Predators*. Edinburgh: Oliver and Boyd.

Metcalfe, N. B. 1984. The effects of mixed-species flocking on the vigilance of shorebirds: Who do they trust? *Anim. Behav.* 32:986–993.

Milinski, M. 1977a. Experiments on the selection by predators against spatial oddity of their prey. *Z. Tierpsychol.* 43:311–325.

Milinski, M. 1977b. Do all members of a swarm suffer the same predation? *Z. Tierpsychol.* 43:373–388.

Milinski, M. 1984. A predator's costs of overcoming the confusion-effect of swarming prey. *Animal Behav.* 32:1157–1162.

Miller, L. and J. Confer. 1982. A study of the feeding of young common terns at one site in Oneida Lake during 1980. *Kingbird* 32:167–172.

Milne, H. 1974. Breeding numbers and reproductive rate of eiders at the Sands of Forvies National Nature Reserve, Scotland. *Ibis.* 1116:135–152.

Mineau, P. and F. Cooke. 1979. Rape in the lesser snow goose. *Behaviour* 70:280–291.

Mock, D. W. 1976a. Social behavior of the boat-billed heron. *Living Bird* 14:185–214.

Mock, D. W. 1976b. Pair-formation displays of the great blue heron. *Wilson Bull.* 88:185–230.

Moller, A. P. 1975. Ynglebestanden of Sandterne *Gelochelidon n. nilotica* Gmel. i 1971 i Europa, Afrika og Vestasien, med et tilgageblik over bestandsaendringer i dette arhundrede. *Dansk Orn. Foren Tidsskr.* 69:1–8.

Moller, A. P. 1981. Breeding cycle of the gull-billed tern (*Gelochelidon nilotica* Gmel.) especially in relation to colony size. *Ardea* 69:193–198.

Montevecchi, W. A. 1977. Predation in a salt marsh laughing gull colony. *Auk* 94:583–585.

Montevecchi, W. A. 1978a. Corvids using objects to displace gulls from nests. *Condor* 80:349.

Montevecchi, W. A. 1978b. Nest site selection and its survival value among laughing gulls. *Behav. Ecol. Sociobiol.* 4:143–161.

Montevecchi, W. A. 1979. Predator-prey interactions between ravens and kittiwakes. *Z. Tierpsychol.* 49:136–141.

Montevecchi, W. A., and J. Wells. 1984. Fledging success of northern gannets from different nest-sites. *Bird Behaviour* 5:90–95.

Moran, G. 1984. Vigilance behavior and alarm calls in a captive group of meerkats, *Suricata suricatta*. *Z. Tierpsychol.* 65:228–240.

Moreau, R. E. 1944. Clutch size: a comparative study, with reference to African birds. *Ibis* 86:286–347.

Morris, R. D. 1986. Seasonal differences in courtship feeding rates of male common terns. *Can. J. Zool.* 64:501–507.
Morris, R. D. and M. J. Bidochka. 1982. Mate guarding in herring gulls. *Colonial Waterbirds* 5:124–130.
Morris, R. D. and R. A. Hunter. 1976. Factors influencing desertion of colony sites by common terns *(Sterna hirundo)*. *Canadian Field Nat.* 90:137–143.
Morris, R. D., I. R. Kirkham, and J. W. Chardine. 1980. Management of a declining common tern colony. *J. Wildlife Manage.* 44:241–245.
Morris, R. D. and D. A. Wiggins. 1986. Ruddy turnstones, great horned owls, and egg loss from common tern clutches. *Wilson Bull.* 98:101–109.
Moynihan, M. 1955. Some aspects of reproductive behavior in the black-headed gull (*Larus ridibundus ridibundus* L.) and related species. *Behaviour* suppl. 4:1–197.
Moynihan, M. 1958. Notes on the behavior of some North American gulls. II: Non-aerial hostile behavior of adults. *Behaviour* 12:95–182.
Moynihan, M. 1959a. A revision of the family Laridae (Aves). *Amer. Museum Novitates* 1928:1–42.
Moynihan, M. 1959b. Notes on the behavior of some North American gulls. IV: The ontogeny of hostile behavior and display patterns. *Behaviour* 14:214–239.
Moynihan, M. 1960. Some adaptations which help to promote gregariousness. *Proc. 12th Intern. Ornithol. Congress.* 12:523–541.
Muller-Schwarze, D. 1968. Circadian rhythyms of activity in the adélie penguin *(Pygoscelis adeliae)* during the austral summer. *Antarct. Res. Ser.* 12:133–149.
Muller-Schwarze, D., and C. Muller-Schwarze. 1973. Differential predation by South Polar skuas in an adélie penguin rookery. *Condor* 75:127–131.
Mumme, R. L., W. D. Koenig and F. A. Pitelka. 1983. Mate guarding in the acorn woodpecker: Within-group reproductive competition in a cooperative breeder. *Anim. Behav.* 31:1094–1106.
Munro, J. A. 1949. Studies of waterfowl in British Columbia. Baldpate. *Can. J. Res. sect. D.* 27:289–294.
Murphy, R. C. 1936. *Oceanic Birds of South America.* 2 vols. New York: American Mus. Nat. History.
Murray, B. G., Jr. 1971. The ecological consequences of interspecific territorial behavior in birds. *Ecology* 52:414–423.
Murray, B. G., Jr. 1979. Population Dynamics: Alternative Models. New York: Academic Press.
Murray, B. G., Jr. 1981. The origins of adaptive interspecific territorialism. *Biol. Review.* 56:1–22.
Murray, B. G., Jr. 1985. Evolution of clutch size in tropical species of birds. *Ornithol. Monographs* 36:505–519.
Murray, B. G., Jr. 1988. Interspecific territoriality in *Acrocephalus:* a critical review. *Ornis Scand.* 19:309–313.
Murton, R. K. 1971a. The significance of a specific search image in the feeding behavior of the wood pigeon. *Behavior* 39:10–42.
Murton, R. K. 1971b. Why do some bird species feed in flocks? *Ibis* 113:534–536.

Myers, J. P. 1978. One deleterious effect of mobbing in the southern lapwing (*Vanellus chilensis*). *Auk* 95:419–420.

Nelson, J. B. 1966. The breeding biology of the gannet *Sula bassana* on the Bass Rock, Scotland. *Ibis* 108:584–626.

Nelson, J. B. 1970. The relationship between behavior and ecology in the Sulidae with reference to other sea birds. *Oceangr. Mar. Biol. Ann. Rev.* 8:501–574.

Nelson, J. B. 1975. The breeding biology of frigatebirds—a comparative review. *Living Bird.* 14:113–155.

Nelson, J. B. 1979. *Seabirds: Their Biology and Behavior.* New York: A and W Publishers.

Nettleship, D. N. 1972. Breeding success of the common puffin *Fratercula arctica L.* on different habitats of Great Island, Newfoundland. *Ecol. Monogr.* 42:239–268.

Nettleship, D. N. and T. R. Birkhead. 1958. *The Atlantic Alcidae.* New York: Academic Press.

Nice, M. M. 1941. The role of territory in bird life. *Amer. Midland Nat.* 26:441–487.

Nice, M. M. and J. Ter Pelkwyk. 1941. Enemy recognition by the song sparrow. *Auk* 58:195–214.

Nicholls, G. H. 1977. Studies of less familiar birds: bridled tern. *Bokmakierie* 29:20–23.

Nickell, W. 1964. Fatal entanglements of herring gulls (*Larus argentatus*) and common terns (*Sterna hirundo*). *Auk* 81:555–556.

Niebuhr, V. 1981. An investigation of courtship feeding in herring gulls *Larus argentatus. Ibis* 123:218–223.

Nisbet, I. C. T. 1973a. Courtship feeding, egg size, and breeding success in common terns. *Nature* 241:141–142.

Nisbet, I. C. T. 1973b. Terns in Massachusetts: Present numbers and historical changes. *Bird Banding* 44:27–55.

Nisbet, I. C. T. 1975. Selective effects of predation in a tern colony. *Condor* 77:221–226.

Nisbet, I. C. T. 1976. Early stages in postfledging dispersal of common terns. *Bird Banding* 47:163–164.

Nisbet, I. C. T. 1977. Courtship-feeding and clutch size in common terns *Sterna hirundo.* In B. Stonehouse and C. M. Perrins, Eds., *Evolutionary Ecology,* 101–109. London: Macmillan.

Nisbet, I. C. T. 1978. Population models for common terns in Massachusetts. *Bird Banding* 49:50–58.

Nisbet, I. C. T. 1981. Behavior of common and roseate terns after trapping. *Colonial Waterbirds* 4:41–46.

Nisbet, I. C. T. 1983a. The status of tern populations in northeasten United States and adjacent Canada: summary and overview. *Colonial Waterbirds* 6:100–103.

Nisbet, I. C. T. 1983b. Territorial feeding by common terns. *Colonial Waterbirds.* 6:64–70.

Nisbet, I. C. T. 1983c. Paralytic shellfish poisoning: Effects on breeding terns. *Condor* 85:338–345.

Nisbet, I. C. T. 1989. Long-term ecological studies of seabirds. *Colonial Waterbirds* 12:143–147.

Nisbet, I. C. T. and M. Cohen 1975. Asynchronous hatching in common and roseate terns, *Sterna hirundo* and *S. dougallii*. *Ibis* 117:374–379.

Nisbet, I. C. T. and W. H. Drury. 1972a. Post-fledging survival in herring gulls in relation to broodsize and date of hatching. *Bird Banding* 43:161–172.

Nisbet, I. C. T. and W. H. Drury. 1972b. Measuring breeding success in common and roseate terns. *Bird Banking* 41:97–106.

Nisbet, I. C. T., and M. J. Welton. 1984. Seasonal variations in breeding success of common terns: Consequences of predation. *Condor* 86:53–60.

Nisbet, I. C. T., J. M. Winchell, and A. E. Heise. 1984. Influence of age on the breeding biology of common terns. *Colonial Waterbirds* 7:117–126.

Nuechterlein, G. L. 1981. "Information parasitism" in mixed colonies of western grebes and Forster's terns. *Anim. Behav.* 29:985–989.

O'Malley, J. B. E. and R. M. Evans. 1983. Kleptoparasitism and associated foraging behaviors in American white pelicans. *Colonial Waterbirds* 6:126–129.

Onno, S. 1967. Nesting colony of the common gull. *Orn. Kogumik.* 4:114–148.

Orians, G. H. 1961a. Social stimulation within blackbird colonies. *Condor* 63:330–337.

Orians, G. H. 1961b. The ecology of blackbird *Agelaius* social systems. *Ecol. Monogr.* 31:285–312.

Orians, G. H. 1969. On the evolution of mating systems in birds and mammals. *Am. Nat.* 103:589–604.

Orians, G. H. 1980. *Some Adaptations of Marsh-Nesting Blackbirds.* Monographs in Population Biology No. 14. Princeton: Princeton University Press.

Orians G. H. and M. F. Willson. 1964. Interspecific territories of birds. *Ecology* 45:736–745.

Palmer, R. S. 1941a. A behavior study of the common tern (*Sterna hirundo hirundo* L.). *Proc. Boston Soc. Nat. Hist.* 42:1–119.

Palmer, R. S. 1941b. 'White-faced' terns. *Auk* 58:164–178.

Paludan, K. 1951. Contributions to the breeding biology of *Larus argentatus* and *Larus fuscus*. *Dansk. Naturhist. Foren Vidensk. Meddel.* 114:1–128.

Panov, E. N. and L. Y. Zikova. 1987. [Influence of ecological and social factors on the reproductive success of the great black-headed gull *(Larus ichthyaetus)*.] *Zool. Zhurn.* 66:883–894.

Parker, G. A. and D. I. Rubenstein. 1981. Role assessment, reserve strategy, and acquisition of information in asymmetric animal conflicts. *Animal Behav.* 29:221–240.

Parkes, K. C., A. Poole, and H. Lapham. 1971. The ruddy turnstone as an egg predator. *Wilson Bulletin* 83:306–307.

Parmalee, D. F. and S. J. Maxson. 1974. The Antarctic Terns of Anvers Island. *Living Bird* 13:233–250.

Parsons, J. 1971. Cannibalism in herring gulls. *Br. Birds.* 64:528–537.

Parsons, J. 1975. Seasonal variation in the breeding success of the herring gull: an experimental approach to pre-fledging success. *J. Animal Ecology* 44:553–573.

Parsons, J. 1976. Factors determining the number and size of eggs laid by the herring gull. *Condor* 78:481–492.

Parsons, K. 1985. Proximate and ultimate effects of weather on two heron species in Massachusetts. Unpubl. Ph.D. diss. New Brunswick, N.J.: Rutgers University.

Partridge, L. 1978. Habitat selection. In J. R. Krebs and N. B. Davies, eds., *Behavioral Ecology: An Evolutionary Approach*, 351–376. Sunderland, Mass.: Sinauer Assoc.

Patterson, I. J. 1965. Timing and spacing of broods in the black-headed gull *Larus ridibundus*. *Ibis* 107:433–459.

Patton, S. R. and W. E. Southern. 1978. The effect of nocturnal red fox predation on the nesting success of colonial gulls. *Proc. Colonial Waterbird Group* 1:91–101.

Payne, R. B. 1969. Breeding seasons and reproductive physiology of tricolored blackbirds and redwing blackbirds. *Univ. Calif. Publ. Zool.* 90:1–137.

Payne, A. P. and H. H. Swanson. 1972. The effect of sex hormones on aggression in the male golden hamster. *Proc. Soc. Endocrinology* 53:11–12.

Payne, R. B. and H. F. Howe. 1976. Kleptoparasitism by gulls of migrating shorebirds. *Wilson Bull.* 88:349–351.

Pearson, T. H. 1968. The feeding biology of sea-bird species breeding on the Farne Islands, Northumberland. *J. Animal. Ecol.* 37:521–551.

Pessino, C. M. 1968. Redwinged blackbird destroys eggs of common and roseate terns. *Auk* 85:513.

Peters, J. L. 1934. *Check-list of Birds of the World*. vol. 2. Cambridge: Harvard University Press.

Petrie, M. 1983. Female moorhens compete for small fat males. *Science* 220:413–415.

Pettingill, O. S., Jr. 1939. History of one hundred nests of Arctic tern. *Auk* 55:420–428.

Phillips, G. C. 1962. Survival value of the white coloration of gulls and other sea birds. PhD. thesis, Oxford University.

Pielou, E. C. 1977. *Mathematical Ecology*. New York: Wiley.

Pienkowski, M. W. 1982. Diet and energy intake of grey and ringed plovers, *Pluvialis squatarola* and *Charadrius hiaticula*, in the nonbreeding season. *J. Zool.* 197:511–549.

Pienkowski, M. W. and P. R. Evans. 1982. Breeding behavior, productivity and survival of colonial and non-colonial shelducks *Tadorna tadorna*. *Ornis Scandinavica* 13:101–116.

Pierotti, R. 1980. Spite and altruism in gulls. *Amer. Natur.* 114:290–300.

Pierotti, R. 1981. Male and female parental roles in the Western Gull under different environmental conditions. *Auk* 98:532–549.

Pierotti, R. and C. A. Bellrose. 1986. Proximate and ultimate causation of egg size and the "third-chick disadvantage" in the Western gull. *Auk* 103:401–407.

Pierotti, R. and E. C. Murphy. 1987. Intergenerational conflicts in gulls. *Animal Behav.* 35:435–444.

Plummer, M. V. 1977. Predation by black rat snakes in bank swallow colonies. *Southwest. Nat.* 22:125–126.

Poole. R. W. 1974. *An Introduction to Quantitative Ecology.* New York: McGraw-Hill.

Post, P. W. and M. Gochfeld. 1979. Recolonization by common terns at Breezy Point, New York. *Proc. Colonial Waterbird Group* 2:128–136.

Post, P. W. and G. S. Raynor. 1964. Recent range expansion of the American oystercatcher into New York. *Wilson Bull.* 76:339–346.

Powell, G. V. N. 1974. Experimental analysis of the social value of flocking by starlings *(Sturnis vulgaris)* in relation to predation and foraging. Anim. Behav. 22:501–505.

Power, H. W. and C. G. P. Doner. 1980. Experiments on cuckoldry in the mountain bluebird. *Am. Nat.* 116:689–704.

Power, H. W., E. Litovich, and M. P. Lombardo. 1981. Male starlings delay incubation to avoid being cuckolded. *Auk* 98:386–389.

Pulliam, R. H. 1973. On the advantage of flocking. *J. Theor. Biol.* 38:419–422.

Quadagno, D. M., H. E. Shubeita, J. Deck, and D. Francoeuer. 1979. The effects of males, athletic activities, and all-female living conditions on the menstrual cycle. Eastern Conf. on Repro. Behav. New Orleans, La.

Quinney, T. E., B. N. Miller and K. R. S. Quinney. 1981. Gulls robbing prey from great blue herons *(Ardea herodias)*. *Can. Field-Naturalist* 95:205–206.

Ramsey, J. J. 1968. Roseate spoonbill chick attacked by ants. *Auk* 85:325.

Randall, R. M., B. M. Randall and T. Erasmus. 1986. Rain-related breeding failures in Jackass penguins. *Gerfaut* 76:281–288.

Rauzon, M. K., C. S. Harrison, and R. B. Clapp. 1984. Breeding biology of the blue-gray noddy. *J. Field Ornithol.* 55:309–321.

Raynor, G. S. 1972. Overland feeding flights by the common tern on Long Island. *Kingbird* 22:63–71.

Regelmann, K. and E. Curio. 1983. Determinants of brood defence in the great tit *Parus major* L. *Behav. Ecol. Sociobiol.* 13:131–145.

Richard, M. H. and R. D. Morris. 1984. An experimental study of nest site selection in common terns. *J. Field. Ornithol.* 55:457–466.

Richter, W. 1983. Hatching asynchrony: The nest failure hypothesis and brood reduction. *Am. Nat.* 120:828–832.

Ricklefs, R. E. 1980. "Watchdog" behaviour observed at the nest of a cooperative breeding bird, the rufous-margined flycatcher *Myiozetetes cayanensis. Ibis* 122:116–118.

Ripley, S. D. 1961. Aggressive neglect as a factor in interspecific competition in birds. *Auk* 78:366–371.

Ripley, B. D. 1981. *Spatial Statistics.* New York: Wiley.

Robertson, C. J. R. 1984. *The Readers' Digest Complete Book of New Zealand Birds.* Sydney, Australia: Readers' Digest.

Robertson, R. J. 1973. Optimal niche space of the red-winged blackbird: Spatial and temporal patterns of nesting activity and success. *Ecology* 54:1085–1093.

Robertson, W. B., Jr. 1964. The terns of the Dry Tortugas. *Bull. Florida State Mus.* 8:1–95.

Robinson, S. K. 1985. Coloniality in the yellow-rumped cacique as a defense against nest predators. *Auk* 102:506–519.

Rockwell, E. D. 1982. Intraspecific food robbing in glaucous-winged gulls. *Wilson Bull.* 94:282–288.

Roskaft, E. and T. Jarvi. 1983. Male plumage colour and mate choice of female pied flycatchers *Ficedula hypoleuca*. *Ibis* 125:396–400.

Rowan, W., K. M. Parker, P. L. Sulman, K. Pearson, E. Isaacs, E. M. Elderson, and M. Tildesley. 1919. On the nest and eggs of the common tern *(S. fluviatilis)*. A cooperative study. *Biometrika* 19:308–354.

Roselaar, C. S. 1985. *Sterna hirundo*, common tern. In S. Cramp, ed., *Handbook of the Birds of Europe, the Middle East, and North Africa*, 71–87. Oxford: Oxford University Press.

Russell, P. A. 1979. Fear-evoking stimuli. In W. Sluckin, ed., *Fear in animals and man*, 86–165. New York: Van Nostrand Reinhold.

Ryder, J. P. 1980. The influences of age on the breeding biology of colonially nesting seabirds. In Behavior of Marine Animals, vol. 4: *Marine Birds* J. Burger, B. L. Olla, and H. E. Winn, eds., 153–168. New York: Plenum.

Ryder, J. P. and P. L. Somppi. 1979. Female-female pairing in Ring-billed Gulls. *Auk* 96:1–5.

Safina, C. 1990a. Foraging habitat partitioning in roseate and common terns. *Auk* 107:351–358.

Safina, C. 1990b. Bluefish mediation of foraging competition between roseate and common terns. *Ecology* 71:1804–1809.

Safina, C., and J. Burger. 1985. Common terns foraging: seasonal trends in prey fish densities, and competition with Bluefish. *Ecol.* 66:1457–1463.

Safina, C. and J. Burger. 1988a. Ecological dynamics among prey fish, bluefish and foraging common terns in a Coastal Atlantic system. In J. Burger, ed., Seabirds and Other Marine Vertebrates: Competition, Predation and Other Interactions, 95–173. New York: Columbia University Press.

Safina, C. and J. Burger. 1988b. Use of sonar for studying foraging ecology of seabirds from a small boat. *Colonial Waterbirds* 11:234–244.

Safina, C. and J. Burger 1988c. Prey dynamics and the breeding phenology of common terns. *Auk* 105:720–726.

Safina, C. and J. Burger. 1989. Inter-annual variation in prey availability for common terns at different stages in their reproductive cycle. *Colonial Waterbirds* 12:37–42.

Safina, C., J. Burger, M. Gochfeld and R. Wagner. 1988. Evidence for food limitation of common and roseate tern reproduction. *Condor* 90:852–859.

Safina, C., R. H. Wagner, D. Witting, and K. Smith. 1990. Dynamics of diets delivered to roseate and common tern chicks. *J. Field Ornithol.* 61:331–338.

Safina, C., D. Witting and K. Smith. 1989. Viability of salt marshes as nesting habitat for common terns in New York. *Condor* 91:571–584.

Saliva, J. and J. Burger. 1989. Effect of experimental manipulation of vegetation density on nest site selection in sooty terns *Sterna fuscata*. *Condor* 91:689–698.

Salt, G. W. and D. E. Willard. 1971. The hunting behavior and success of Forster's tern. *Ecology* 52:989–998.

Sargeant, A. B., S. H. Allen, and R. T. Eberhardt. 1984. Red fox predation on breeding ducks in midcontinent North America. *Wildlife Monographs*. 89:1–41.

SAS. 1982. *User's Guide: Statistics*. Cary, N. C.: Statistical Analysis Institute.

Schaller, G. B. 1972. *The Serengeti Lion*. Chicago: University of Chicago Press.

Schnell, G. D. 1970. A phenetic study of the suborder Lari (Aves). II: Phenograms, discussion, and conclusions. *Syst. Zool.* 19:264–302.

Schnell, G. D., B. L. Woods, and B. J. Ploger. 1983. Brown pelican foraging success and kleptoparasitism by laughing gulls. *Auk* 100:636–644.

Schoen, R. B. and R. D. Morris. 1984. Nest spacing, colony location, and breeding success in herring gulls. *Wilson Bull.* 96:483–488.

Schoener, T. W. 1974. Resource partitioning in ecological communities. *Science* 185:27–38.

Schreiber, R. W. and N. P. Ashmole. 1970. Sea-bird breeding seasons on Christmas Island, Pacific Ocean. *Ibis* 112:363–394.

Sealy, S. G. 1973. Interspecific feeding assemblages of marine birds off British Columbia. *Auk* 90:796–802.

Searcy, W. A. 1984. Song repertoire size and female preferences in song sparrows. *Behav. Ecol. Sociobiol.* 14:281–286.

Serventy, D. L., V. Serventy, and J. Warham. 1971. *The Handbook of Australian Sea-Birds*, Sydney, Australia: A. H. and A. W. Reed.

Severinghaus, L. L. 1982. Nest site selection by the common tern *Sterna hirundo* on Oneida Lake, N.Y. *Colonial Waterbirds* 5:11–18.

Shalter, M. D. 1978a. Studies of mobbing behavior abound. *J. Ornith.* 119:462–463.

Shalter, M. D. 1978b. Mobbing in the pied flycatcher: Effect of experiencing a live owl on responses to a stuffed facsimile. Z. Tierpsychol. 47:173–179.

Shields, W. M. 1984. Barn swallow mobbing: self-defence, collateral kin defence, group defence, or parental care. *Anim. Behav.* 32:132–148.

Shields, W. M. and J. R. Crook. 1987. Barn swallow coloniality: a net cost for group breeding in the Adirondacks? *Ecology* 68:1373–1386.

Shields, W. M., J. R. Crook, M. L. Hebblethwaite, and S. S. Wiles-Ehmann. 1988. Ideal free coloniality in the swallows. In C. N. S. Slobodchikoff, ed., *The Ecology of Social Behavior:* 189–228. San Diego: Academic Press.

Shugart, G. W. and W. C. Scharf. 1983. Common terns in the northern Great Lakes: current status and population trends. *J. Field Ornithol.* 54:160–169.

Sibley, C. G. and J. E. Ahlquist. 1972. A comparative study of the egg white proteins of non-passerine birds. Bulletin 39, Peabody Museum of Natural History, Yale University, New Haven.

Sibley, C. G., J. E. Ahlquist, and B. L. Monroe, Jr. 1988. A classification of the living birds of the world based on DNA–DNA hybridization studies. *Auk* 105:409–424.

Sick, H. 1965. Breeding sites of *Sterna eurygnatha* and other sea birds off the Brazilian coast. *Auk* 82:507–508.

Siegel, S. 1956. Nonparametric statistics for the behavioral sciences. New York: McGraw-Hill.

Siegel-Causey, D. and G. L. Hunt, Jr. 1981. Colonial defense behavior in double-crested and pelagic cormorants. *Auk* 98:522–531.

Siegel-Causey, D. and S. P. Kharitonov. 1990. The evolution of coloniality. In D. M. Power, ed., *Current Ornithology*, vol. 7. New York: Plenum.

Siegfried, W. R. 1972. Ring-billed gulls robbing lesser scaup of food. *Can. Field-Naturalist* 86:86.

Siegfried, W. R. and L. R. Underhill. 1975. Flocking as an anti-predator strategy in doves. *Anim. Behav.* 23:504–508.

Sladen, W. J. L. 1958. The Pygoscelid Penguins. *Falkland Is. Depend. Surv. Sci. Rep.* 17:1–97.

Slobodkin, L. B. 1968. How to be a predator. *Amer. Zoologist* 8:43–51.

Smith, N. G. 1969. Provoked release of mobbing—a hunting technique of *Micrastur* falcons. *Ibis* 111:241–243.

Smith, N. G. 1972. Migrations of the day-flying moth *Urania* in Central and South America. *Carib J. Science* 12:45–58.

Smith, S. M. 1980. Demand behavior: A new interpretation of courtship feeding. *Condor* 82:291–295.

Smith, S. and E. Hosking. 1955. Birds fighting: Experimental studies of the aggressive displays of some birds. London: Faber and Faber.

Smythe, N. 1970. On the existence of "pursuit invitation" signals in mammals. *Amer. Nat.* 104:491–494.

Snapp, B. D. 1976. Colonial breeding in the barn swallow *Hirundo rustica* and its adaptive significance. *Condor* 78:471–480.

Sokal, R. R. and F. J. Rohlf. 1982. *Biometry*. San Francisco: W. H. Freeman.

Soper, M. F. 1969. Kermadec Islands expedition reports—the Grey Ternlet. *Notornis* 16:75–80.

Southern, H. N. 1938. Posturing and related activities of the common tern (*Sterna h. hirundo* L.) *Proc. Zool. Soc. London* 108:423–431.

Southern, L. K. and W. E. Southern. 1979. Absence of nocturnal predator defense mechanisms in breeding gulls. *Proc. Colonial Waterbird Group*, 2:157–162.

Southern, W. E. 1974. Copulatory wing-flapping: A synchronizing stimulus for nesting ring-billed gulls. *Bird Banding* 45:210–216.

Southern, W. E. 1977. Colony selection and colony site tenacity in ring-billed gulls at a stable colony. *Auk* 94:469–478.

Southern, W. E., S. R. Patton, L. K. Southern, and L. A. Hanners. 1985. Effects of nine years of fox predation on two species of breeding gulls. *Auk* 102:827–833.

Spaans, A. L. 1971. On the feeding ecology of the herring gull *Larus argentatus* Pont. in the northern part of the Netherlands. *Ardea* 59:73–188.

Spaans, M. J. and A. L. Spaans. 1975. Enkele gegevens over de broedbiologie ban de Zilbermeeuw *Larus argentatus* op Terscheling. *Limosa* 48:1–39.

Spurr, E. B. 1974. Individual differences in aggressiveness of adélie penguins. Anim. Behav. 22:611–616.

Spurr, E. B. 1975. Breeding of the adélie penguin *Pygoscelis adeliae* at Cape Bird. *Ibis* 117:324–338.

Stelfox, H. A. and G. J. Brewster. 1979. Colonial-nesting herring gulls and common terns in northeastern Saskatchewan. *Canadian Field-Naturalist* 93:132–138.

Stonehouse, B. 1956. The king penguin of South Georgia. *Nature* 178:1424–1426.

Storey, A. E. 1987a. Characteristics of successful nest sites for marsh-nesting common terns. *Can. J. Zool.* 65:1411–1416.

Storey, A. E. 1987b. Adaptations for marsh nesting in common and Forster's terns. *Can. J. Zool.* 65:1417–1420.

Taylor, I. R. 1979. The kleptoparasitic behaviour of the artic skua *Stercorarius parasiticus* with three species of tern. *Ibis* 121:274–282.

Teague, G. W. 1955. Observaciones sobre las aves indigenas y migratorias del orden Charadriiformes (chorlos, gaviotas, gaviotines y sus congeneres) que frecuentan las costas y esteros del litoral del Uruguay. *Comunic. Zoolog. Museo Historia Natural Montevideo* 72:1–58.

Tenaza, R. 1971. Behavior and nesting success relative to nest location in adélie penguins *Pygoscelis adeliae*. Condor 73:81–92.

Thompson, D. B. A. 1983. Prey assessment by plovers (Charadriidae): net rate of energy intake and vulnerability to kleptoparasites. *Anim. Behav.* 31:1226–1236.

Tinbergen, N. 1953. *The Herring Gull's World*. London: Collin's New Naturalist.

Tinbergen, N. 1956. On the functions of territory in gulls. *Ibis* 98:401–411.

Tinbergen, N. 1959. Comparative studies of the behavior of gulls Laridae. A Progress Report. *Behaviour* 15:1–70.

Tinbergen, N. 1963. On adaptive radiation in gulls *Tribe Larini*. *Zool. Bonner. Beitrage* 39:209–223.

Tinbergen, N. 1967. Adaptive features of the black-headed gull, *Larus ridibundus* L. *Proc. 14 Intl. Ornith. Congr.* 14:43–59.

Tinbergen, N., M. Impekoven, and D. Frank. 1967. An experiment on spacing out as a defence against predation. *Behaviour* 28:307–321.

Tornkovich, P. C. 1986. [Material on the biology of the ivory gull on Graham Bell Island (Franz Joseph Land)]. In V. D. Illitchev, ed., *[Current Problems In Ornithology]*. Moscow: Academy of Sciences of USSR.

Trivelpiece, W. and N. J. Volkman. 1979. Nest-site competition between adélie and chinstrap penguins: An ecological interpretation. *Auk* 96:675–681.

Trivers, R. L. 1972. Parental investment and sexual selection. In B. Campbell, ed., *Sexual Selection and Descent of Man*: 136–179. Chicago: Aldine.

Trull, P. 1983. Shorebirds and noodles. *American Birds* 37:268–269.

Tullock, G. 1979. On the adaptive significance of territoriality: Comment. *Amer. Nat.* 113:772–775.

Underwood, R. C. 1982. Vigilance behavior in grazing African antelopes. *Behaviour* 79:81–107.

Urban, E. K., C. H. Fry, and G. S. Keith. 1986. *The Birds of Africa*, vol. 2. New York: Academic Press.

Vader, W. 1979a. Fiskemaker raner beitende laFiskemaker raner beitende lapps-pover. *Fauna* 32:62–65.

Vader, W. 1979b. En makelig rodnebbternae. *Fauna* 32:34–35.

Van Gelder, R. G. 1984. The mammals of the State of New Jersey: A preliminary annotated list. New Jersey Audubon Soc. occas. paper no. 143.

Veen, J. 1977. Functional and causal aspects of nest distribution in colonies of the sandwich tern. *Behav. Suppl.* 20:1–193.

Verbeek, N. A. M. 1979. Some aspects of the breeding biology and behavior of the Great Black-backed Gull. *Wilson Bull.* 91:575–582.

Verbeek, N. A. M. and J. L. Morgan. 1978. River otter predation on glaucous-winged gulls on Mandarte Island, British Columbia. *Murrelet* 59:92–95.

Vermeer K. 1963. The breeding ecology of the glaucous-winged gull, *(Larus glaucescens)* on Mandarte Island, Brit. Col. occ. paper Brit. Col. Prov. Mus. 13:1–104.

Verner, J. 1964. Evolution of polygamy in the long-billed marsh wren. *Evolution* 18:252–261.

Verner, J., and G. H. Engelsen. 1970. Territories, multiple nest building, and polygyny in the long-billed marsh wren. *Auk* 87:557–567.

Veitch, C. R. 1985. Methods of eradicating feral cats from offshore islands in New Zealand. *ICBP Tech.Publ.* 3:125–142.

Vieth, W., E. Curio, and E. Ulrich. 1980. The adaptive significance of avian mobbing. III: Cultural transmission of enemy recognition in blackbirds: Cross species tutoring and properties of learning. *Anim. Behav.* 28:1217–1228.

Voous, K. H. 1963. Tern colonies in Aruba, Curaçao, and Bonaire, South Carribean Sea. Proc. 13th Int. Ornithol. Congress: 1214–1216.

Waltz, E. C. 1982. Resource characteristics and the evolution of information centers. *Am. Nat.* 119:73–90.

Waltz, E. C. 1983. On tolerating followers in information-centers, with comments on testing the adaptive significance of coloniality. *Colonial Waterbirds* 6:31–36.

Ward, L. D. and J. Burger. 1980. Survival of herring gull and domestic chicken embryos after simulated flooding. *Condor* 82:142–148.

Ward, P. 1965. Feeding ecology of the black-faced dioch *Quelea quelea* in Nigeria. *Ibis* 107:533–536.

Ward, P., and A. Zahavi. 1973. The importance of certain assemblages of birds as "information centres" for food finding. *Ibis* 115:517–534.

Watson, G. E. 1975. *Birds of the Antarctic and Sub-Antarctic.* Washington D.C.: Smithsonian Institution.

Weidmann, U. 1956. Observations and experiments on egg-laying in the black-headed gull *Larus ridibundus*. *Br. J. Anim. Behav.* 4:150–161.

Weller, M. W. 1959. Parasitic egg laying in the redhead *(Aythya americana)* and other North American Anatidae. *Ecol. Monogr.* 19:333–365.

Werschkul, D. F. 1982a. Parental investment: influence of nest guarding by male little blue herons *Florida caerulea*. *Ibis* 124:343–347.

Werschkul, D. F. 1982b. Nesting ecology of the little blue heron: promiscuous behavior. *Condor* 84:381–384.

Wesolowski, T., E. Glazewska, L. Glazewski, E. Hejnowicz, B. Nawrocka, P. Nawrocki, and K. Okonska. 1985. Size, habitat distribution, and site turnover of gull and tern colonies on the middle Vistula. *Acta Ornithologica* 21:45–67.

Wetmore, A., and B. H. Swales. 1931. The birds of Haiti and the Dominican Republic. *U.S. Natl. Mus. Bull.* 155:1–483.

White, M. G., and J. W. H. Conroy. 1975. Aspects of competition between Pygoscelid penguins at Signey Island, South Orkney Islands. *Ibis* 117:371–373.

Wickler, W. 1985. Coordination of vigilance in bird groups. The "Watchman's Song" hypothesis. *Z. Tierpsychol.* 69:250–253.

Wiggins, D. A. and R. D. Morris. 1986. Criteria for female choice of males: courtship feeding and parental care in common terns. *American Naturalist* 128:126–129.

Wiggins, D. A. and R. D. Morris. 1987. Parental care of the common tern *Sterna hirundo. Ibis* 129:533–540.

Wiggins, D. A. and R. D. Morris. 1988. Courtship and copulatory behaviour in the common tern *Sterna hirundo. Ornis Scand.* 19:163–165.

Wiggins, D. A., R. D. Morris, I. C. T. Nisbet, and T. W. Custer. 1984. Occurrence and timing of second clutches in common terns. *Auk* 101:281–287.

Wiklund, C. G. 1982. Fieldfare *(Turdus pilaris)* breeding success in relation to colony size, nest position, and association with merlins *(Falco columbarius). Behav. Ecol. Sociobiol.* 11:165–172.

Wiklund, C. G. and M. Andersson. 1980. Nest predation selects for colonial breeding among fieldfaress *Turdus pilaris. Ibis* 122:363–366.

Williams, A. J. 1974. Site preferences and interspecific competition among guillemots *Uria aalge* (L.) and *Uria lomvia* (L.) *Ornis. Scand.* 5:113–121.

Williams, G. C. 1966. *Adaptation and Natural Selection.* Princeton, N.J.: Princeton University Press.

Wilson, E. O. 1975. *Sociobiology: The New Synthesis.* Cambridge, Mass.: Harvard University Press.

Wilkinson, G. S. and G. M. English-Loeb. 1982. Predation and coloniality in cliff swallows *(Petrochelidon pyrrhonota). Auk* 99:459–467.

Windsor, D., and S. T. Emlen. 1975. Predator-prey interactions of adult and prefledging bank swallows and American kestrels. *Condor* 77:359–361.

Wishart, R. A. 1983. Pairing chronology and mate selection in the American wigeon *(Anas americana). Canadian J. of Zool.* 61:1733–1743.

Wittenberger, J. F. and G. L. Hunt, Jr. 1985. The adaptive significance of coloniality in birds. In D. S. Farner, J. R. King, and K. C. Parkes, eds., *Avian Biology,* 3:1–79. New York: Academic Press.

Woodward, P. W. 1972. The natural history of Kure Atoll, northwestern Hawaiian Islands. *Atoll Res. Bull.* 164:1–318.

Yasukawa, K. 1981. Male quality and female choice of mate in the red-winged blackbird *(Agelaius phoeniceus). Ecology* 62:922–929.

Yom-Tov, Y. 1975. Synchronization of breeding and intraspecific interference in the carrion crow. *Auk* 92:778–785.

Ytreberdg, N. 1956. Contribution to the breeding biology of the black-headed gull (*Larus ridibundus* L.) in Norway. *Nytt Mag. Zool.* (Oslo), 4:5–106.

Zahavi, A. 1975. Mate selection—A selection for a handicap. *J. Theor. Biol.* 53:205–214.

Zubakin, V. A. 1973. Cannibalism in gull-billed tern (*Gelochelidon nilotica*). *Zool. Zhurn* 52:32–36.

Zubakin, V. A. 1985. [Parallelism in the evolution of coloniality in birds]. In V. A. Zubakin, V. V. Ivanitsky, E. E. Stotskaya, and S. P. Kharitonov, eds., *[Theoretical aspects of coloniality in birds]*, 42–47. Moscow: Moscow Society for Nature Study.

Zubakin, V. A. and Avdanim, V. O. 1982. [Peculiarities of colonial nexting in *Rhodostethia rosea*] *Bull. Moscow Univ.* 12:1754–1756.

Zusi, R. L. 1958. Laughing gull takes fish from black skimmer. *Condor* 60:67–68.

Zusi, R. L. 1962. Structural adaptations of the head and neck in the black skimmer. Proc. Nuttall Ornithol. Club, no. 3. Cambridge, Mass.

Taxonomic Index

Subject Index

Developmental defects, 318, 320